MANUFACTURING TIME

A GUILFORD SERIES

Perspectives on Economic Change

Editors

MERIC S. GERTLER
University of Toronto

PETER DICKEN
University of Manchester

Manufacturing Time:
Global Competition
in the Watch Industry, 1795–2000
AMY K. GLASMEIER

The Regional World:
Territorial Development in a Global Economy
MICHAEL STORPER

Lean and Mean:
Why Large Corporations Will Continue
to Dominate the Global Economy
BENNETT HARRISON

Spaces of Globalization:
Reasserting the Power of the Local
KEVIN R. COX, *Editor*

The Golden Age Illusion:
Rethinking Postwar Capitalism
MICHAEL J. WEBBER and DAVID L. RIGBY

Work-*Place*:
The Social Regulation of Labor Markets
JAMIE PECK

Restructuring for Innovation:
The Remaking of the U.S. Semiconductor Industry
DAVID P. ANGEL

Trading Industries, Trading Regions:
International Trade, American Industry,
and Regional Economic Development
HELZI NOPONEN, JULIE GRAHAM, and ANN R. MARKUSEN, Editors

Manufacturing Time

*Global Competition
in the Watch Industry, 1795–2000*

Amy K. Glasmeier

THE GUILFORD PRESS
New York   London

© 2000 The Guilford Press
A Division of Guilford Publications, Inc.
72 Spring Street, New York, NY 10012
www.guilford.com

All rights reserved

No part of this book may be reproduced, translated, stored in a retrieval system, or transmitted, in any form or by any means, electronic, mechanical, photocopying, microfilming, recording, or otherwise, without written permission from the Publisher.

Printed in the United States of America

This book is printed on acid-free paper.

Last digit is print number: 9 8 7 6 5 4 3 2 1

Library of Congress Cataloging-in-Publication Data
Glasmeier, Amy.
   Manufacturing time: global competition in the watch industry, 1795–2000 / Amy K. Glasmeier.
     p. cm. — (Perspectives on economic change)
   Includes bibliographical references and index.
   ISBN 1-57230-589-4
   1. Clock and watch industry—History. I. Title. II. Series.

HD9999.C582 G53 2000
338.4′7681114′09—dc21
                                                               00-042997

Prologue

In 1989, while a professor at the University of Texas, I was approached by a man from the Swiss region known as the Jura, who asked a deceptively simple question: "How can one of the world's preeminent regions of industrial precision reposition itself in the face of technological change?" This question, posed seven years after the introduction of the personal computer, was not trivial. Indeed, wouldn't the Jura, once the world's center for precision machining, be well poised to participate in the microelectronics revolution?

The answer seemed to lie within the 150-year-old struggle for watch industry supremacy. To understand what has happened during this period within the firms themselves, I needed to examine how industry leadership was transferred among widely disparate locations, production systems, and industrial cultures. While I pursued this investigation, it became evident that the answer to industry success and lost fortunes lay in a complex intersection between the characteristics and qualities of the product and the larger environment in which it was produced.

There are many examples of failed regional economies and industrial production systems throughout history. Lancashire, England, and the textile industry; Liverpool and shipbuilding; the Connecticut Valley in the United States and the brass metal industry; Pittsburgh and steel; Detroit and autos; Houston and oil; Los Angeles and aerospace; and countless others. These cases clearly show that the nature of regional economic decline and renewal is not homogeneous. These places did not all experience regional transformation in the same way—some places failed and then succeeded again, some not only succeeded but went on to greater heights, and some failed miserably and permanently. Confusing matters further, in some cases it was the industrial base that failed, while in others it was the regional context that declined. Thus even the object of investigation was not equivalent across cases. Digging further back in time, I discovered one factor that unified these experiences: their decline did not occur in isolation, but was the result of the interaction between these centers of industrial activity and others also vying at the same time for the leadership position. Thus, my purpose in this book has been to integrate internal and external events into an understanding of how one industry rose and declined among very different circumstances, settings, and cultures.

Understanding why one competitor succeeds and another fails requires knowledge of how each competitor is situated in real time. Why one region succeeds at one point in time is not always immediately apparent—and perhaps is only evident with hindsight, no matter how detailed our contemporary understanding of its current trajectory. The industrial trials and tribulations recounted in this study span a long enough period of economic history and include a great enough variety of countries, cultures, and technological transformations, to isolate short-term events from the effects of longer-term and more systemic conditions that alter the fortunes of important participants.

The book has relevance for numerous audiences, including economic geographers, economic historians, regional planners, political scientists, economists, business professionals, and policy analysts. For economic geographers and economic historians, the book provides a detailed account of the evolution of industrial cultures and systems of production and explains how these structures of social relations are altered by the effects of technological change, economic crisis, and geopolitical identities. Regional planners may find the book helpful in identifying the importance of local institutions as vehicles of change. Political scientists will find that knowledge of the historical evolution of the industry contributes to our understanding of how national trade politics and policy are shaped by local industrial constituents. The book points out the fragility of industrial coalitions that come together and break apart depending on the convergence of forces endogenous to the industry itself as well as those that are larger, exogenous shaping events.

The book also contributes to a growing body of literature on evolutionary economics that takes as its point of departure the idea that industries are not comprised of a group of homogeneous firms but rather are composed of an array of actors with more or less capability to influence their own environments. Here is a detailed look at the role of path dependency, demonstrating the high degree of interaction between the meaning and effect of path dependency and other shaping events. The history of the global watch industry fills in the details of path dependency's genesis, its potential durability, and its contingency. Business professionals will find the story of organizational evolution relevant to an understanding of missteps made in the face of high-quality strategic information. Why firms fail to act in the face of strategic knowledge has much to do with corporate culture, and a corporation's ability and willingness to learn "outside the box." Finally, policy analysts, especially those interested in economic development, may find some value in the way I have sorted local conditions from larger forces of change. The extent that public officials can influence the trajectory of a regional economy is directly related to the level of respect given to their understanding of economic processes and business conditions by industry leaders. Being able to walk in someone else's shoes can foster understanding of the extent that businesses are truly willing to partner with local officials, citizens, and other relevant actors to chart a path leading to regional wealth creation and community economic stability.

The Swiss watch region no longer reigns supreme over the world watch industry. The greatest loss to the region is the degradation of skill, which occurred during the period of most serious economic crisis. Could individual firms and the industrial complex as a whole have jumped technological paradigms from micromechanical to microelectronic and resumed global leadership, based on the evidence presented in this book? No. But a single answer cannot be found that conclusively demonstrates why the region's firms failed to make this transition successfully.

Examining an individual industry across time and space uncovers the multifold interventions and interactions that occur and influence the ability of local actors to respond in the face of needed change. Some situations require only marginal adjustments in local behavior, while in other cases the magnitude of change required is monumental. Proximity to the stimulus for change turns out to be as important as learning by doing and learning by emulation. Ironically, however, the sources of magnitudinal change are rarely proximate to all sites of interest; this fact, above all others, demonstrates that "local embeddedness" can limit needed change because the meaning of change cannot be interpreted within existing systems of learning and information dissemination if their origin is outside the local sphere of influence. "Falling behind" is not the same as "falling out" of the race, as the case of the Swiss watch industry clearly shows. But falling behind increases the likelihood of becoming a marginal player over time, which may ultimately mean forfeiting all but the memory of once being great. Only time will tell.

Acknowledgments

Given the time it took for me to write this book, there are many people who have contributed to it, and to whom I owe great appreciation.

My editors, including Series Editors Peter Dicken and Meric S. Gertler, have provided guidance and continuous support throughout this project. Peter Dicken read the manuscript more than once and made excellent suggestions along the way about how to make it better. Peter Wissoker, acquisitions editor at The Guilford Press and a thoughtful manuscript mentor, also read an early version of the work and provided very detailed comments, which proved essential to moving the project forward. Guilford's editorial staff, including William Meyer and Philip Holthaus, are stupendous and worked exceptionally hard to make sure that the text was as clear as possible. In the end, however, none of the editors or reviewers is responsible for errors in the text. Omissions and oversights are my responsibility alone.

This project began while I was teaching at the University of Texas at Austin. The chairman of the Department, Terry Kahn, gave me tremendous encouragement and support, without which the project could never have begun. There were many students who helped with the initial investigation, providing research and translation assistance. These include Rolf Pendall, Bettina Brunner, Kelley Furgeson, Achmed Ktioui, Amy Kays, and Jeff Thompson. Rolf in particular started the ball rolling and dug up significant historical sources in the project's early stages, and Amy prepared untold numbers of drafts of reports as the project took shape. Ann Groschitch provided translation assistance in Washington, DC, along with Graeme Burt, Keith Rokoske, and Colin Polske at Penn State. Alex Ebaugh contributed several summers sorting through historical documents to compile the history of the U.S. electronic industry's foray into watch manufacturing. Jeanna Gatten did a meticulous job on the tables and charts, and Erin Heithoff did wonderful work reconstructing the maps. Finally, Stephen Belcher, Assistant Professor at Penn State, also provided editorial and translation assistance.

A number of professionals and individuals related to the industry provided generously of their time. Michael Harrold kindly provided permission to publish the charts and tables from his excellent history of the U.S. watch industry. Noriuki Sugiura assisted in countless ways during my trips to Japan. In particular, he accompanied me to firm interviews and took me to Nagano Prefecture, where Seiko Watch is located. Wonmin Higuichi provided great company during our visits to Seiko in Tokyo. Daniel Chuen Chee Wo pointed out the meaning of Seiko's strategy in the company's early attempts to access the U.S. market. Dr. Kam-Hon Lee helped me acquire key technical documents in Hong Kong along with Dr. Charles Ko, who helped set up appointments with key members of the Hong Kong watch industry.

Without question, this project would never have happened had it not been for a fateful meeting with Luc Tissot in Jürgen Schmandt's office at the University of Texas at Austin. Mr. Tissot posed the initial question about the fate of the region, and in the early life of the project provided funding for the investigations of the industry in Hong Kong, Switzerland, and Japan. Mr. Tissot was an essential part of this project, providing access to members of the Swiss watch industry and contributing untold numbers of hours to interviews about his experiences there. He accompanied me on the first trips to Hong Kong and Japan and provided a unique view onto the world of the Swiss industry coming of age. He also provided many of the insights needed to understand the chain of events that led to the crisis in the industry in the early 1980s. It is safe to say that this book would never have been written without him.

Several people at the Tissot Economic Foundation in Le Locle, Switzerland, were exceptionally helpful. Wendy Tiallard, a dear friend, provided translation and research assistance, and was a wonderful companion on our many adventures in the Swiss countryside; Jocelyn Tissot provided administrative support and encouragement; and Pierre Rossel was instrumental in helping me understand aspects of Swiss culture.

Countless others in Switzerland gave freely of their time as I worked my way through the history of the Swiss watch industry and its evolution in the Jura region. Claude Michaud of INSEAD provided strategic insights about the origins of the crisis in the industry. Denis Miallat of the University of Neuchâtel was kind enough to meet with me on several occasions and provided documents he had written on the watch industry over a 20-year period. Carolyn Veya shared her first-hand experiences as a woman working in the industry and traveled with me on several interviews to talk with workers and their families. Industry contacts were essential. René Retornaz, former president of the Watch Federation and its chief trade negotiator for more than 40 years, provided excellent insights into the environment faced by Swiss industrialists over the postwar period. T. Raja was helpful in providing industry statistics in a pre-web world. François Jequier shared his family's history in the watch industry. Thierry Kneuss, secrétaire fédératif of the watch industry workers' union, Fédération Suisse des Travailleurs de la Métallurgie et de l'Horlogerie, pointed out the importance of labor peace in the evolution of the Swiss watch industry. Charles Dubois provided invaluable insights about the Swiss industry's initial foray into Hong Kong. Raymond Béra talked at length about the Swiss marketing strategy in Asia. Francis Châtelain shared with me the metal watchband and case he made for the first electronic digital watch introduced by Texas Instruments. An untold number of company executives and their workers kindly provided me with an understanding of their labor processes and strategic problems. This information proved essential to my understanding of the industry and demonstrated that the skills of the region were its enduring competitive advantage.

During academic year 1998–1999, while I was on sabbatical at the Department of Urban Studies and Planning at MIT, Karen Polenske and Barbara Baran were gracious hosts. Being at MIT gave me access to Harvard's amazing library. Richard Samuels, Professor of Political Science at MIT, pointed me in the direction of the World War II Bombing Survey, which I needed to confirm the location of precision machining in Japan during the war. With this clue, I was able to find the missing piece to the puzzle of how the Japanese watch industry experienced its remarkable postwar recovery. There are also countless others who have contributed to my thinking on the world watch industry, and to them I owe a debt of thanks.

I received considerable financial support from a number of institutions to complete this project. The Tissot Foundation provided travel and support funds during the critical first years. The University of Texas at Austin provided research support for graduate assistance. The Guilford Press provided a travel advance to return to Asia in the early 1990s. MIT and Penn State provided resources for a sabbatical leave. The Pennsylvania State University College of Earth and Mineral Sciences Wilson Fund for international travel provided funds to return to Hong Kong and Japan in the late 1990s. Special funds also were provided by the Department of Geography for the production of the maps.

The Institute for Policy Research and Evaluation, in particular, Michelle Aungst, Patti Doroschenko, and Lee Carpenter, worked miracles with early versions of the text. Lee's critical editing and Patti and Michelle's magic text processing and formatting brought the manuscript down the home stretch. This book would not have seen the light of day without their hard work. My thanks also go to Dr. Irwin Feller, who allowed the text to be developed in the Institute. The Department of Geography at Penn State also provided assistance. Special thanks to Kathy Sherman, Karen Royer, and Rosie Long for shepherding the book to its final destination.

I owe many of my friends a wealth of thanks for quietly and supportively listening to me all these years. Special thanks to Jim Schoch, who helped me develop the arguments and understand the bigger picture, and to Barbara Baran, who lovingly allowed us to yak until all hours of the night trying to figure out the story. Mike Conroy deserves thanks for reading the text when I was the most anxious about its form. JK and Brenda Bohlke watched over several summers as the text took shape. David Dibiase and Cindy Brewer listened patiently to the story of watches too many times to count. Erica Schoenberger and Flavia Martinelli offered untold amounts of encouragement in the book's final stages. Students in my Geography 505 seminar—Larry Wood, Kelly Prince, Michael Glass, Christy Jocoy, Jenn Melbye, and Tracey Farrigan—had the patience and good humor to read the base texts that are referenced in this book without so much as a hint of complaint. And my family, who put up with many missed weekends, nights, and even vacations—they know their love and generosity are greatly appreciated. Without them, this book could never have been written. This book is dedicated to them and to my mother, the late Nancy Foster Glasmeier.

List of Tables and Figures

Tables

TABLE 5.1.	Watchmakers in Neuchâtel, 1752–1812	96
TABLE 5.2.	Gender Breakdown, Watch Industry of 1870	102
TABLE 5.3.	Watch Movements, 1864–1887	103
TABLE 6.1.	Total Output, Waltham and Elgin Corporations, 1880	119
TABLE 6.2.	Production of Watches, Jeweled and Dollar, United States, 1880–1930	125
TABLE 7.1.	Swiss Watch Exports to the United States, 1864–1882	132
TABLE 7.2.	Watch Industry Employment in the Swiss Jura, 1870–1920	133
TABLE 7.3.	Production of Movements, Waltham Watch Company, 1926–1935	142
TABLE 7.4.	Watches, Movements, and Materials Imported into the United States, 1912–1926	147
TABLE 7.5.	Number of Persons Employed in Swiss Watchmaking Industry by Size of Enterprise, 1929	150
TABLE 7.6.	Imports of Swiss Watches and Movements, 1934–1942	153
TABLE 7.7.	U.S.–Swiss Trade during World War II	153
TABLE 8.1.	The Japanese Watch Industry: Production, Shipments, and Stocks of Wristwatches during the Postwar Period	169
TABLE 8.2.	The Japanese Watch Industry: Watch and Clock Factory Shipments, 1951–1959	170
TABLE 8.3.	The Japanese Watch Industry: The Record of Export and Import of Watches	171
TABLE 8.4.	Evolution of Production and Exports of Watches and Clocks	174
TABLE 8.5.	Per-Unit Value of Watch Movements Imported into the United States	176

TABLE 9.1.	Swiss Watch Exports to the United States	196
TABLE 9.2.	Ten Principal Customer Nations of the Swiss Watch Industry, 1950, 1955, and 1960	198
TABLE 9.3.	Number of Firms and Workers in the Swiss Watch Industry, 1960–1980	200
TABLE 10.1.	Watch Exports, Entrepôt Trade, and Imports from 1968 to 1991	220
TABLE 10.2.	Watch Manufacturer Numbers and Employees	221
TABLE 10.3.	Main International Markets and Exports, 1960–1991	222
TABLE 10.4.	Exports and Imports in 1950s and 1960s	224
TABLE 10.5.	Watch Product, 1960–1991	227

Figures

FIGURE 4.1.	Mining and metal industries—about 1725 (Liverpool, England region).	74
FIGURE 5.1.	The Swiss watch region, 1920.	95
FIGURE 6.1.	American watch factories.	109
FIGURE 7.1.	Watch company population, 1860–1930.	135
FIGURE 7.2.	Yearly watch production, jeweled versus unjeweled, 1860–1930.	135
FIGURE 8.1.	Distribution map of major watch and clock factories in Japan.	167
FIGURE 10.1.	Sketch map of cities and provinces of China.	217

Contents

Prologue — v
Acknowledgments — ix
List of Tables and Figures — xiii

1 ■ From Keeping Time to Keeping Pace — 1
History Revisited, 5
Regions and Industries Linked, 6
Production Organization and the Fate of Regions, 8
Taking the Long View, 10
Layout of the Book, 13

2 ■ The Need for and Constraints on Change — 15
Technological Change, 17
Economic Crises, 22
War and Geopolitical Conflicts, 31
Summary, 36

3 ■ The Organizational Development of the World Watch Industry — 40
The Organizational Structure of Watchmaking in the Craft Era, 41
The Organization of the Watch Industry in the Factory Era, 45
The Emergence of Managerial Capitalism and Factories, 49
The Rise of Collective Capitalism, 53
Organizational Culture and Identity, 58
Regional Culture and Identity, 62
Summary, 63

xv

4 ■ The Burden of Being First: The Foundations of Watchmaking and Timekeeping Technology and Britain's Ascent and Decline as the Dominant Watch Manufacturing Region 64

Timekeeping in History, 65
What's All the Fuss About?: Precision, 66
History of Watch Technology, 66
Early Antecendents of the Watch Industry, 69
The Burden of Being First, 70
The Long Road Downhill, 85
Summary, 86

5 ■ Why Switzerland?: The Rise of the Jura System of Watch Manufacturing 88

The Emergent Systems: Switzerland and the United States, 91
Early History of the Swiss Watch Industry: One Nation's Loss Is Another One's Gain, 92
Mechanization and Toolmaking, 98
Creating a Competitive and Cooperative Environment, 99
From Craft Production to Protofactories, 101
Summary, 105

6 ■ The American System of Watch Manufacturing 107

Early History, 107
The Waltham Watch Company, 112
The 1860s–1880s: New Stimuli and New Entrants, 115
Sowing the Seeds of Their Own Demise, 119
Making Watches for the Masses, 121
The Hamilton Watch Company of Lancaster, Pennsylvania, 127
Summary, 128

7 ■ More Than One Way to Win a War 130

The Swiss Watch Industry Reemerges as a Major Industrial Competitor, 131
The Jura Industry's Past First Haunts, Then Saves It, 132
Costs of Redirection, 136
Time on Your Hands, 141
The U.S. Industry after the Turn of the Century, 144
Hamilton after the Turn of the Century, 145
World War I and Its Aftermath, 145
The Interwar Years: Optimism Gives Way to Decline, 148
The Statut de l'Horlogerie and the Codification of the Swiss System, 150
Summary, 152

8 ■ An Unexpected Competitor 155

The Meiji Era and the Revival of Engagement
   with the Outside World, 157
What Did They Know and When Did They Know It?, 160
A Roundabout Way into Watches, 161
Hattori Seiko, 164
World War II and Relocation Out of Tokyo, 166
Summary, 173

9 ■ Only the Young Survive: The U.S. Watch Industry 178
between the World Wars and after World War II

The U.S. Industry after World War I, 180
Hamilton from the 1920s to the 1960s, 182
Bulova from the 1950s to the 1980s, 186
Timex from the 1940s to the 1970s, 189
Abandoning Industry Regulation:
   Instituting Industrial Change, 193
Summary, 201

10 ■ Going Electronic, Moving to Hong Kong 203

The Challenge of Quartz: Now Anyone
   Could Compete, 205
Digital Delay, 206
The U.S. Strategy for Digitals, 208
Which Way to Watch a Watch?, 210
Watchmaking in Hong Kong, 221
The Quartz Revolution and Its Link
   to the Electronics Industry, 225
Organization of the Hong Kong Watch Industry, 229
Problems Facing the Industry, 234
Future Prospects, 238
Summary, 240

11 ■ Can One Man Save an Industry? 242

Given Up for Dead, 243
The Antecedents of the Swatch Watch, 245
The Emergence of Swatch, 246
The Internationalization of Watch Manufacturing, 247
Reorganization à la Japanese, 247
New Markets, 249
The 1990s and Beyond, 250
Summary, 257

12 ■ Success Goes to the Nimble, Regardless of Size 260
Distribution and Market Segmentation, 262
The Role of Skill, 263
Institutional Arthritis Sets In, 264
Industry Structure, Regional Performance, and Enduring Competitive Advantage, 266
Endogenous Attributes and the Weight of Exogenous Events, 266
Conclusions, 273

Notes	275
References	281
Index	297
About the Author	311

MANUFACTURING TIME

1 From Keeping Time to Keeping Pace

In 1969, on the eve of the electronics revolution, the Fédération de l'Industrie Horlogère Suisse (FH; Federation of the Swiss Watch Industry) commissioned a marketing study by the American marketing guru Daniel Yankelovich to ascertain the probable size of the future electronic watch market. Earlier, in 1966, a consumer survey had already indicated "considerable interest in a new electronic watch. Customers saw increased accuracy, novelty, and a good product for presents" (Yankelovich, 1969, p. 3). By 1969, the interest had grown further, and the Swiss were feeling pressure to respond. Société Generale de l'Horlogèrie Suisse SA (ASUAG; the acronym is derived from the German translation: Allgemeine Schweizerische Uhrenindustrie AG), the big Swiss watch parts producer, had grudgingly just given in to the licensing of the tuning-fork technology developed by Max Hetzel, a former Swiss watchmaker who had emigrated to the United States after being rebuffed by Swiss mechanical watch manufacturers. Hetzel sold his tuning-fork watch under the trade name "Accutron." Hetzel's monopoly on the product enabled him to set prices very high; indeed, some believed them to be set unfairly high. Nonetheless, the 1969 Yankelovich study argued that

> despite Accutron's high prices, limited retailer support (because of less than "Keystone" margins), repair problems and poor styling, the trade agrees that the Accutron has made a remarkable market penetration. In our judgment, this performance is caused by the existence of an unusually strong current consumer demand and an accelerating potential demand for an Accutron type of watch—specifically a tuning-fork watch. Accordingly, most retailers believe the entire watch market can be expanded by the growth in emphasis on electronic watches because of large numbers of non-electronic watches that will become obsolete for the consumer. The trade feels the market expansion will be caused by three factors: 1. A higher retail price substitution sale for automatic wind watches; 2. Consumers will be buying

watches at a faster rate than they expected because of availability of a new type of watch as opposed to the availability of only conventional watches; 3. Consumer demand for tuning-fork watches will be accelerated by its image of modernity and identity with "space-age technology." (p. 6)

Following this assessment, Yankelovich went on to say that "in our judgment, the demand for the tuning-fork movement watch from the trade and the consumer is growing at a rapid pace. The Swiss could be left behind unless they are prepared to meet the demands of aggressive retailers and potential competitors" (p. 7).

By the late 1960s, the big question for the Swiss was not whether to go electronic but which of Switzerland's major watch company brands would be the first to introduce a tuning-fork watch. The decision was not a trivial one. The Swiss were already facing growing market upheaval in their established market segments. Timex, a U.S. watch manufacturer, had recently come out with an electronic watch priced at $40 U.S. Demand for the watch was building, though slowly because habitual buyers of conventional Timex watches were confused by this watch's relatively high price (the conventional Timex watch cost less than $20). The Swiss could not see any profit in chasing this low-end market segment with a new technology. For the Swiss, "new" meant complicated and therefore expensive; consequently, the Swiss typically introduced new technology in watch brands at the top of the price pyramid.

Simultaneously, U.S. electronic component producers were also pressuring the Swiss as they moved rapidly into downstream markets in search of outlets for the huge component surpluses being churned out by their new factories. From the beginning of the microelectronics era, firms moved quickly down the production process learning curve, such that product cycles were being eclipsed in a matter of months. In the precomputer era, there was little more than missiles and pocket calculators to absorb output. For electronics producers, the watch market seemed to be a sure bet, with huge potential.

In the past, the Swiss strategy had been to bring a new technology to market through a high-priced brand. They expected to collect a premium price because of the "innovative nature" of the new product. This time, however, the Swiss did not control the market and thus their entry strategy could not be dictated solely by their past successes. This was the first time since the middle of the 19th century that the Swiss had not dictated market performance.

But what could have happened, and should have happened, did not happen. Yankelovich clearly foresaw two important trends that would govern the electronic watch age. First, the demand would be huge, but only for a reasonably priced product that could be purchased by the masses. Demand for electronic watches would ultimately displace the consumption of traditional mechanical watches—the lifeblood of the Swiss industry. Second, if the Swiss didn't make the right move at the right time, Japanese, not U.S., producers would eventually take control of the electronic watch market. Yankelovich sharply reported:

> In our opinion, the Japanese could pose a formidable threat to the Swiss, unless the Swiss move ahead in introducing the electronic watch in the United States. Several large retailers report that they are looking toward Japanese suppliers for electric/electronic watches. They admit they would rather sell a Swiss tuning-fork movement watch, but they are unwilling to wait an inordinate length of time for the Swiss to make it available. (p. 8)

Presaging the next technological innovation, Yankelovich pointed out: "The concept of the quartz crystal watch is favorably received by retailers who know of it. This innovation will put the Swiss in the enviable position of being able to offer a *complete* electronic watch line (tuning-fork and quartz crystal movements) in a wide price range" (p. 8).

Why was such a clear message ignored by the Swiss? Past experience suggests that just before major technological change, weak signals portending the tranformation are coming from many directions. But, because there are so many signals, all with varying amplitudes, their meaning is often confusing and conflict-filled, and therefore their significance is muted. In a highly contested industry, many voices are important. In the watch industry, the Swiss were burdened by all of them: markets, producers, competitors, new technologists, macroeconomists. Admittedly, the most immediately important message, that was coming from retailers, was highly mixed—this source of market intelligence was sending confounding and very self-interested signals. Standard Swiss watch markets were of a split opinion concerning the new technology. Mass retailers of medium- and high-priced merchandise, such as department stores, thought the market would be bullish for electronics. They wanted to get a product on the market fast in order to move quickly into the profitable (for them) realm of private-label production and distribution. As Yankelovich reported, "In the judgment of some aggressive, large mass retailers and merchandisers, if the Swiss do not provide this type of electronic watch in the near future—others will" (p. 7).

In the traditional jewelry market, on the other hand, jewelers were, as Yankelovich noted, antagonistic toward tuning-fork technology because Hetzel had eaten into their profits and because his company provided very poor repair service—historically, a key to customer satisfaction and loyalty in the jewelry trade. Retailers' short-term perspectives aside, Yankelovich's report was clear: bring a product to market that people can afford. The Swiss, who already dominated the mechanical watch end of the industry, would miss the moment, someone else would prevail, and inaction would become the watchword.

Indeed, the Swiss were stopped in their tracks. Unable to go forward in selecting who would bring an electronic watch to market because of interorganizational rivalries and a historically constructed production paradigm that linked precision with cost, Swiss reaction time to the new challenge was fatally slow. In the absence of a new consensus about the nature of the competitive environment, they could only look backward and see the world according to their historical experience. Reminiscences about an earlier, happier, and more civilized (translate into controlled) time filled the boardrooms of the major Swiss watch firms. The practice of searching for answers to competitive threats based on past experience is a common and sometimes fatal flaw in organization learning, which is history-based and reflective of past experience.

Hampered by the ideology of precision manufacturing, which demanded a high-priced introduction, and lacking the strategic intelligence to know that just around the corner the mass market of the electronics age loomed, the Swiss system simply stalled. In two years the Swiss share of the U.S. market fell precipitously from 80 percent to 20 percent. Firms were hemorrhaging. Retailers were returning cartons of products unopened. Inventory was piling up and had to be written off in million-dollar lots. Hidden assets evaporated overnight. Meanwhile, new product development stood still in the face of the pending technological revolution. Running for cover became the order of the day.

Ultimately, the tuning fork was not the straw that broke the Swiss lock on the world watch industry. It was simply an intermediate irritant. Indeed, tuning-fork technology still required considerable machining during parts manufacturing and was, comparatively speaking, very complex. It would never be the foundation of a cheap watch that the common man or woman could afford. Had the crisis been the result of the tuning fork, the Swiss might have been able to respond because the technological trajectory of this intermediate technology was actually complementary with mechanical watch technology. Even Yankelovich misforecast the tsunami that would be unleashed with the advent of the quartz crystal, the technology that took the entire watch industry by storm in a matter of a few years (Jequier, 1991). This new technology was nothing like that associated with mechanical watches; the centuries-old link between the tradition of precision and mechanics was broken forever. Although an inkling of what could happen was evident in 1966, the weak signals for evolution and then for revolution, while heard, went unheeded despite the market intelligence available.

Why couldn't the world's leader in watch manufacturing see the writing on the wall? How could a wholly new product emerge out of the United States, a country whose domestic watch industry had been seriously compromised and almost killed off by government control over procurement during World War II? How, ultimately, did a country like Japan, which for all intents and purposes lacked a watch industry at the turn of the century, rise up and dominate the industry worldwide only 70 years later? And then, with an equally swift shift, how did a small island country like Hong Kong, with a marginal history of industrialization, become the world's center for the watch industry by the end of the 20th century?

History Revisited

The striking thing about these questions is that with some slight recasting they could describe underlying facets of the world watch industry as it went through various crises and episodic events over a period of some 200 years. Of particular significance were the 1830s, when, due to myopia, the British yielded market control to the Swiss, and in another 30 years saw the Swiss shut out of America both in terms of access and technology, allowing a nascent domestic industry to take control of the home market. At another turning point, the 1920s, the introduction of the wristwatch brought similar crises and challenges for the Swiss; this time it was the Americans who stumbled by failing to move quickly into wristwatch production after World War I.

A sequence of additional developments and challenges was played out during and after the Great Depression and World War II that again altered the competitive circumstances facing the watch industry. And these developments led in turn to the great crisis resulting from the microelectronics revolution of the 1970s. Who would have guessed that an entirely new economic dogma espoused by the Japanese would serve as the foundation for the emergence of a small number of firms capable of supplying an unprecedented number of watches for the global market? While not wishing to succumb to the temptation of saying everything in history repeats itself, I must nonetheless note that there are reoccurring themes that help us to understand how industries and regions evolve through time.

History is not repeated, at least not with the same participants, order of events, and pattern of outcomes. And yet, reactions to deflations and depressions, wars, the emergence of new competitors, and technological change are in part shaped by past reactions to similar circumstances. (Gourevitch, in *Politics in Hard Times* [1986], now in a 1996 second edition, makes a similar argument about the political response to economic crises.) As time proceeds, if the original participants remain in the game, their reactions are shaped by past experiences, past paradigms, and past conventions (Choi, 1993; Gourevitch, 1986, p. 33; North, 1990). Hence, to the extent that history is repeated, it is not so much that the events are similar as much as it is that the actors respond in ways that are cumulative and historically fashioned (Fuellhart, 1998; Gourevitch, 1986). Past patterns are called up to contend with the immediacies of change.

This book traces the developmental experience of the watch industry over its life, starting in the late 17th century. My purpose in pursuing such a long industrial history is to investigate the ways in which culture, institutions, organizations, and actors interact with and respond to change-inducing events. In some cases these events represent crises, in others, episodic junctures that are both endogenous and exogenous in origin and that call forth degrees of rupture with past practice. I am particularly interested in understanding systemic reactions to the effects of technological change in terms of how the magnitude of change is interpreted, translated, and then reacted to by different industrial cultures and systems, through the mediation of organizations and organizational behavior.

Regions and Industries Linked

This project began in response to a simple question: Why did the Jura-based Swiss watch industry collapse in the 1970s in the face of discontinuous technological change? Previous periods of significant technological change and unexpected events, though threatening, had been overcome in the preceding 200 years. Why, then, did the introduction of microelectronics and the subsequent manufacture of quartz-based watches have such a profound effect on the world's premier watch industry? Why didn't the same technological thrust wreak havoc on Japan, the emerging leader of the industry? At first blush, the answer to this question is that the Swiss simply misjudged the situation, a fate reminiscent of many industry leaders that lose their way (Grove, 1996). But, as my introductory story suggests, such a conclusion is unsatisfying because it can be reduced to a simple tautology: "Leaders of industry and leading industrial regions inevitably decline—it's just a matter of time" (a point implied in Storper and Walker, 1989). This answer is unsatisfying because it is both too simple and wrong.

That all industries and regions ultimately decline is a trivial statement, one that wholly undervalues and virtually dismisses the underlying conditions that precipitated the crisis and the ultimate response to it. This usurpation of events leaves the mistaken impression that there is something pat, rote, repetitive, and inevitable about the process of change. It also succumbs to the mistaken impression that the process of economic development somehow leads to equilibrium. In fact, what is valuable about investigating an industry and the regions in which an industry embeds itself over a long period of time is the ability to see and separate out those factors that are enduring contributors to change versus those that are fleeting; those that are linked by path dependency versus those that are more deliberate and independent in outcome; those that are unique to culture versus those that are transcendent of the society in which they unfold. A long and detailed historical examination also reveals the importance of the interaction among events, events that when experienced in isolation produce one outcome but that when experienced in conjunction with other sources of change produce a very different outcome.

The answer is also wrong because it denies actors the ability to function autonomously, albeit encumbered by but not completely disabled by past experience. There are many examples that illustrate how deliberate actions and coping strategies enabled particular firms in an industry or a location to better manage the effects of either internally or externally induced change. There also are many instances where for peculiar cultural and ideological reasons some firms, industries, and places were more capable of coping with change because of their belief in organized, deliberate, purposeful response (e.g., the Japanese vs. the Americans). Inevitability presumes that agents are incapable of deliberate actions in the face of pressures and unexpected events. Inevitability also presumes that there are no differences in the reaction patterns of agents, in this case, firms in industries and the places in which firms reside. While past patterns of behavior do in fact govern much of what we too often simplistically see as the result of purposeful actions, nonetheless organizations and individuals can and do purposely weigh the consequences of different strategies and plans and, depending on their degree of self-consciousness, can and do modify the future. In part, what explains self-consciousness is idiosyncrasy, but there is a growing recognition of the importance of culture (Schoenberger, 1997), institutions (North, 1990), ideology (Samuels, 1994), and paradigms (Choi, 1993), which open new avenues of understanding and which unburden us of the need to draw quick conclusions about complex processes. Complexity can in fact help clarify what is often simplified to the point of conjecture or, worse, insignificance.

Production Organization and the Fate of Regions

Another purpose I had in writing this book was to respond to the euphoria unleashed by the proposition that in the last 20 years American Fordism, a complex institutional system of rigid mass supply and mass demand dominated by large firms, has reached its apex and has been eclipsed by something that has come to be known as post-Fordism, flexible specialization, and its variants. The new industrial paradigm offers an alternative model of organization based around small firms that, presumably because of their flexibility and implied inventiveness, are viewed as nimble competitors in an increasingly globalizing world. In this "brave new world" large firms have become dinosaurs whose bulky, overweight, and sluggish forms put them in danger of extinction. While the intense scholarly enthusiasm of the early 1980s for the passage across a "new industrial divide" has waned somewhat (thanks to a growing body of empirical work that demonstrates the contingent nature of this form of industrial success), nonetheless today there is still a great deal of loose speculation about the merits of a world that has moved away from mass production, toward a more tailored, customized system of wealth creation.

The watch industry provides a fascinating lens through which to examine many of the hypothesized relationships and outcomes implied by this new paradigm. This is so because the industry has passed through several rounds of mass and custom market structure, during which it has taken on different organizational structures dictated by the prevailing form of competition. Perhaps the two most important realms in which this book contributes to larger debates about the meaning, significance, and implications of models of capitalist development are the recognition of the strategic learning process which reveals the powerful shaping force of history and historical experience, and the distinct differences between large and small firms and their abilities to acquire new information. Secondarily, this book probes the extent of cross-fertilization and contacts, which have historically existed among emergent competitors, and has helped shape the strategic trajectory of new industrial constellations.

Heralding the benefits of small firm clusters as fonts of innovation, numerous authors (Piore and Sabel, 1984; Saxenian, 1990; Scott, 1988; Storper, 1997) have ascribed (and to a far lesser extent empirically verified) a special quality of the process by which firms learn to act strategically in a localized industry context. Missing from this reporting is a recognition of the effects of change on small firm clusters and industrial systems more generally. In fact, much of the reporting surrounding the degree of small firm resiliency ignores the fact of change. A growing body of increasingly critical evidence points out the peculiar vulnerabilities of small firm complexes in periods of rapid and unpredictable change in technology and markets (Glasmeier, 1991; Varaldo and Ferrucci, 1996). The watch industry allows us to explore the effects of change in various dimensions and clarifies the circumstances in which small firm complexes cope with unforeseen events and deliberate about their options in periods of technological and broader economic change.

The second realm to which this book contributes is an understanding of the process of economic development. Over the last 25 years there has been a strong interest in explaining the nature and tendencies of the increasingly global economy. Theories of the 1980s tended toward binary distinctions regarding the path of development of first and third world countries. The most pronounced theoretical claim, variously described as the *new international division of labor*, suggested that first world castoffs were the basis of spatial decentralization and therefore of third world industrialization (Frobel, Heinrichs, and Kreye, 1980). This characterization essentially relegated less developed countries to a position of permanent inferiority. In response to the rise of the newly industrializing countries (NICs), scholars sought additional explanations for current trends. Alternative perspectives based on the product cycle suggested that even if places were initial sites of low-skilled activity, eventually even this backward position could be surmounted through the accumulation of assets associated with increasingly more sophisticated activities (Norton and Rees, 1978). Later theoretical propositions attempting to explain how seemingly regressive countries were able to "leapfrog" from a backward to a modern state emphasized the role of human capital and explicit gestures on the part of national governments to target strategic industries. Protection for infant industries, closed markets, and high-cost domestic consumption provided the conditions industries needed to flourish and become vigorous international competitors (Amsden, 1989).

Still another perspective attempts to explain territorial development with the argument that industries "are able to create a productive capacity that did not exist before, often without very much regard to the previous conditions of the places in which they are situated. To a large degree, they provide their own impulses toward development, endogenously, in place" (Storper and Walker, 1989, p. 128). But suppose that in the life of a single industry, aspects of all of these perspectives apply. Hence, it might not be necessary to have spatial diffusion through the transmission of capital assets across space. Instead, it might be possible for diffusion to be carried forth by people rather than by branch plants. Then what appears to be wholly new manifestations of development are really just a human capital–based spatial continuum (Hoke, 1990). Earlier attempts to locate a single theoretical explanation for patterns of development have often included dismissing previous explanations that are historically constructed and still operative. In addition, there has been an unfortunate tendency to ignore attributes of the industrial development of different eras in the attempt to explain contemporary circumstances. The practice of downplaying both the experience of industrial development in different countries at different times and the degree to which they overlap avoids the complex and contingent reality of the situation.

Taking the Long View

Industrial and regional changes are processes that respond to internal and external stimuli. In both cases such powerful forces as the development of new technologies, financial crises, economic ideology, and major geopolitical events (e.g., wars) strongly influence and in many instances dictate the direction and rate of change. The way change is manifested depends upon the dominant mode or style that characterizes the production system. The ways in which independent and interdependent realms of industry and region interact with and are influenced by technological change, financial crises, ideological predispositions, and geopolitical events are the subjects of this book.

Over the last 200 years technological change has greatly influenced the spatial distribution of economic activity and the dominant mode of industrial organization. Technological change takes several forms, including both incremental and radical, the emergence of new technology systems, and the evolution of technoeconomic paradigms (Freeman and Perez, 1988). The strength and effect of these stimulants of change have dramatically different impacts on industrial systems. They are heightened, muted, and in some cases mediated by other changes operating in and shaping the environment. The severity of change is filtered through other developments occurring in the society and the economy at large which may have little or nothing to do with the specific technological change confronting the industry in question (Dosi and Orsenigo, 1988).

Financial crises, another stimulant of change, place pressures on firms and limit, restrict, or alter the ability of actors to move and operate within the environment. Financial crises also impact the flexibility with which firms can act in response to coterminous changes in the environment. Although the experiences of individual countries vary to some degree, based on the level of economic development of the particular country at the time of the crisis, nonetheless certain economic and institutional characteristics are associated with a crisis. According to Gourevitch (1986), these characteristics include "a major downturn in a regular investment/business cycle, a major change in the geographic distribution of production, and a significant growth of new products and productive processes. These three properties operate at an international level that deeply implicates domestic economies and so link conflicts over industrial responses within each country to international trends" (p. 20). Gourevitch's insightful exposition of the effects of economic crises on politics and policy has important applications to our concern about how industries and regions respond to economic crises. Economic crises result in declines in demand, which precipitate layoffs, force the redirection of resources, and in some instances result in opportunities lost.

The ways in which the effects of crises manifest themselves are dictated by and are historically contingent upon the dominant institutional character of the industrial system resident in a specific country at a single moment in time (Lazonick, 1991). In this context, *institutional character* is defined as the means of production control, production financing, and production coordination that yield a system of wealth creation through private enterprise. Part of the institutional context is the role played by the nation-state in ensuring the efficacy and sustainability of the dominant mode of coordination. Drawing from Lazonick's (1991) portrayal of the historic development of the industrial systems of Great Britain, the United States, and Japan, it is possible to show how "the very different strategies and structures of business organizations that brought first Britain and then the United States to international dominance persist as impediments to organizational and technological transformation in subsequent eras when more powerful modes of business organization had appeared on the international scene" (p. 23).

Two other perspectives are important to any explanation of industrial and regional change. The first, the role of industrial culture, becomes critical in explaining how, in the presence of high-quality and exacting information actors (in this case, firms) in an industry find it difficult to respond. In contradistinction to the neoclassical view that suggests that problems faced by industries are simply the result of incorrect prices, which once adjusted yield a balanced outcome, the cultural theory of the firm espoused by Schoenberger (1997) suggests that the failure to act comes about in situations where firms face "very deep challenges to their experience and understanding of time, space, and competition" (p. 211). As she says, "These processes, in effect, provide underpinnings of a geographically and historically specific but large-scale cultural crisis of the firm" (p. 211). Of major importance when considering the role of culture in shaping the reaction space in which firms operate is a recognition that an inability to act permeates capitalist organizations to their very lowest levels, and is transmitted through the surrounding society. As Schoenberger suggests, the prevailing industrial ideology of the day can be so pervasive and persistent that it limits the ability of even a region's citizens to see the need for change.

The second perspective, the economic ideology surrounding the militarization of technology, has profound implications for industries that create goods that are dual use: military and civilian. In consumer goods–based industries, market competition determines the trajectory and success of firms. In contrast, industries that produce products with unusual technological capabilities, which are subject to dual uses, such as watches, carry an added burden of operating within a realm marked distinctly by the ideological predisposition of the nation-state and its attitude toward the interdependence of military and civilian technology. Different attitudes regarding the role of technology in national security can result in the compartmentalization of knowledge development that distinguishes between specific wartime purposes and civilian purposes. How a nation chooses to use its industrial base in wartime efforts, on the one hand, while allowing the market to prevail as the determinant of technology development during peace time, on the other hand, has major implications for the transition between two very different periods of demand, and ultimately for the decision-making process that governs choices among the multiple technological trajectories that occur inside firms.

It is easy to say that all industry leaders and leading industrial regions ultimately fail; it is far harder to explain the range of options that govern the long-term outcome of industrial and regional development. Throughout the last 200 years, many turning points have been reached where some industrial leaders have fallen back while others have moved forward. Equally probable are cases of surprising resiliency in the face of change, where failed competitors have leapt forward after a period of serious decline to overtake challengers. These conditions lead to many questions:

- How do we explain such circumstances?
- How do we separate short-term episodic events and their influences from long-term structural change?
- How do we distinguish between durable competencies and ephemeral conditions?
- How do we explain the convergence of technique in a world that is cast as being comprised of leaders and laggards?

The history of the world watch industry presents an important, highly detailed, and animated case in which to consider these important issues.

Layout of the Book

In the next 11 chapters I attempt to capture the historical development of the world watch industry, beginning when the British were dominant in the mid-18th century. The book is organized in three parts. Chapters 2 and 3 lay out a set of five factors that have had a major influence on the competitive experiences, strategies, and major actors in the industry over the last 150 years. The history of the early industry in Britain begins in Chapter 4, first by introducing the watch itself and then by reviewing the early life of the industry. Chapter 5 introduces the Swiss watch industry and attempts to answer the question of how a poor, landlocked country came to dominate the world watch industry by the mid-19th century. Chapter 6 recounts the rise of the "American system" of watch manufacturing and documents how it came to life and swiftly challenged the Swiss position as worldwide industry leader. Chapter 7 chronicles the inexorable decline of the U.S. industry, which was brought to its knees by cutthroat competition within the domestic industry and by the techno-military ideology in which the industry was embedded.

Chapter 8 introduces the most important challenger to Swiss industry leadership in the 20th century: the Japanese watch industry. Through an examination of the history of Japan's 20th-century industrialization, I suggest how this war-torn economy was able to rise to prominence in a technology-based industry in a matter of a few decades, and to overtake the former world leader. Chapter 9 brings back into focus the Swiss industry, which throughout much of the 20th century held its position as the world's leader in watch manufacturing through clever styling, excellent marketing, and great flexibility. This position was first shaken by the Japanese in the late 1960s, and then throttled by the next major competitor on the scene, Hong Kong. The most dramatic shift in the world watch industry in the last 100 years was the introduction of the electronic watch.

Chapter 10 recounts the worldwide competition that ensued as all major watch- producing countries struggled first to cope with change and then to dominate the resulting product innovation. A new competitor emerged victorious: Hong Kong rose and swiftly dominated the low end of the watch market. How this happened and its effect on the rest of the industry forms the conclusions to Chapter 10. In Chapter 11 I return to the Swiss industry, which fought to regain a position in the low end of the industry through a clever marketing ploy based on prior investments in innovation that had been shelved for lack of an industry consensus about marketing strategy. This chapter raises the question of whether a vertically disintegrated system of production can ultimately retain market leadership squeezed on the one hand by a large low-wage, mass-producing system, such as that in Hong Kong, and on the other hand a highly mechanized, fully automated system of production, such as that in Japan. Chapter 12 brings together the long history of the industry and attempts to draw lessons from the many experiences that have unfolded over the last 250 years of watch manufacturing.

2 *The Need for and Constraints on Change*

Why do firms respond to crises as they do? Over the last decade many writers have attempted to answer this deceptively simple question. Some have concluded that firms' responses to change are based on their organizational culture (Saxenian, 1994; Schoenberger, 1997). Others suggest that the context in which the culture is created and embedded determines a firm's reaction to unexpected events (Saxenian, 1994; Gertler, 1993; Gabher, 1993). Further, according to the business strategy literature, firms respond to crises and unexpected events based on their size, sector, management structure, and whether they have an overall strategy or just operate on automatic pilot, careening from crisis to crisis (Fuellhart and Glasmeier, 1999). These perspectives help us understand the range of actions business organizations have taken in the face of needed change. But they provide little guidance about the range of possible actions firms could have taken given the environmental context. In other words, what was the backdrop in which change needed to occur and to what extent were firms constrained in their options for change?

When considering how firms might behave in a crisis, we usually assume that they will respond with a strategy. However, choice implies forethought and purposeful action in which, as Peter Gourevitch (1986) suggests, "we may find neither consciousness nor coherence"; nonetheless, to make sense of how events subsequently unfold, Gourevitch says, "we must use both notions" (p. 35). Explaining why particular outcomes occur in the historical development of an industry and its firms requires at least two intermediate steps. The first involves identifying the sources of change—both within and without the industry. Second, the prevailing industry organizational structure that represents the dominant industrial culture (no matter how loosely defined) must be isolated in order to explicate a firm's range of possible reactions. Mediating this process of give-and-take is the intensity of the stimulus for change. In some cases the intensity of affect associated with or embodied in a stimulus of change is very weak. Thus it has only limited consequences for an industry or an industrial region or culture. An example might be the setting of a standard that clarifies the function or performance of a product. Such an event may be temporarily burdensome, but usually requires only incremental adjustment to existing operations. Conversely, sources of change can be so intense that there is virtually no way to return to the status quo. In this case, adjustment requires a radical and systemic departure from past patterns of behavior. An example of this might be a technological discontinuity, which essentially makes a whole generation of sunk capital obsolete and fundamentally challenges former ways of doing things. What intensifies or moderates the effects of change are the stimulus for change and the degree of departure from past experience the deviation implies. A further moderating influence is the extent to which sources of change unfold in a singular fashion or are unleashed like an avalanche carrying away everything in their path. In the face of a cascading series of coterminous events, no single response usually proves sufficient to avert chaos.

The goal of this book is to examine how and why the development of a single industry took the form it did over time and space and how the ensuing degrees of change that confronted competitors in the watch industry both permitted and limited their ability to respond to threats. Comparisons across countries require a set of events, factors, and influences to be made apparent in such a way as to allow comparisons among the experiences of individual participants. First we need to define the basis upon which we interpret the degree of possible reactions a firm or set of firms might have had, separate from the dominant conditions of the day. In other words, what are the key elements that encourage or constrain an industry's firms (and, where appropriate, its industrial culture) from responding to crises and unexpected events? These constraints can be endogenous (specific to the industry in location) and/or exogenous (extraterritorial and structural).

Drawing inductively from the experience of industrial development of the last 250 years, in this and the next chapter I examine six constraints that influenced the responses from firms and industrial systems: technological change, economic crisis, war, technomilitary ideology, industrial paradigm, and dominant industrial culture. These six exogenous and endogenous factors are moderated or filtered through the effects of time and the relative position of countries in the evolving global economy. Each section sets forth a set of conditions that help interpret the significance of these different factors, and then offers a synthesis of their effects on the world watch industry.

There is no necessary order to the factors I consider important to interpreting the various responses of industrial systems to change. However, over the course of the last 150 years, changes in technology—incremental, radical, and ultimately system transforming—have had truly profound effects on and altered the options faced by national industries. Thus, I begin by reviewing the ways in which differing degrees of technological change influence the range of options faced by firms. My particular interest here is in the spatial manifestation of technological change, whether the effect of change is proximate to the industry in question or has a cascading effect across the entire industrial system.

Technological Change

In a useful taxonomy of innovations, Chris Freeman and Carlota Perez (1988) highlight four distinct patterns of technological change: incremental, radical, technology system transforming, and technoeconomic paradigm shifts. What distinguishes these experiences is the degree of change implied and the proximity of the effect of change—Did the change occur within the industry in question, was it peripheral to it, or did it originate in a wholly different industrial setting? To what extent was the resulting change complementary to existing practice or reflective of a complete discontinuity? Freeman and Perez's typology can be summarized as follows (pp. 45–47):

1. Incremental innovations occur almost constantly depending on "demand pressures, sociocultural factors, technological opportunities and trajectories." This type of innovation arises out of existing industrial practices usually as a result of some problem that requires immediate solution. While important in more effectively accomplishing the task at hand, incremental innovations do not alter the course of an industry's trajectory, but rather assist in the steady increase in productivity.

2. Radical innovations are discontinuous events, which in recent times are the result of deliberate R&D efforts. Radical innovations are inconstant through space and time, but usually precipitate new products and markets and ensuing capital investments. In the past they have occurred in conjunction with changes in products, processes, and organizational innovations. While they bring about dramatic change in the immediate surrounding environment, the effect is not pervasive throughout the economy unless a swarm of radical innovations occur simultaneously.

3. Technology system-transforming innovative change has an effect on more than the immediate environment and represents the conver-

gence of incremental and radical innovations combined with organizational and managerial changes. System-transforming innovative change brings together alterations in sets of interrelated industries and represents changes in a family of economic activities.

4. Technoeconomic paradigm shifts are of such major consequence that they alter the entire economy. So extensive are the effects of this transforming influence that whole new technology systems are the result. So pervasive are these shifts that whole new families of services, institutions (including distribution channels), human capital resources, and transactional environments unfold.

Of critical importance to this story, given its emphasis on how different production systems and industrial cultures respond to change, is the degree to which a geographic effect is associated with different forms of technological change. In the case of incremental and radical technological change, Freeman and Perez (1988) argue that the presumed embeddedness of the innovation most seriously influences the immediate environment. Simultaneously, the spatial effect is presumably quite pronounced because the proximate locality is the likely primary beneficiary of the development. This locality is the font and the recipient of change.

The spatial effects of technology system-transforming innovative change are not as clear-cut. The effect of a system-transforming change can be quite far-reaching, impacting residual, even peripheral, realms because of their industrial history and economic base. In fact, there is good reason to suspect that such overarching changes would have localized impacts first, given embedded and tacit knowledge, but that longer term effects would be more far-reaching (i.e., would influence more than the surrounding vicinity). The source of a techno–economic paradigm shift, however, may or may not unfold proximate to the site of its effect. What makes this latter category of change so powerful is its pervasiveness; thus its effect can be quite unpredictable. In some cases, the changes could be quite remarkable in their positive implications, while in other instances, the changes wrought could be quite debilitating.

While Freeman and Perez (1988) clearly distinguish the implied spatial relationships between the first two forms of change, the third and fourth sources have less obvious geographies. Moreover, as technology system-transforming changes and paradigm shifts occur, both contain multiple components and can have multiple effects on nonproximate environments. Thus, conceivably, their locational impacts can be quite distant from and therefore totally unexpected in intent compared with the point of their origin. Of particular relevance to this book is the summary offered by Freeman and Perez as a way of implementing their taxonomy:

New technoeconomic paradigms develop initially within the old, showing its decisive advantages during the downswing phase of the previous Kondratiev cycle. However, it becomes established as a dominant technological regime only after a crisis of structural adjustment, involving deep social and institutional changes, as well as the replacement of the motive branches of the economy. (p. 47)

They go on to suggest that the diffusion of a paradigm shift affects all facets of the economy, including best practice forms of organizational structure; skills sets of work forces; mix of products; locational patterns of investment; opportunities for small firms; strategic positions of large firms; and patterns of consumption.

Of major importance in understanding how paradigm shifts will affect existing systems of wealth-creating activities is that their greatest effects are unknowable because the production system in which they might most effectively be embedded has yet to develop (Dosi and Orsenigo, 1988). Take, for example, microelectronics and the final market of computers. Microelectronics were created long before computers came into being to serve as products that depend on them. Other sectors of the economy were the initial market targets, since computers employing microelectronics had not been developed. In this instance, tangential industries were more susceptible to the application of the new technology because they functioned in already established product areas with all the necessary links to the market, such as through distribution channels. In this case, such a product domain is an identifiable terrain that can be violated. In other words, this kind of transformation can be imagined, whereas for sectors of the future their transformation cannot be readily comprehended (e.g., watches and calculators vs. personal computers and hand-held communication devices) (Dosi and Orsenigo, 1988). Also, and significantly, at the latter stages of a product's life it is peculiarly susceptible to incremental adjustments through the application of wholly tangential developments (e.g., radios and microelectronics vs. vacuum tubes). As Freeman and Perez (1988) comment: "We are referring to a combination of interrelated product and process, technical, organizational, and managerial innovations, embodying a quantum jump in potential productivity for all or most of the economy and opening up an unusually wide range of investment and profit opportunities. Such a paradigm change implies a unique combination of decisive and technical advantages" (p. 47).

Of interest to us here is the extent to which the intensity of such a change has an original spatial effect that "leads off" and then is followed by the complementary effect of the transformation in other locations, and/or whether it is solely established settings that experience the full force of the new development. Given the importance of receptivity and the presumed nascent nature of what will one day become the rightful domain in which a technological innovation resides, it is conceivable that tangential, easily exploitable sectoral settings are readily bombarded by the effects of the innovation, long before the ultimately fundamental technology takes shape in entirely new product domains (Dosi and Orsenigo, 1988).

Let's switch gears now and suppose that the paradigm-shifting innovation is close to the heart of certain residual industries. How, then, does its introduction change the course of events? As Freeman and Perez (1988) suggest, "By paradigm change we mean precisely a radical transformation of the prevailing engineering and managerial common sense for best productivity and most profitable practice, which is applicable in almost any industry" (p. 48). In effect, Freeman and Perez are suggesting a complete upheaval and transformation of the prevailing means of doing business; there is no turning back—the future will not be similar to the past. This explanation is followed by precise conditions: the transformation produces or yields perceived low and rapidly falling relative costs. Simply stated, the outcome is far less costly than prevailing alternatives. Due to the nature of the transformative innovation, there will be an increasing abundance of supply that will drive down and maintain lower prices; abundance is both a reflection of immediate demand and a forecast regarding demand in the foreseeable future (Freeman and Perez, 1988). Thus, the forceful nature of the change is permanent and in the medium term inexhaustible.

But does such a transformation in all of its complexity come from within or from outside the established system? Freeman and Perez (1988) suggest that it occurs when "the existing paradigm or regime is almost exhausted; ... old trajectories show persistent limits to growth and future profits are seriously threatened such that the high risks and costs of trying new technologies appear as clearly justified (Arthur, 1988). And it is only after many of these trials have been obviously successful that further applications become easier and less risky investment choices" (p. 49). Choi (1993) suggests that paradigmatic upheaval is a long process of realization often led by deviant behavior that is first rejected by the dominant system, but that ultimately serves as the basis for dramatic change. Thus, in the case of industrial paradigmatic shifts, the source for change could easily arise from within the system itself, at first be rejected, and ultimately form some aspect of the stimulus for change. As the introduction to this book suggests, the invention of the tuning fork was rejected by Swiss watch manufacturers and initially viewed as an unnecessary irritant introduced by a deviant. It was not the source of a paradigm shift but simply a precursor of what was yet to come.

Of major importance given both of these perspectives is the extent to which the manifestation of paradigmatic changes must take place within the realm in which the system-transforming effect is ultimately realized. In other words, there is no necessary relationship between the geography of cause and effect. This is quite contrary to historical notions of technology diffusion, which lack sensitivity to the broad range of technological changes that can and do unfold, and the cross-sector, cross-fertilization that might occur in tandem with the development of a new innovation.

In trying to situate the effect of a paradigm shift, one must identify the prerequisite conditions leading up to pervasive acceptance of the new system. Basing their conclusions on long wave research, some critics argue that the penetration of new technologies is most effective or is subject to the least resistance during periods of economic decline when sticking with the status quo has a relatively low payoff and when the status quo represents the greatest mismatch between poor present and better future states. Freeman and Perez (1988) suggest that a paradigm shift requires "a full scale re-accommodation of social behavior and institutions to suit the requirements and the potential of a shift which has already taken place to a considerable extent in some areas of the technoeconomic sphere" (p. 59). In other words, there is often a larger trend, which precedes the widespread adoption of the new trajectory, which has to be achieved. This process of achievement occurs as a result of political search, experimentation, and adaptation.

Thus, radical technoeconomic departures from the status quo are likely to build in intensity and effectiveness, and to be accompanied by a growing and increasingly widespread deterioration in prevailing conditions. Persistent conditions cannot be unassailable; in fact, their declining credibility ultimately enables a new regime to take hold (Choi, 1993).

The presence of new technology is a profound influence on the ability of industries and industrial regions to cope with change. The intensity and speed of change has important implications for coping. The changes that are most readily aligned with the status quo will have a resounding effect within the immediate vicinity of change. Clearly, many of the small-scale but critically important innovations in watch materials, machine manufacture, and product design first had a concentrated effect on the Swiss watch industry and then radiated out to other countries' watch complexes after 1850. The Swiss ability to regain leadership several times during the 1860–1950 period was tied to the region's continuing generation of incremental innovations, which had profound effects on the vertically and horizontally fragmented industrial system. The more far-reaching and episodic changes may have little to do with the targeted outcome, and may therefore exhibit few system-adjusting characteristics (Lundvall, 1988).

Certainly, the locus of change cannot be taken for granted. The emergence of the microelectronics revolution and the introduction of quartz technology occurred far away from the Swiss watch industry. The initial effects of these paradigm-shifting innovations were not felt in the Swiss watch industry, but instead in distant U.S. electronics firms searching for a downstream market in which to embed the latest semiconductor and quartz crystal technologies. Simultaneously, and half-way around the world, Japanese watch firms after World War II—unlike the Swiss—could take up this innovation because it was derived from the existing base of knowledge and was thus neither earth-shattering nor culturally displacing. Thus the Swiss could contend with change in a highly effective manner when it originated within the Swiss industrial complex and use it as a basis for reasserting leadership; but the more radical transformation required in the face of a paradigm shift took more time and money and required serious retooling to cope. The most profound systemic changes are not likely to be local in origin. As to the impact paradigmatic changes have on the international locus of technology leadership, Freeman and Perez (1988) add: "Since the achievement of a 'good match' is a conflict-ridden process and proceeds very unevenly in differing national political and cultural contexts, this may exert a considerable influence on the changing pattern of international leadership and international patterns of diffusion" (p. 60).

Economic Crises

Economic crises have profound impacts on countries, industries, and communities. The means by which they play out and are ultimately resolved have lasting effect on previous social, economic, political, and institutional relationships. Simultaneously, the buildups leading to economic crises are rarely short-lived and can usually be traced to long and often drawn-out changes that unfold over a period of years (see Landes, 1969, p. 236, for a discussion of the events leading to the 1873 financial decline). Over the last 150 years, a series of major economic adjustments have altered the relationships among workers, employers, industries, and national governments. In some cases changes already have a foothold in industries on the eve of a downturn, while in other cases a downturn precipitates changes in an industry—particularly changes in the nature of organizational relationships. A brief review of these crises provides a framework in which to consider how different industrial systems might have been constrained to respond.

The Crisis of 1873–1896

No one likes surprises, least of all whole populations within countries used to constant economic growth over a long period. As David Landes (1969) notes in describing the deep recession of the late 19th century, "The years 1873 to 1896 seemed to many contemporaries a startling departure from historical experience. Prices fell unevenly, sporadically, but inexorably through crisis and boom—an average of about one-third on all commodities. It was the most dramatic deflation in the memory of man" (p. 231).

The 1873–1896 era was a period of great economic instability. As a result of the increasingly widespread effects of the Industrial Revolution, the accelerated diffusion of machine manufacture throughout the 19th century resulted in ever-increasing efficiency and cost reductions across industries. Prices declined and competition intensified (Gourevitch, 1986). Trade restrictions, held in check during the 1850s and early 1860s by economic growth, proliferated in a new wave of protectionism in most industrial economies, save Britain. In the previous two decades the expansion of free trade was seen to benefit everyone. Technological change applied to all industry, producing increasing profits and prosperity (Gourevitch, 1986). First Britain, then Germany pursued policies of free trade. This optimism was short-lived, however. As Gourevitch reports:

> After more than 20 years of boom, and about ten years of liberal trade, troubles began to arise which in their turn would last two decades or more. The hard times that began in 1873 derived in a direct way from the good times that preceded them. Technology and investment created overcapacity, and eventually the familiar logic of investment cycles caught up with the domestic and international economies. In that logic the race to make profit by investing in new plants so squeezes the profit margin that investment stops, and with it comes a falloff in demand which leaves all those factories with a large capacity to produce but without buyers for their products. (p. 72)

In previous periods of recession and depression, economic leaders such as Britain could retard economic decline through investments abroad and imports. But, with the internationalization of the economy and the growing pervasiveness of technological innovations and economies of scale in manufacturing and agriculture, the effects of the economic crisis of the 1870s could not be mitigated by the actions of a single nation (Sandberg, 1974).

Responses to the effect of the growing spatial division of labor generally came down to two options: free trade or protection. As Peter Gourevitch points out in his book *Politics in Hard Times* (1986), the two options would have had decidedly different effects. Free trade would have required increased mobility of labor and capital in order to allow the effects of comparative advantage to define different countries' production profiles. The alternative, protection, would have restricted cheaper goods and hopefully allowed impacted industries time to restructure and establish a better competitive footing.

A variety of positions on trade were taken by Western European countries: Britain, having the most to gain from free trade, resisted protection. Germany, having just realized the benefits of its late industrialization, strongly resisted the inflow of goods. The smaller countries, deeply dependent on trade, could do little more than stand on the sidelines while the titans divvied up the market. As for the United States, there was widespread clamoring for the protection of industry. In fact, rates of protection experienced in the 1870s were really just a continuation of rising tariffs that began with the Morrill Tariff Act of 1864, which were designed to generate revenues in support of the Civil War effort. During the early 1870s, tariff increases prevailed. Low tariffs were only kept in place on raw resources, thus reducing the material input cost to U.S. manufacturers. But on the whole, and in particular for imported high value-added manufactured goods, tariff rates hovered at 40 percent (Shannon, 1967).

Not all countries experienced the economic crisis similarly. Japan, after almost two centuries of seclusion, entered the latter part of the 19th century with an obsessive desire to reverse the backward state of its economy and bring the nation into parallel with the Western countries. Restricted from raising tariffs above 5 percent, Japan was vulnerable to the effects of falling prices on commodity and consumer goods in the West. The declining price of cotton virtually decimated the Japanese cotton industry; thereafter Japanese manufacturers turned toward silk production with a vengeance to take up the slack (Allen, 1928). But the declining price of Western manufactured goods worked to the advantage of the Japanese, who were focused on catching up with the West. Technology was imported and foreign workers hired to facilitate technology transfer; their acquisition was no doubt enhanced by the falling prices of goods (Odagiri and Goto, 1993). Capital investment made by the government supported the development of new business ventures. Thus, the economic crisis of the late 19th century, while problematic, had a far less devastating effect on Japan than on her Western industrial rivals. Japan's march toward modernization conveniently coincided with a period of deflation and the stagnation of her rivals' economies.[1]

The effect of the 1870s–1890s crisis was very much felt by industries and communities in which manufacturing plants and agricultural commodities generated tradable goods. The political response by continental

European countries and the United States to changing international economic conditions was predominantly to protect the gains that had been made by domestic industries as they attempted to "catch up" with Britain's industries or received favorable treatment under laws that supported infant industries. Corporate interests in contradistinction with proprietary capitalists worried about their exposure given the rising level of capital in more capital-intensive industries (Gourevitch, 1986). Halfway around the world, Japan, unable to close its borders to trade, adapted to this and in some cases abandoned budding industries, while benefiting from low-priced strategic commodities that were essential to playing catchup.

The late-19th-century economic crisis precipitated differential responses in support of a country's watch industry. Despite some calls for protection, in Great Britain the overwhelming direction of the economy was toward free trade. Thus British-made watches were not accorded effective protection against the onslaught of the Swiss or the U.S. watch industries. While the British government could try to keep contraband goods out, this had little effect on preserving overseas markets. The United States, which deliberately used tariffs on virtually all industries as a means of financing the national government, quite comfortably used trade barriers to protect its infant watch industry. It was obvious to all after the worlds fair in Philadelphia that watch manufacturing capability was a key element of a nation's industrial prowess and military might. Japan could neither use tariffs to restrict market access nor protect its infant industries. Indeed, its actions during the crisis were quite the opposite. Given an undervalued currency, a deflationary cycle yielding cheap commodities, and a relative lock on Asia's growing markets, Japan used the economic crisis at the end of the last century to buy new technology from abroad to build up basic industries needed for later industrialization.

The Crisis of the 1920s and the Great Depression

There are a multitude of explanations for the economic crisis of the late 1920s and the Great Depression. European countries experienced the 1920s in shades of similarity. Britain suffered from declines in labor-intensive sectors that faced international competition. Unemployment was uneven but never fell to prewar levels. Several important industries were confronted with major dislocation and decline after years of neglect and low capital investment. Labor disputes ensued as livelihoods collapsed under the weight of economic instability. Twinned with these structural problems were the long-term effects of World War I, during which industries were commandeered for the war effort and then left to recover on their own amid chaotic and increasingly trade-sensitive markets.

Germany followed a similar path, though the political fallout of the financial crash was far more extreme. Germany's postwar shortage of capital, initially filled by U.S. investments, became critical in the late 1920s, when U.S. investors became more and more vulnerable to declines in the domestic economy. German firms trying to hold on had to cut payrolls drastically and had to liquidate their assets in a desperate effort to remain in business.

Since the time of the Great Depression the United States accounted for approximately half of the industrial world's output, its role in the crash cannot be underestimated (Kenwood and Loughheed, 1992). The search for single explanations of the United States's role in the decline have busied analysts for years, but important to us here is the effect of the crash on industries, whole regions, and labor. In the United States, industries with greater international exposure battled with domestic industries that were aligned with labor unions and citizens groups to protect themselves against foreign imports (Gourevitch, 1986). The United States, by restricting access to its huge market, wreaked havoc on the smaller, export-dependent European countries. In addition to trade-sensitive industries, luxury goods industries also experienced serious declines in demand. First, wage cuts were tried, with some industries imposing cuts of 30 percent or more. But eventually cuts proved insufficient to rectify the free fall of decline in demand and major job loss resulted (Moore, 1945). Durable consumer goods were also affected as demand peaked and markets became saturated. Decline in demand was attributed to the growing maldistribution of gains from productivity increases which "tended to favor profits over wages" (Kenwood and Loughheed, 1992, p. 225).

Japan did not escape unscathed from the world financial crisis. From the beginning of the 20th century, Japan had embarked upon an explicit program to diversify her economy and free it from its total dependence on textile exports. By the eve of the Great Depression, the country had successfully shifted its output away from textiles, which comprised more than 50 percent of all traded goods in the 1910s but amounted to less than 30 percent by the early 1930s. The metal goods and food products industries took up the slack (Lockwood, 1954). When the global depression hit, Japan, like other countries, found the U.S. market closed off to its trade. In Japan, this took a heavy toll, given that the United States and China accounted for more than half the country's exports. The effect of the Great Depression was minimally muted by the fact that Japan's economy was dualistic. A significant portion of the country's workers were employed in traditional industries producing for local consumption;

these industries and workers were less affected by international trade. But even the movement toward diversification and the reliance on a dualistic economy proved insufficient to protect Japanese workers and firms from international economic decline. Faced with the inevitable need to cut wages, employers increased working hours and eventually were forced to let workers go (Allen, 1940). The major traded goods sectors such as silk were devastated, forcing a massive reallocation of productive factors toward other industries. Monetary changes also were required; Japan was eventually forced to devalue her currency in order to achieve a reasonable balance of payments. This devaluation proved fortuitous inasmuch as in subsequent years a low-value currency allowed the country to capture new markets through the sale of low-cost goods.

The Swiss, much like firms in the United States, were seriously harmed by the decline in global demand for exports. On the eve of the Great Depression, the Swiss watch industry was more capital-intensive and hence had much more to lose than only two decades earlier. The costly move into wristwatches that was needed to retake the American market put many family firms' fortunes at risk. The intensive effects of the Great Depression led to industry-wide regulation as a means of coping with economic crisis. Such avenues were not available to American watch firms, which were still recovering from the effects of World War I. As for the Japanese, the winds of war were building as the global economic crisis wreaked havoc on Japan's planned industrial ascendancy. Industries were selected and foreign exchange was metered out in line with the national goal to become an economic powerhouse. Watches were a privileged product, not only for their functionality but also for the competency small-scale precision manufacturing prowess contributed to the production of war materiel.

The Great Depression brought ruin to many people, firms, and communities through its pervasive nature, including massive wage reductions, a complete end of capital investment in established lines of industries, the closure of traditional markets, the destruction of labor relations, protectionist sentiments that closed off primary markets, and the pileup of inventory. No region went unharmed. Hairline cracks in industries and the industrial relationships within them turned into mile-wide fissures of discontent and asset devaluation as the global economy stagnated.

The Crisis of 1929–1949:
The Post-Depression and World War II Compromises

While the phenomenon of war is treated here separately as an influence on the evolution of industries, the period following the Great Depression is not easily separated from the effects of World War II. The political crisis created by the buildup to and the waging of World War II had profound effects on the world economy and the major national interests of nations on both sides. Of particular importance to our story is the extent to which the devastation of the Great Depression, which was followed by the war buildup, precipitated social and economic relations of a specific form. As Gourevitch (1986) suggests, "Labor accepted capitalist management of an economy run on the basis of market incentives (even nationalized industries) in exchange for a welfare system, high wages, employment-oriented macroeconomic policy, and constitutional protection of organizing rights" (p. 29). Business, too, accepted these compromises in return for a relatively stable environment, including growing free trade. In the three macroregions of study (the United States, Japan, and Switzerland), industrial capitalists sought to control the amplitude of uncertainty and unbridled competition by forming various versions of cartels. In some cases the gesture was quite powerful in its effect, while in others it was unsuccessful in providing a stable environment. The crisis really unfolded as countries' economies were pushed and in some cases commandeered into supporting the war effort, pulling industries off course as the need for military materiel grew.

As for Japan, the post-Depression years reflected a growing tension between the civilian and the military economies. As C. G. Allen noted in 1940, "Throughout 1929–1937 development of the industrial system was profoundly affected by the efforts of powerful groups in Japan (especially the Army) to direct the country's resources along lines determined by strategic and military considerations, and to impose over many kinds of enterprise a national control with the same ends in view" (p. 70). Coupled with this growing tension was the effect of recent past productivity, which led to increasing investments that positioned many Japanese industries to become global leaders (Lockwood, 1954; Suzuki, 1994). This was true for strategic industries, such as metals and chemicals, but also for consumer goods industries, such as textiles (Odagiri and Goto, 1993). It is correct to say that by the early 1930s, despite the Great Depression, Japanese industries were well on their way to becoming strong international competitors with technological parity in a number of vital industries; thus, even though the war devastated the physical infrastructure of the country, the lessons previously learned could and were easily reactivated after the war.

Thus, the interwar years were characterized by a long period of slow recovery and marked economic instability. The effects of the Great Depression took a decade to slough off. In some sectors, the depressing conditions were never overcome. The World War II effort distorted the experience of the major watchmaking regions' industries. The Swiss, having grown rich from wartime production, missed the boat in terms of technological change stimulated by the war effort, particularly in the areas of radar, microelectronics, and computers. Japan, in contrast, despite suffering extraordinary economic devastation, retained, and in some realms even strengthened, its ability to work in advanced areas of electrical and mechanical product development. Japan's major watch firms had moved their production operations out of Tokyo, thus preserving manufacturing capability. The U.S. industry, commandeered for the war effort and enticed by cost-plus contracts to produce for various branches of the military, woke the day after the war ended to find an abrupt end to orders for war materiel. Firms had gained considerable know-how from the science needed to fuel the war effort. But with virtually no assistance in the difficult adjustment process back to civilian production priorities, the leading American firms found it difficult to capitalize on their newfound knowledge and competency. Calls for industry protection in the form of increased tariffs to break the lock on the U.S. market gained by the Swiss during the war provided a momentary and largely illusory respite from unrelenting competition.

The 1950s: Rebuilding after the War

World War II gave everyone involved some things they had not planned for: inflation, material shortages, redirected industries, and pent-up consumer demand. As they came out of the war, the major Western powers exhibited an intense desire to move quickly toward a new future by putting the war behind them. Many things can be said about this era, but aside from the Korean War it was not characterized by crisis. These years were unique in providing a period of recovery for the Western world through the rebuilding of war-torn economies, and for Japan through contracts to supply material goods for its part in the Korean War. For European countries involved in but not mauled by World War II, these years were a period of relative calm. Markets were largely captive, global demand was booming, and few worrisome threats appeared on the horizon.

In these years, Western trading partnerships were being extended into Asia. America in particular was doling out economic development in exchange for democratic vows (Samuels, 1994). Japan, once an unruly warring competitor, now appeared as a tame ally, becoming a military materiel depot and the closest captive geopolitical partner of the United States. The importance of trade policy as security policy began to take shape at this time (Glasmeier, Thompson, and Kays, 1993).

Although it had suffered profound devastation, Europe made a surprisingly rapid recovery. Many countries outside the direct line of fire enjoyed bountiful growth as the war-torn economies of the primary aggressors were rebuilt. While it is incorrect to say that the situation returned to prewar conditions, the lack of uncertainty and fear made the 1950s a decade to remember.

Devastation was everywhere in Japan, but so was assistance in recovery efforts (Redford, 1980). The United States had a tremendous stake in Japan, and sought to stabilize the region at all costs. Significant technology transfer helped reinvigorate the economy and rekindle the formation of a technology-based economy. Rehabilitation of some industries occurred, while others benefited from the tremendous learning that had taken place during the buildup associated with World War II. Many strategic industries got their start as part of the war effort, while others received sharp infusions of capital (Ohawa and Rosovsky, 1973). The war had clearly decimated some industries, particularly those located in Tokyo, but others removed from the site of extreme battle to the countryside were well positioned to take up consumer production anew, after the armistice was signed.

The 1970s: The Oil Crisis, Inflation, and Monetary Instability

"The golden years," like all good things, eventually came to an end. In previous decades, a long, slow buildup eventually unleashed a new crisis, but the 1970s were different. Almost overnight the cost basis of business worldwide virtually doubled as the price of oil doubled, redoubled, and doubled again. The resulting economic onslaught was compressed and surgical in effect, as the developed nations' economies contracted in response to this supply-side inflationary shock (Brenner, 1998). Other contributing factors were equally sharp in their effect: high interest rates, declining productivity, increasing foreign trade competition, rising wage levels, and, most specifically, falling rates of profit.

The United States had caught a serious cold, and the rest of the world, except for Japan and Germany, was on the verge of pneumonia. Japan and Germany were alarmingly nimble competitors that had begun to press the U.S. hegemon to large effect (Brenner, 1998). Germany produced products that were a good value, while Japan's undervalued currency made everything that it produced seem like it was on sale. The United States, mired in a banking crisis—the result of the halcyon days of the postwar run-up of stock and real estate values—could do nothing more creative than set up a protectionist wall in reaction to the rise of new agile competitors. Switzerland, at the time the most widely recognized haven for international capital, had to revalue its currency in response to the huge inflow of petrodollars seeking a safe harbor during a period of unprecedented inflation. What was good for the Swiss banks was devastating for Swiss manufacturing industries, particularly in export sectors like watches.

For places outside, but within the influence of, America's orbit and increasingly that of the global economy, attempts to defend domestic currencies proved challenging. Traded goods were even harder to defend as new low-cost enclaves emerged as formidable competitors. Being low cost was paired more and more often with high skills (Brenner, 1998). Industrialized economies were increasingly confronting adversaries with strong state support, a trained work force, and high levels of productivity. Decades of success in delivering a high and rising standard of living suddenly became a nemesis. Being on top meant having everything to lose.

While economic crises arise for a variety untold sets of reasons, their outcomes are remarkably similar in effect: displacement, new competitors, new systems, and new geographies of competition. Tracing out the experiences of the last 100 years in an admittedly abbreviated form suggests that your starting point is important and that distinct institutional and cultural proclivities have strongly influenced your experiences along the way.

War and Geopolitical Conflicts

In virtually any text written about industrial history, the role of war as driver or source of industrial change punctuates the progression of events. Wars and war-associated activities alter pathways of trade, speed technological change and technology development, impact the existence and structure of institutions, change the allocation of natural resources, and in many cases shift the spatial distribution of economic activity. War is about combatants, those seeking an end to the altercation and those promoting it, struggling over territory and resources (Sandholtz et al., 1992). Given the usually numerous parties to such conflicts, the effect of war impacts a country's national economic system in multifold ways depending on its geopolitical stance in the conflict (belligerent, non-belligerent, neutral) and its previous state of development.

The Effect of War on Trade

Over the last 150 years, five war-related events have had major impacts on the spatial distribution of economic activities. Four of the five were international in nature: World War I, World War II, the Korean War, and the Viet Nam War. All transformed economic activity. In contrast, the effects of domestically based wars, such as the U.S. Civil War, are seen predominantly in their effect on trade. In the case of the Civil War, two factors were paramount in transforming the international competitive terrain: the restriction of imported manufactured goods into the U.S. market, and the huge stimulus provided to domestic industries that produced goods for the military. Regarding the first, tariff restrictions already in place were strengthened and in some cases raised by the federal government as it sought to increase revenue to finance the war. Average tariffs were 40 percent, and some imported goods were banned outright. In the second, specialty materials and goods, including cotton fabrics, weapons, explosives, and other materiel, experienced huge increases in demand, fostering the development of domestic industries to supply such goods.

The Effect of War on Technological Progress

As more parties to a conflict emerge, the effect of war on industrial change takes on a peculiar pattern of behavior depending on the geopolitical position of the combatants and neutral parties. The strategic nature of an industry in a combatant's war effort can strongly influence the speed and direction of technological change within that industry. Noted economic historian David Landes (1969) describes the effect of war on European industrial development:

> The effect of war on technological progress and economic development is a moot subject, and economic historians have not found it difficult to adduce evidence in support of both the optimistic and pessimistic point of view. The explanation of these contradictory data is simple enough: the fact is that war serves both to promote and to impede innovation and growth, and there is no a priori reason to assume that the balance will fall on one side or the other of the ledger. Nor do we have at the moment the techniques or information needed to establish a balance sheet for any given conflict; the less so, as the problem is complicated by the difficulty of disentangling war from the many other forces influencing the economic conjuncture in a given place and time. In effect, one has to conjecture what the economy might have done had peace prevailed. (pp. 421–422)

Of peculiar importance in the case of watches is both the technology itself and the role chief competitors played in the major wars of the 20th century (Gordon and Dangerfield, 1947). The key producers played many roles, shaped by their geopolitical positions in the conflict. In the United States and Great Britain, industries were commandeered in support of the war effort. In Japan, an industry was totally redirected for the war effort. Despite the fact that the industries in all three countries were unilaterally captured in support of the war effort, the effect of this change was quite different in all three cases. A fourth key player, Switzerland, was a neutral; although its industry was not expropriated in the service of her national government, nonetheless its juxtaposition between the Allies and the Axis powers had an unprecedented influence on the industry's development during World War II. As one Swiss watchmaker said, "We sold to both sides. To the British we sold parts for navigational systems for British warplanes; to the Germans we sold fuses for bombs. To America we sold watches, as their industry was focused on the war effort. Being neutral, we had no choice. We ran way over normal capacity. It was an unusual time; fortunes were made by many" (L. Tissot, personal communication, 1989).

The effects of expansion for war on an industry depend sharply on the extent to which the good pressed into service is essential in the execution of the struggle. Again, Landes (1969) provides poignant insight into how war changes the fate of industry:

> The stimulatory effect of war takes the form of sharply increased demand for certain goods and services, a demand that presses against severe constraints of supply—shortages of labor, real capital, and raw materials. Those of the older industries producing "nonessential" consumer goods are often compelled, either by fiat of the state or resource bottlenecks, to curtail output and investment. Those of the older industries producing for the war effort are in a better position to profit from heightened demand; but their expansion does not necessarily entail advances in technique or improvements in equipment. On the contrary, their first recourse will be to such unemployed capacity as may have existed before the increase in demand. This may take the form of stand-by machines of somewhat lower efficiency or even obsolete equipment resurrected for the emergency; so that the war may actually promote a kind of technological retrogression. Even when new plant is required and the resources are made available, the gain in productivity may not be so great as it would ordinarily; for time is short, and the quickest solution to may problems is the tried-and-true technique. In general, expansion of output in these older branches will take the form of capital widening rather than deepening. (p. 422)

These gains, however limited or contrived they might be, must in the end be juxtaposed with the dislocations and effects of reconversion to peacetime use, which can, depending on the length of the war, be crippling. Reconversion in the absence of governmental assistance, as Ann Markusen and Joel Yudken (1992) argue in their book about dismantling the cold war economy, may be unattainable given the cultural and interinstitutional changes that often occur in commercial firms called into military service. Problems of reconversion arise because labor skills, marketing, distribution, and materials acquisition are all affected during a buildup, so much so that it proves difficult to return to prewar conditions (Markusen and Yudken, 1992). The ability to return to normalcy is to some degree governed by the technomilitary ideology of a country.

At the same time Landes (1969) makes a second point worth underscoring here. During times of war, given expensive and uncertain choices of technology and the implied scale of investments that need to be made over a short period of time, capitalists may not make major capital investments despite what appears to be the long-run productivity-increasing benefits of such efforts. As Wolfram Fischer (1991) comments regarding the resistance of early entrepreneurs to the adoption of new techniques,

> Early entrepreneurs did calculate. Their calculations may have been rudimentary by twentieth-century standards, but they calculated the costs of their investments (including interest rates), of raw materials and wages, and they calculated probable returns. The commercially oriented entrepreneur would soon find out that he could do without greater investments if he saved on costs, used written off machinery until its physical death, employed less or cheaper labor (in this case women), and bought his raw materials at other places or switched over to more lucrative product lines. Fairly often he would open up new establishments in regions of cheap labor and waterpower, or even close down a factory when costs ran too high. (p. 147)

This point, referenced at length, is all the more acute in the case of war. Partly because of the lack of additional equipment to bring into operation (either because of shortages or restrictions on the use of resources) and partly because of the incredible uncertainty of a war effort, capitalists are perhaps even less inclined to make new investments during such periods of heightened insecurity. The upside of this forethought is the low risk of capital; the downside is that competitors, given different initial conditions, may have been forced to make such investments. Thus once the war is over, the penny-wise entrepreneur can turn out to be the pound-foolish businessman if his competitor leaps ahead with new, productive technology that enables lower costs.

The Role of Technomilitary Ideology on Industrial Change

While war dramatically directs and redirects the flow of resources and goods among parties to the conflict, the ideological basis of national perceptions of the relationship between economy and security shapes the intellectual basis of industrial problem solving.

"Technomilitary ideology" describes a nation's system of beliefs and its understanding of the relationship between military interests and the evolution and development of a national industrial system of wealth creation. In the last 20 years Japan's ascendancy as a world economic power has precipitated significant research aimed at understanding the difference between American and Japanese perceptions of technology and security and how this has influenced the technological rivalry between the two countries. As Richard Samuels (1994) writes in *Rich Nation, Strong Army*, an economic history of the Japanese military–industrial culture, "The institutional development of whole economies (and thereby the trajectories of innovation and growth) depends on the way technology is understood strategically and the role it plays ideologically" (p. 3). In contrast to the Japanese, who made no distinction between national defense objectives and economic interests,

> Cold War America defended territory but acted as if its economic interests were a subordinate component of national security. . . . The Japanese were never taught to make such a distinction between national defense objectives and economic interests and indeed reject such arguments as naive wherever they are presented. . . . Two different ways of approaching civilian and military technology development thus grew from the divergent ideas that the United States and Japan had about national interests. . . . In the United States during the Cold War, private research and development grew considerably smaller relative to public R&D. In Japan, the opposite was true. . . . The Japanese may have demonstrated, like the Venetians and the Dutch before them, that butter is as likely as guns to make a nation strong and, further, that nations cannot be strong without advanced technology. . . . In essence, the Japanese story is one in which ideology and institutions are linked, shaping strategic choices based on different conceptions of national interests than are widely accepted in the United States. (pp. 3–4)

Thus when it came to industrial reconversion after the wars of the 20th century, U.S. firms were left to figure out how to manage and finance the process of change. Military R&D, which had been so central to military production, shaped domestic R&D activity. The search for and selection of solutions to production problems took on a decidedly complex form.

In the case of watches, competitors could be and were in decidedly different positions after World War II. The Swiss, for example, had little trouble reconverting given their maintenance of civilian production during the wars; their military markets demanded it. No markets had been lost; in fact, the U.S. market in both World Wars had been retaken from U.S. firms preoccupied with military production. At the same time, there was no technology push from the war effort to propel the Swiss into new technological realms that might have led to the development of new products. Capital investment was kept to a minimum and women were employed to assemble products. Thus, in some unusual sense, being neutral denied the Swiss the benefits of technological developments from World War II.

The same was not the case in Japan. Japanese firms integrated military production with consumer-based production spheres. Thus, conversion to prewar conditions was managed literally overnight. More importantly, as technological change began to press upon the watch industry and new technology paths had to be chosen, the Japanese were not divided about the direction of change because competitors were few and conceptions of industry needs and military needs were one and the same. Thus, at the end of the war, new developments were quite naturally embedded in the postwar reemergence of the industry.

On the other hand, in the early days of the digital age in the United States, the competing interests of the consumer and the military–industrial sectors strained and eventually elongated the process of technology selection, thereby obscuring the direction the watch industry would ultimately take (Numagami, 1996).

Perhaps the more complex intersection ultimately existed between the Swiss position of neutrality and the lack of national direction or encouragement toward the pursuit of a new technology that was highly linked to the strategic concerns of competitor states. In other words, given that the impetus for the electronics era originated outside the specific need to tell time, the lack of any ideological predilection toward either the integration of military and civilian activities, or a single-minded emphasis on technological mastery for the sake of strategic superiority, left the Swiss out in the cold.

The point here has been that wars act as focusing devices that shift resources in multiple directions, sometimes leading to technological advances, but sometimes limiting industrial change. In the case of advanced technology sectors, linked as they often are to military prowess, the ideological belief system surrounding the relationship between the military and industry might strongly influences domestic industries' response to both the demands of war and the return to civilian production once hostilities conclude. What appears at first blush to be a failure of technology selection obscures the more fundamental source of the problem: the motivational goal of achieving a new technological capability transcendent of a specific industrial application.

Summary

Why does the same industry take different forms in different countries and economies? How does the moment in time in which an industry emerges affect its relative position in the evolving global competition? How does the original system of organization ultimately influence the long-term pattern of industrial development? What are the privileges and penalties that accompany the selection of a competitive strategy? Does the selection of competitive structure predestine the response options firms have at their disposal? Can a new competitor emerge from thin air or must there be preconditions that presage the longer term outcome?

While in the previous discussions of exogenous factors I have emphasized discontinuous events, I now wish to turn toward a discussion of endogenous conditions that evolve in conjunction with the emergence or the evolution of new participants in a global industry. For a host of reasons—including resource distribution, geography, international access, stage of development, and culture—industries do not take shape in the same way or at the same time in all places. Nonetheless, there are preconditions which, if unmet, are likely to thwart or retard development (Landes, 1969). Of primary importance are the standard of living and the stock of accumulated wealth and institutional capacity that undergird a country's prospective development. Over a 200-year history of industrialization, initial conditions strongly suggest the importance of the relative placement of different countries through time (Maddison, 1991; Williamson, 1991). This latter distinction is particularly important when discussing the evolution of an industry and the emergence of new competitors that eventually overtake the status quo. While a considerable degree of surprise is accorded the emergence of Japan and the NICs of the last 20 years, their ascendance cannot be viewed as unprecedented, as some authors suggest. To the contrary, as with all of the participants in the world watch industry, nascent or latent tendencies existed long before the emergence of the industry, presaging its development. Just as an example, more than 300 years ago, Japan was noted for its exquisite ability to work at small scale and with fine attention to detail in the construction of automata, small objects, powered by micromechanics. The ability to work with high precision at small scale is the requisite skill of the world's centers of watch manufacturing. Thus the ability to perform and the necessary skill predate the emergence of the Japanese watch industry.

Preconditions become all the more important when a country comes of age and industrializes. There is considerable evidence of and interest in how backward or latecomer regions and countries race ahead of preexisting leaders. Alexander Gershenkron[2] made the point that while late developers face many difficulties, nonetheless there are substantial benefits to being second (Gerschenkron, 1962)—particularly in terms of learning from the successes and failures of one's predecessors and to exploit an accumulated pool of knowledge and technique (Landes, 1991). But, just as there are benefits, so there are costs. Landes and others note that coming second requires more resources, more focused effort, and more control over actions and outcomes. If the capitalist class is incapable of organizing itself to bring about development, the state must be still stronger in its efforts. Clearly, Gerschenkron left many ill-defined points in his argument, such as questions about scale, scope, and institutional prerequisites. Nor did he mention what might trigger the conditions by which leapfrogging could occur. Of interest to us is the extent to which an event such as war and the technoeconomic paradigm of a country focuses resources around key strategic technologies, becoming a basis, for example, of Japan's push forward in the development of quartz technology.

Although this book's focus is the evolution of the watch industry, beginning with the British industry, nonetheless the antecedents of the structural form of the industry in Britain date back to the beginning of the industry and the development of the personal timekeeper. Watch manufacturing evolved through five distinct stages: craft; putting out, or commissioned labor; protofactory; mass production; and vertical integration. Each of these stages corresponds to a particular labor process, material acquisition and distribution system, mode of production, institutional structure and mode of coordination, and methodology of technological change. Given the array of associated factors that must align with a mode of production organization to facilitate effective coping and change, it is likely that lags in one could in fact become the impediment to the realization of another's potential (Sabel and Zeitlin, 1997; Scranton, 1997). For example, rigidities in a distribution system may thwart the realization of gains from technological innovation. At the same time, a lag in the constellation of factors forces firms to make adjustments that may be viewed as irrational initially but that can become defining moments for an entirely new trajectory within an industry. Many examples exist where the more deliberative Japanese took a wait-and-see attitude, which produced just enough of a lag that it allowed other components of a manufacturing process to catch up and come into sync with others (Numagami, 1996). It is important to recognize that the level of complexity in and the multifold forms of coordination that are necessary for textbook market-based solutions to unfold are an admitted rarity in the history of industry (see Fischer, 1991, for important examples of why waiting makes sense).

Like other industries (Sabel and Zeitlin, 1997), the history of the watch industry is anything but an example of pure forms of flexibility or mass production. Indeed, as Sabel and Zeitlin (1997) point out, after years of rigid adherence to the world as one immutable form of production organization or another, they too have come to believe that

> the experience of fragility and mutability, which seems so novel and disorienting today has been, in fact, the definitive experience of the economic actors in many sectors, countries and epochs in the history of industrial capitalism. Precisely because they [firms] have been aware of the complex

dependence of every form of economic organization on multiple shifting background conditions, they have constantly experimented with institutional designs that until recently would have been judged economic solecisms. For the same reasons, they have rarely interpreted economic and technological progress as continual and ineluctable progression towards a single set of practices that in their self-perfection would ultimately pass into a sphere of transhistorical permanency. But this double perception of mutability and fragility, we will argue further, has not led them to exalt catch as catch-can muddling through as the organizing principle of reflection and action. What we find instead is an extraordinarily judicious, well-informed and continuing debate within firms, and between them and public authorities, as to the appropriate responses to an economy whose future is uncertain, but whose boundary conditions at least in the middle term are taken to be clear. (p. 3)

Competition takes place in a context in which deviations from the immutable are commonplace and may in fact be the status quo. Thus, to understand why a specific situation works out the way it does requires a complex understanding of the various constraints and contingencies that an industry and firms within it face at any point in time.

3 The Organizational Development of the World Watch Industry

The world watch industry has progressed through several distinct eras of industrial development since the 17th century, when regional industrial fortunes were tied to craft-based organizations and guilds. As the industry progressed and changed from a craft-based system reliant on the skills of masters and their helpers to one more fragmented and decentralized around a system of putting out, or commissioned labor, new social structures and industrial cultures emerged. With the eventual collapse of putting out and the formation of an urban protofactory system and then a more formal factory-based system, social relations and industrial structure once again changed. Now the industry relied on and was structured around greater task specialization and deskilling of the work process. By the middle of the 19th century, yet another evolutionary step was taken. The production process changed and the scale of operation was modified toward higher levels of vertical integration. In response to the ensuing structural changes, major producers of watches responded incrementally, taking on those facets of required change that best fit their existing mode of operation. This incrementalist tendency led to both favorable and problematic outcomes, depending on the level of change required. The more profound the effects of a called-for change, the greater its challenge to the status quo. As this chapter suggests, organizational learning is highly incremental and history-dependent. The greater the difference between new and past patterns and practices, the more difficult change becomes. The deeper the cultural norms, the more resistant to change they are.

The Organizational Structure of Watchmaking in the Craft Era

As with other manufactured luxury goods, the production of watches in the 15th century unfolded within a system of handicraft manufacture under the direction of master craftsmen. An exceedingly clever and dexterous person, the master was "assisted by one or more journeymen or apprentices" (Landes, 1969, p. 43). This system of relative individual independence did not remain immutable, however, as Landes notes:

> Fairly early, however, as far back as the 13th century, this independence broke down in many areas, and the artisan found himself bound to the merchant who supplied the raw materials and sold his finished work. This subordination of the producer to the intermediary ([or, less often] of weak producers to strong ones) was a consequence of the growth of the market. Whereas once the artisan worked for a local clientele, a small but fairly stable group that was bound to him personally as well as by pecuniary interest, he now came to depend on sales through middlemen in distant, competitive markets. (p. 43)

The regulation of this form of production rested with the *guilds*, production organizations that regulated skills and access to skill, and to a much lesser extent access to markets. By restricting access to skills, craftsmen were able to preserve their markets and the public credibility of their products. Guilds acted as brakes on the expansion of output and thus overproduction. In later periods, guilds also thwarted technological change, a mistake that ultimately led to their own demise.

The movement from craft production to the putting-out system was an outgrowth of the merchants' growing control over producers. The expansion of the market and the lack of access to and familiarity with its extensiveness placed artisans at a decided disadvantage vis-à-vis merchants. Market reach and knowledge of distant tastes shifted the advantage to merchants who could accumulate demand and periodically fill it (Glasmeier, 1990). Writing of this time of transition, Landes (1969) details the advantages accorded the merchants:

> Only the merchant could respond to the ebb and flow of demand, calling for changes in the nature of the final product to meet consumer tastes, recruiting additional labor when necessary, supplying tools as well as materials to potential artisans. It was largely in this way that the rural population was drawn into the productive circuit. Very early, urban merchants came to realize that the countryside was a reservoir of cheap labor; peasants eager to eke out the meager income of the land by working in the off-season, wives and children with free time to prepare the man's work and assist him in his

> task. And though the country weaver, nail maker, or cutler was less skilled than the guildsman or journeyman of the town, he was less expensive, for the marginal utility of his free time was, initially, at least, low, and his agricultural resources, however modest, enabled him to get by on that much less additional income. Furthermore, rural putting out was free of guild restrictions on the nature of the product, the techniques of manufacture, and the size of enterprise. (p. 44)

This new putting-out system was marked by substantial flexibility and innovation, as Landes (1969) points out in describing the woolen industry of Britain in the late 1700s: "In particular it was free to develop cheaper fabrics, perhaps less sturdy than the traditional broadcloths and stuffs, but usable and often more comfortable. This freedom to adjust and innovate is particularly important in light industry, where resources and similar material considerations often are less important as locational factors than entrepreneurship" (p. 45).

Putting out, while highly efficient in an era with few economic alternatives, grew less and less viable as an organizing system over time. As workers grew wealthier and additional avenues of income creation emerged, merchants found it increasingly difficult to control output and quality. It became harder to meet increased demand when workers began to assert more control over their lives and the labor process. The need to find an alternative, more controllable system of manufacture grew overwhelming, as markets grew and demand burgeoned. In Britain, the geographic range of the putting-out system had reached its limits. Transport costs became a real impediment to profits.

Lacking control of workers' efforts and reaching the limits of geographic expansion, employers slowly but steadily shifted to a new system in which machines began to replace human labor. Machines enabled employers to organize labor within specific locales and structures, thereby providing them with more control over their workers. This transformation began and grew strongest in Britain, particularly in textiles. A lag in implementing the new system on the Continent later proved to be central to the emergence of the Swiss as watch industry leaders in the 1800s.[1] Unfortunately, the British watch industry failed to heed the wider trend toward standardization and mechanization unfolding in such industries as textiles and machinery, while still contending with higher labor costs compared with other continental countries. As is often the case, the transmission of a new innovation is neither total nor taken up by everyone at the same time. Industry leaders often have little incentive to follow the innovation frontier, given the builtup competency and nonproduction-related relationships in distribution and supply channels already in place. The lag made inevitable by sunk costs that are never truly written off becomes a weight that can in certain circumstances drag down the industry leader.

The movement to a factory-based system of manufacture was neither rapid nor total. The slow demise of the putting-out system provides one explanation for the emergence of factories: the need to control labor and the production process. Volumes have been written on this subject, from a wide variety of perspectives and with an equally broad number of emphases. What is important to note here is the way the movement from a more decentralized to a more centralized system removed control of production from workers and turned it over to the capitalist. Of additional importance is the recognition that the process of change did not occur in the same way simultaneously in all relevant locations. Nor did it occur comprehensively even within a single location, as later discussion points out. In fact, as Landes (1969) notes, the toll associated with industrialization was significant:

> On the entrepreneurial side, they [factories] necessitated a sharp redistribution of investment and concomitant revision of the concept of risk. Where before, almost all the costs of manufacture had been variable—raw materials and labor primarily—more and more would now have to be sunk in fixed plant. The flexibility of the older system had been very advantageous to the entrepreneur: in times of depression, he was able to halt production at little cost, resuming work only when and in so far as conditions made advisable. Now he was to be a prisoner of his investment, a situation that many of the traditional merchant-manufacturers found very hard, even impossible, to accept. (p. 42)

Other factors necessarily had to accompany such a risky and expensive strategy. Labor costs could be an impediment. Labor supply surely was an absolute constraint. A rising standard of living and a growing equalization of income could act as the necessary stimuli to entice entrepreneurs to make needed investments. Obviously, depending on the significance of each of these factors, a businessman might hesitate to make the investment in a factory.

The ways in which mechanization unfolded across industrial sectors are also relevant to our discussion. Why some sectors mechanized early and rapidly while others resisted machine support can be reduced to several intervening factors. The nature of the good largely determined its susceptibility to mechanization. This attribute went hand in hand with the extent of the market for the good. If the good was a high-value/low-volume one and of a luxurious nature, labor costs were relatively insignificant and thus a minimal stimulus to change, unless, of course, the transition to machines was twinned with a movement toward reaching a broader market with a more widely disseminated good. In this case, shifts in the market could require change simply to hold a longer term position in the market. Institutional factors could also serve as impediments to change. The number of possible explanations supporting the lack of investment are certainly sufficient to demonstrate that the prospect of a smooth functioning factor market acting as a stimulus for change was contingent on elements other than the short-term costs of production.

What, for example, defines the nature of a good in terms of its target market? Unless constrained by a technological impediment, the choice of market position is usually quite subjective. A good is defined as luxurious primarily by the capitalist who deems that it should be made, and secondarily by other organizations such as guilds or distribution channels that have reason to resist change and uphold a specific market position. Thus, the nature of the good and its market destiny are to some extent flexible and largely related to the entrepreneur's own assessment of the market. Technological impediments in a labor-intensive industry are few, except for worker skill. The value of skill can be withheld, but only for so long, given that a laborer's wage is a function of an individual's output. In view of the prospect for standardization of a product's market and the eventual cheapening and mass production of a good, why did some industries such as watches resist or arrive late to mechanization? Why didn't the British watch industry follow in lockstep with other industries?

Factor cost differences are one constraint, but they are only binding when and if another competitor already exists or enters the market. A new competitor has two primary means of altering the competitive terrain: by redefining the market for the product (in itself a powerful competitive threat) and by pursuing the competition with a significantly new technique. Indeed, given the penchant for mechanization in many facets of British industry, it is surprising that the world's leader in watches did not follow other British industries and pursue a model of mechanization. In the face of an alternative strategy, what other forms of resistance are relevant? Landes (1969) inventories many other relevant constraints, ranging from capital availability and cost to the scale and complexity of technology. He finds all but one of these to be necessary but insufficient explanations of diffusion: the one is the qualitative receptivity of an industry to mechanization. The test case was cotton, a latecomer in textile development, which rose to prominence and preeminence even as many of the original technological inventions were targeted toward problems of wool. Here I wish to underscore the idea that in the metalworking industries an abundance of problems associated with material inputs existed, particularly those relating to consistency of performance, which may have colored watch manufacturers' willingness and ability to pursue mechanization.

No, the answer to the British watch manufacturers' resistance to mechanization in the face of a new competitor rests with a complex interplay of time and perceived threat. Constraints within the industry in the form of guild restrictions placed heavy burdens on the British industry. But even assuming that this burden was relaxed, the British perspective on the product was one of a high value-added good, produced exclusively for the gentry and to a lesser extent for the growing middle class. Equally important, however, was the fact that the competition, in this case the Swiss, were not themselves pursuing mechanization. Thus, the introduction of a new price point in the absence of a new means of organization was not perceived as a sufficient threat to combine interests and change the way business was done.

The Organization of the Watch Industry in the Factory Era

One of my significant concerns in this book is to explain how the transition to factory-based production altered the relative roles of labor and capital. During the craft era workers exhibited some control over the labor process (declining through time). This had important implications for the volume of output and therefore for the size of the markets that could be served, the level and source of innovations, and the source and control of power. The need to harness these capabilities into a more centralized system of command and control had major implications for the manner in which industrial systems evolved. As Lazonick (1991) points out, the transition to the era of proprietary capitalism brought the integration of asset ownership with managerial control, and with this integration an internalization of aspects of the craft system. In particular, in order for firms to "generate [by the standards of the time] high-quality products at low unit costs" (p. 25), firms relied on an "ample supply of highly skilled and well-disciplined labor. Senior workers—who eventually came to be known collectively as the 'aristocracy of labor'—not only provided their own skills to the building and operating of machinery but also recruited junior workers whom they trained and supervised on the shop floor" (p. 25).

This arrangement had much merit. Fixed costs were relatively low, the demands placed on the state for training were muted, and the production and control of the labor process was ensured through what could be described as a system of hereditary transfer of opportunity. Senior workers trained junior workers, passing on their skills and know-how.

The heavy reliance on skilled labor had obvious spatial implications.

Firms were more or less bound to a place. This dependence on labor is one of the prime determinants of spatially bound production systems aptly captured by such well-known observers as Marshall (1997). The external economies within such a complex were substantial. Materials, labor, and services (distribution) were sufficiently honed that, as Lazonick (1991) points out, spatial proximity "enabled work in progress to flow through its vertically specialized branches, and to market its output" (p. 26). These regional concentrations facilitated high levels of specialization. Consequently, employers' ongoing reliance on skilled labor to organize work on the shop floor minimized the need for business firms to invest in the development of managerial structures and organizational capabilities (p. 26). It also meant that pressures to cut costs were resolved through market transactions rather than through internal restructuring.

The system was self-perpetuating. Because proprietors had access to this complementary system of transactions, they had no incentive to invest in or to develop complex managerial structures. Instead, in the early phases of factory industrialization, proprietors tended to own single establishments and rely on the market for the provision of necessary inputs. Firms tended to remain small. There was a continuous flow of new establishments, formed to fill niches. The manufacturing process was inherently labor-intensive and reliant on a cadre of skilled and unskilled workers to manipulate the raw material into final, finished form. The advantages of this supply system, combined with the benefits of a labor force that was self-reproducing, resulted in low fixed costs. Although problems of coordination would eventually become disabling, nonetheless Britain built a globally competitive industry in the early 19th century based on just such a system of organization.

What unseated this remarkably flexible system was the emergence of mass production and the manufacture of volumes of products, all of more or less the same quality and capability. In the early era of mass production, the "American system" became the model to emulate. Unfolding in fits and starts, this system has been characterized as "the sequential series of operations carried out on successive special-purpose machines that produce interchangeable parts" (Ferguson, quoted in Hounshell, 1984, p. 15). Implicit in this description is the move away from a dependence on skilled labor, a system based upon decentralized and fragmented production, and highly variable, relatively high-cost goods (Hounshell, 1984). Costs were not the only concern. Equally important was similarity of parts, interchangeability, and rapid production. Like armaments, watches were produced in an otherwise highly decentralized manner where parts were similar but certainly not interchangeable, and where master craftsmen and finishers were required during the final assembly process.

With the emergence of the American system, originally in the production of armaments, entrepreneurs in related fields turned their attention toward the application of the same principles to their industries. There was considerable debate when special purpose machines were first used as to whether they would yield a final product that was less expensive and more accurate. Other expectations also were challenged as the first examples of the so-called American system took shape. Of particular importance was the rancorous debate about interchangeability and whether it could ever be achieved. Up until that time, parts were manufactured in a decentralized fashion and then fitted together by a skilled worker who would file or otherwise "finish" the pieces and then assemble them by hand. Rough-hewn parts were the standard of the day. Interchangeability was not inextricably linked to whether the product was manufactured by machine. Indeed, as Hounshell (1984) rightly points out, interchangeability could be created by hand-manufacturing methods as well. Thus the two processes, production by machine and interchangable parts, while in some ways linked, are not one and the same nor locked in a necessarily dependent relationship. This is an important point, particularly in the case of watches, which evolved in different places at different times based upon partial integration of such ideas and capacities.

Of particular historical importance to the history of watch manufacturing is the well-documented fact that many machine producers and manufacturers made excessive claims both about the ability to produce at mass volumes and the ability to generate interchangeable parts. Even the reputations of such heralded inventors as Eli Whitney, one father of the American system, have been challenged recently by historians who believe that the true achievement of the American system took much longer, and was far less pervasive and less successful, than the pundits and entrepreneurs of the day claimed (Hounshell, 1984). If claims really were exaggerated and if capabilities were only haltingly achieved, then the standards used to judge competitive capabilities would be off. Even more important is the possibility that the application of an ideal in an incomplete manner could actually form the basis of longer term problems, when latecomers actually *did* implement the ideal. Thus, mechanized production could have been implemented, but the advantages of such a high-volume system, in the absence of both precision machining and an ability to assemble parts in large volume accurately and in a timely fashion, yielded very high fixed costs and serious problems of inventory control. Hounshell (1984) successfully demonstrated the problems these issues presented to the armaments industry. Watchmaking, given the small scale and high levels of precision it involves, could easily have succumbed to problems of partial adoption of an ideal type.[2]

Two other points associated with the early implementation of the American system are critical. Both relate to the role of skilled labor in high-intensity assembly industries. The first point refers to the importance of skilled workers in the manufacture and maintenance of machine tools. Regarding the armaments industry and the Harpers Ferry establishment, Hounshell (1984) notes that John H. Hall, a manager and early advocate of interchangeability, which he implemented at the Harpers Ferry Armory, annually employed as many skilled machinists as he did armorers (Smith, cited in Hounshell, 1984, p. 43). In an attempt to generalize, Hounshell notes, "But as later experience would show, such expenditures and allocations of labor resources were characteristic of manufacture under the American system" (p. 43). Thus, rather than getting away from or out from underneath a dependence on skilled workers, the American system actually created an insatiable demand for skilled labor, hence making the management of the production process precarious. This leads us to the second point: How did the factory owners manage this labor relationship?

In the case of the American system, the pursuit of precision took a peculiar path. While history shows that the pursuit of machine manufacture took precedence over the pursuit of precision, one without the other would create an enduring problem: too much intermediate output, too easily flawed, and yet very expensive to rectify. A century would pass before this major flaw/oversight in the mental construction of the American system would be called into question. The advent of an inventory system based around "just in time" versus "just in case" delivery verified the ignored importance of precision as the primary goal of this new method of manufacturing. Make no mistake—it was not that this problem was ignored. Rather, it was aided and abetted by the nature of the system of production resident in the earliest production plants. It was in the machine tool shops and prototype facilities inside factories that the talents and imaginations of America's greatest early machinists were employed—places where their power, influence, and creativity created a venue, a sheltered and highly dependent space in which to operate and explore. As Hounshell (1984) ably notes in discussing the armaments industry, "Each skilled worker contracted with Colt to use his shop space (Elisha Root), power, machine tools, and materials to produce a particular part or to manage a particular operation for a set piece rate. The contractor agreed to hire, pay, and manage his own workmen. The contractors acted like foremen but had additional administrative functions" (pp. 49–50). Thus a social system was embedded inside the nascent factory system of production, and this social system would define the American system of mass production.

The legacy becomes all the more important when it is considered in light of the organizational implications of moving to mass scale. It would be another 40 years before trial and error ushered in the complete model of America's own unique system of production. And, like manufacturing itself, which requires a period of experimentation and prototyping before the bugs are worked out of the initial design, the interim prototyping period in which production organizations were taking shape was manifested in partial forms of the final system that became the American system of mass production. The interim products, in this case the organizational forms leading up to the vertically integrated firm, embraced a range of capabilities and sensibilities. As Hounshell (1984) notes in discussing the failure of the British to adopt the American system, British manufacturers did not wish to follow the American practices that Anderson, Colt, Nasmyth, and others had declared universally applicable. But manufacturers in the United States did not follow them all, either. Indeed, when Anderson toured Derringer's pistol factory in Philadelphia, he was astonished to find that traditional hand methods were still in use (Hounshell, 1984, p. 64).

It is for history to determine how these incomplete forms of a dominant system fared over time. In some instances (and there are many examples), first-mover advantages in fact become liabilities, even at the small scale of operation exhibited by a cluster of firms in a single industry. As we will see in the case of the Waltham Watch Company, the early American watch manufacturing leader, the partial application of the evolving technique of American-style mass production had serious liabilities. Later entrants pursued strategies that built on and overcame the mistakes evident in the Waltham Company's organizational system. Barriers to entry are high for firms entering an industry for the first time, but their ability to overcome mistakes made by the leading firm while simultaneously finding new methods of production that provide advantages can generate competitive success. The rivalry between the Ford Motor Corporation and General Motors (GM) in the 1920s proved to be the basis of a permanent advantage for GM.

The Emergence of Managerial Capitalism and Factories

The transition to machine-based, vertically integrated, large-scale industrial manufacturing occurred over a long period of time. Metal-cutting machines were in use for the rough shaping of parts as early as the middle of the 18th century. However, it was not until the middle of the 19th century that this technological and organizational innovation was widely implemented. The movement toward this form of production was neither pervasive nor persistent. There were many obstacles to overcome, not the least being the construction of the necessary machinery and resolution of the inherent conflicts presented by machine-based output to established labor relations and hence the production system itself (Lazonick, 1991). Ultimately, the general movement to embrace the American system caused turmoil throughout the entire economic system as machine-made parts and large-scale assembly yielded an unprecedented volume of goods. Many problems were associated with the flow of goods, and these problems spilled over into the acquisition of consistent inputs and, equally important, the development of an effective means of distribution. Goods were no longer produced with a proximate market in mind. Indeed, the international supply of foreign markets became an increasingly important part of firm strategy. To produce at large scale necessarily meant distribution at large scale, ushering in the move toward managerial capitalism.

The transition to machine-based manufacturing and the ensuing complementary innovations in scale and scope of production, distribution, and management of the modern industrial enterprise arose rather gradually starting in the 1870s (Chandler, 1991). In *Strategy and Structure*, Alfred Chandler characterizes the rise of the multiunit, capital-intensive corporation as the result of three investments: production large enough to utilize economies of scale and economies of scope associated with new technological innovations; marketing and distribution structures large enough to sell the goods produced by the new processes of production in the volume in which they were made; and a managerial hierarchy to manage and coordinate the operations of large-scale production and distribution and to allocate resources for future production and distribution (p. 432). That this new paradigm first took hold primarily in the United States and Germany has much to do with the relatively late industrialization of these two countries compared with Britain, and hence to the absence of a preexisting labor process and managerial system to overcome. In the United States, the shortage of skilled labor placed a premium on developing machine-based alternatives to industrial production. Preceding the Civil War and strongly afterward, U.S. firms in a wide range of industries pursued mechanical solutions to the expansion of production to achieve economies of scale and serve burgeoning domestic and international markets.

Within this new institution, many functions were carried out. As Chandler (1991) notes, this new institutional structure "was a governance structure" (p. 433). The rise of the managerial corporation brought with it the integration of multiple functions, multiple units, multiple products, and all the attendant physical and human resources. Although each unit had independent capabilities to execute and manage the enterprise, coordination was offered and accomplished by middle managers, who were themselves managed by top officials of the corporation. Quite strikingly, the emergence of this mode of organization and coordination arose rather quickly at the turn of the century, most prominently in a subset of all the industries comprising the U.S. economy. These industries tended to be sectorally concentrated, consisting of firms that were vertically integrated within the same industry group, and characterized by high-volume, capital-intensive production processes. In other words, what linked these firms were product characteristics that yielded the need for a uniform managerial structure designed to confront and control problems of production and distribution while benefiting from the potential associated with specific sectoral characteristics.

Of particular relevance to our discussion here is Chandler's (1991) comparison of labor- and capital-intensive industrial systems. In the case of labor-intensive systems, adding new machines increased output. In contrast, in more capital-intensive industries, output was increased by rationalizing the process, including rearranging the physical flow of materials, improving machinery, consolidating activities within the same facility, and increasing the use of mechanical power sources to push the process of production through the factory. This dichotomy, though helping schematically to distinguish between the two very different systems of coordination and rationalization, overstates the distinction between different industries and the extent to which one type of sector necessarily leads to a certain selection of coordination outcome over the other.

Many industries did not lend themselves completely to the achievement of the three divine characteristics of a Chandlerian future: scale and scope in production, distribution, and future R&D investments. Some industries, most especially the labor-intensive industries, as Chandler points out, were of insufficient size to engender large production facilities where the scale of distribution would warrant the development of product-specific skills and facilities; and were characterized by manufacturing that was comprised of relatively simple tasks, which eliminated the need for organizational hierarchy. As Chandler (1991) presciently notes,

> In many labor-intensive industries, the large integrated firm had few advantages. Indeed, it often had competitive disadvantages, for size in production facilities not only failed to bring lower costs but also made the firm even less flexible in meeting changes in demand. Even where scale and scope and the integration of production and distribution brought cost reductions, they were rarely sharp enough to permit a small number of firms to dominate the industry. In these labor-intensive industries the competition remained on the basis of price and the ability to move quickly to meet changing demand. (p. 453)

But there were also industries, for example, textiles and precision instruments, that could be considered hybrids or variants on Chandler's themes. Interesting in their own right, they evolved somewhat toward a Chandlerian system. Other industries, such as the watch industry, struggled to blend the benefits of vertical integration into a competitive advantage. Great challenges attended achieving efficiency and developing effective production organizations.

Of similar relevance is the case of firms that formed in an earlier era in which the dominant mode of production was far more labor-intensive, but which through time was subject to the benefits of scale and scope as described by Chandler. The transition from one mode of coordination to another went smoothly in some cases, while in other cases the process of change could be considered incomplete. Watches represent an example in which the effectiveness of the transition related directly to the complexity of the underlying product, the rapidity of change within the product, and the preexistence of coordination mechanisms, particularly in the area of distribution, that caused a halting adoption of managerial innovations. Chandler (1991) makes this point and goes on to say, "If the cost advantages of the economies of scale and scope can account for when and where the modern industrial enterprise made its appearance, they cannot explain the second investment essential to its initial growth, that is, the investment in a national and often international marketing and distributing organization" (p. 440).

Another attribute of importance here is the extent to which the labor process could be made to conform to or be supplanted by machine manufacture and process standardization. The most effective implementation of managerial capitalism occurred in those industries in which the labor process could be transformed through the application of capital equipment, thus leading to the specialization and routinization of manufacturing processes. To the extent that this transformation unfolded incompletely, either because of the skill requirements necessitated at various stages of the production process, or because of the historical organization of the labor process governed by preexisting labor relations and an existing organizational hierarchy, only partial benefits attributable to scale and scope resulted. As with other periods of transition through different industrial eras, partial adoption could be both costly and unpredictable in effect. Overhead associated with capital investments would be huge, while the ineffective application of specialization could yield volatile labor relations and a loss of control over the labor process.

The capital-intensive system reaps the benefits of its condition by turning high fixed costs into low variable costs. Accomplishing this requires vertical integration of technologically separable processes. As

Lazonick (1991) comments, "The ability to capture the benefits of indivisibility requires controlling up and downstream elements. But these gains are even uncertain and depend on an expectation either of capturing additional market share or 'for others' the lower bound set by first movers" (p. 81).

In the case of firms in the U.S. watch industry, the ability to control inputs and market access was not completely within their grasp. Historically, marketing was carried out by wholesalers (a system borrowed from the British) and by the third parties that produced and sold watch cases. Furthermore, because of the need to produce both for the high end—to prove the technical competency necessary to attract investment capital—and for the affordable end of the industry, American watch firms, and their leader, Waltham, could neither benefit from external economies and horizontal fragmentation, nor could the mass-scale producer benefit completely from vertical integration and therefore control of market access and the manufacture of critical parts. This problem most seriously disadvantaged Waltham, the first mover.

Putting this problem aside, time becomes the firm's worst enemy, as Lazonick (1991, p. 133) notes. Even if a firm is capable of designing a production system to operate at optimal scale, it may take years before the innovation is successful. Such a lag impedes achievement of reasonable rates of return.

Thus it is no surprise that the sectors and firms most successful in executing the full complement of attributes of Chandler's modern corporation were those that benefited from scale and scope linked to standardization of the production process. At the same time, however, there were other prerequisites to effective implementation, particularly related to market control. To support the huge capital costs required to achieve economies of scale, it was essential that a relatively few players exercise market power. The emergence of oligopolies in such sectors as oil, steel, and heavy machinery went hand in hand with the rise of the modern corporation. Control was both seized and achieved through investments in up- and downstream linkages, a capability that most smaller, more consumer-oriented goods producers could not achieve.

The Rise of Collective Capitalism

According to Lazonick (1991), the model of market coordination embodied in managerial capitalism remained an effective organizing system until the 1960s. The lack of cutthroat international competition, and therefore ongoing access to stable markets, buoyant growth, and rising standards of living, provided hospitable conditions for the cultivation and persistence of the large, vertically integrated corporation. The seeds of turmoil that would eventually destabilize this system of production, consumption, and market regulation were sown much earlier in the century, however, but lay dormant until after World War II. The challenge to this dominant system arose, ironically, as a new competitor, in this case Japan, coupled long-standing and underappreciated competency in the production of goods with state-supported and yet private sector–led manufacture of new products capable of directly competing in the world's most sophisticated industries and its most developed markets (Conroy and Glasmeier, 1995). The national pursuit of both import substitution and export-led development policies provided infant industries with the protection they required to achieve economies of scale and hence low overall costs. Accustomed to developing technological competency through technology transfer, the Japanese successfully adopted and then modified major innovations, primarily from the United States, to take the lead in key consumer and producer products over the postwar period.

The emergence of the Japanese as effective competitors in many consumer and producer product markets dates back to before World War II. In the last three decades of the 19th century, Japan had begun to develop the competency to produce tradable goods, primarily through the purchase of foreign know-how and technology. The Western nations' lack of interest in and understanding of Japan's capabilities at the beginning of the century is in part explained by the fact that Japan concentrated initially on trade within Asia rather than with the West; as a consequence, there was little visible irritation in Western markets.[3] By the 1920s, however, in several key sectors, Japan was second if not first in the world in terms of exports (Glasmeier et al., 1993; Yoffie, 1990). Although increasingly cut off from international trade outlets starting in the 1930s, Japan continued to build up its productive capacity until just before the declaration of war: this time the emphasis was placed on heavy and chemical industries.

Although seemingly virtually destroyed by the bombing raids that occurred during World War II, the Japanese economy snapped back and was exporting goods to the United States by the middle of the 1950s. From that point forward, and for the next 20 years, Japanese investment in productive capacity averaged 14 percent a year (McCraw, 1986; Rosovsky, 1961). This unprecedented level of investment led to the development of a production system of significant capacity and technical competency twinned with an organizational system that differed quite substantially from that of the era of managerial capitalism.

The emergence of collective capitalism as a distinct mode of organization is both a creature of history and a more modern institutional form. At the end of World War II, Japan had little choice but to pursue an export-led development strategy (McCraw, 1986; Samuels, 1994). The country lacked resources, and had a small land base and a large population. In order to stave off poverty and economic malaise, the nation had to develop export competencies in a variety of sectors. For reasons that can only be understood retrospectively, the Japanese did not choose to pursue products that were labor-intensive despite its large labor force; rather, it moved rapidly into capital-intensive production as part of its economic recovery in the 1950s.

The second major stimulant in the formation of collective capitalism was in part a holdover of the structure of the national economy that had developed since the turn of the century. Japan's economy is comprised of interlocking business organizations that enjoy both vertical and (in many instances) horizontal affiliations. Furthermore, the system of integration reaches down to the level of select blue-collar workers in first-tier firms who are offered lifetime employment in return for a work life dedicated to the firm. Guaranteed their jobs, workers willingly accept technological change as a part of labor–management relations. According to Lazonick (1991), "The extent of collectivization of interests under Japanese capitalism contrasts with the more limited planned coordination of the specialized division of labor under U.S. managerial capitalism and the virtual lack of planned coordination that existed during the days of British proprietary capitalism" (p. 38).

Although the Japanese system has many unique facets, of particular importance is the role of *enterprise groups*—networks of firms—that form the modern-day extensions of the *zaibatsu*, the family-controlled businesses that led the development of the Japanese economy from the end of the last century until the Occupation after World War II. These enterprise groups consist of a central business firm around which radiate vast numbers of both affiliated and nonaffiliated subcontractors. This disintegrated system allows the core firm to enjoy the benefits of vertical integration, especially in terms of financing, technology transfer, and internal markets, without the disadvantages of ownership and bureaucratic control. The core firm is able to ensure quality, quantity, and timely and effective delivery of parts and services due to long-term contracts and investments made by the core firm into subcontracted operations (Lazonick, 1991).

Of critical importance is this system's effect on investment decision making, particularly in the area of technology development. Economist and business strategist Michael Best (1990) points out that the "new competition," in this case the Japanese, has been misunderstood. In his view,

> The successful Japanese firm has combined Schumpeter (creative destruction and an emphasis on continuous organizational innovation) and Pennrose (emphasizing team work as a means of developing distinctive pro-

ductive services), and thereby altered the notion of entrepreneurship from "big ideas by an individual" to a social process of learning within which individual contributions can come from the bottom up, as well from specialist staff. (p. 138)

Blending insights from Lazonick and Best suggests that by pursuing consensual decision making, the Japanese lead firm gains from the experience and know-how of key managers and employees while being able to identify bottlenecks and potential obstacles to strategic decisions. The ability to tap potential sources of innovation across the firm and to act purposefully in this endeavor as a legitimate pursuit of the firm raises the potential for continual innovation at both product and process levels. According to Best (1990), consensus decision making

> when operating effectively, so that individual managers cannot circumvent the group process, ... eliminates competing centers of decision-making power within the organization that might otherwise undertake investments that have conflicting objectives. In effect, investment strategy and managerial structure are organizationally integrated. The lack of rigidities between the layers of transactions within the firm (ex. R&D, product design, product planning, production, and distribution) creates a context in which product development becomes [quoting Imai], "a dynamic and continuous process of adaptation to changes in the environment." (p. 155)

Continuous reflective innovation can occur because (1) design teams are given broad mandates and then allowed to structure their own activities and work; (2) the phases of production are approached simultaneously rather than in a linear fashion, thus promoting identification of problems early in the production chain; and (3) continuous learning is built into new product development because the process of collective decision making facilitates and requires communication across disciplinary boundaries. New information is constantly filtering into situations of decision making, thereby enhancing reflective self-conscious learning (Glasmeier, Fuellhart, Feller, and Mark, 1996).

Lifetime employment, integral to this system, helps build consensual decision making within the firm. Workers who do not fear losing their jobs are more likely to follow the directional lead of managers and at the same time to point out instances where management decision making is not likely to lead to effective outcomes. This mutual trust is imperative given the high cost and high risk associated with R&D investments in new product and process developments. In describing the returns to the firm of lifetime employment and decentralized decision making, Lazonick (1991) suggests that "in a corporate environment, economic success depends on not only individual initiative but also cooperative effort, and collective rewards may provide the appropriate incentive mechanism" (p. 40).

There is considerable debate regarding the extent to which the rise of collective capitalism reflects purely state-led capitalism or whether the relationship between capital and the state is more complex. In particular, Samuels, writing about the development of the Japanese economy in the 20th century, argues that politics in Japan is a politics of reciprocal consent:

> By reciprocity, I imply that jurisdiction can belong to private firms as well as to the state. Control is mutually constrained. In exchange for the use of public resources, private industry grants the state some jurisdiction over industrial structure in the "national interest." Business enjoys privilege, systematic inclusion in the policy process, access to public goods, and rights to self-regulation. It reciprocates by agreeing to state jurisdiction in the definitions of market structure and by participating in the distribution of benefits. (Samuels, quoted in Hart, 1994, p. 124)

Of interest here is the extent to which national resources and the political power of the state have been used to structure the terms of competition in Japan to ensure cooperation across firms to the extent of regulating foreign exchange as a means of reducing interfirm competition, of providing preferential tax credits and export financing to encourage international trade, and by creating consortia to pursue innovations that require large-scale investment in both capital and human resources.

In spite of the frequent portrayal of the Japanese system of collective capitalism as one in which monolithic companies benefit from government policies, not all firms are members of a *kieratsu* (business group) nor have all firms been the recipients of national subsidies and preferential treatment. The Sony and Honda Corporations are frequently cited as breaking with the Ministry of Trade and Industry and essentially charting a course independent of national administrative policy. Like these two, Seiko, the largest producer of watches in Japan, also broke with national market regulation, and has long followed a policy of independence. Transcending the strict boundaries of national government direction, Seiko has benefited from the consensual nature of Japanese business relations and has capitalized on national investments in the creation of a highly skilled and literate work force. The availability of patient long-term capital has been critical in allowing firms to pursue risky investments in next-generation technology.

The transitions between different modes of market regulation are fuzzy at best. Most historians make sharp delineations between them, presumably for convenience. In so doing, however, they leave the nature of partial adoption unexamined. And yet, given that change is a process that occurs over time and space, transition and intermediate conditions are essential components in understanding how both firms and regions react to change. Coming of age varies for everyone and everything; when you were born and the dominant social and ideological conditions of the day have much to say about who you ultimately become. The same is true for firms and regions. Thus, in characterizing the options organizations have within their grasp, of primary importance is the need to recognize the role played by timing in the construction of degrees of freedom to act.

Throughout the life of an industry, then, which spans both time and space, different conventions and rules of behavior form the basis for collective action. Organizational behavior and prevailing institutional norms structure the range of possibilities firms and regions have at their disposal when they are faced with the need to change. Shared understandings arise at a much deeper level, however. Even as the range of options is to some degree predetermined by the prevailing mode of organizational behavior of the day, nonetheless the execution of the task, its actual implementation, is embedded in a specific context. Success within the task environment is mediated through the prevailing, highly localized culture. Responsiveness to change in some cases is facilitated and in other cases thwarted by the prevailing belief system, which transcends the organization of production. I now turn to my last topic of discussion in this chapter, the role of regional/industrial culture in the process of economic change.

Organizational Culture and Identity

In the previous chapter, I suggested that a variety of factors act as mediating influences on the ability of firms within industries and regional sites of industrialization to respond to changes in their environment. Some factors, such as wars and economic crises, stimulate change: they alter the context in which existing industrial structures must operate. Other influences, such as the dominant mode of industrial organization, in effect condition norms of behavior or trajectories of response. Ideology shapes perceptions of what can and should be. I now examine how shared understanding acts at a very microlevel to shape possible patterns of response. In effect, *culture* in this discussion refers to the rules and practices, identities and aspirations, of individuals at the most intimate level. To some extent my interest is in the culture of firms, organizations of individuals that take on an identity and hence are able to undertake collective and purposeful action. Simultaneously, my interest is in the ways culture itself is found within people, which in turn is manifested in regions.

In this way, I hope to bring geography into the discussion, not as a physical place per se, but rather as a reservoir of experiences and actions that in some cases and at different times act as a passive agent, and at others act as a dynamic influence or force in the face of change. Regional identity transcends the identity of the individual and yet is reciprocal over time and can influence the persistence and transmission of individual identify.

In a wide-ranging and insightful exposition of the role played by culture in the ability of firms to act, Schoenberger (1997) presents a detailed account of the nature of corporate culture, both in terms of its popular understanding, particularly within the confines of the business press, and in terms of a more far-reaching delineation of the creation of culture inside the firm. The central question attempted in her study is this: Why, in the face of excellent information, do firms fail to act, even when inaction places the firm in serious, even perilous, jeopardy?

In response to the business disasters of the 1980s, it has become quite fashionable to place the blame for economic upheaval on failings of business culture. Critics argue that culture produces myopia, which limits businesses' understanding of changes occurring in the environment. Much of the more popular work in the business press treats culture as a given and speaks of it in normative fashion (see Glasmeier et al., 1993, for different views of the role of culture in the modern corporation). While these approaches suggest the role culture should play in the modern business organization, I am more interested in where culture comes from, the role it plays within the firm, and how it is maintained. My interest extends even further, for I seek to situate culture in a regional context and to understand under specific circumstances how geographic proximity helps to retain and reinforce identity.

What is business culture? According to Best (1990),

> Culture implies the existence of mutual understandings that allow individuals to transcend the individualism of economic man, where self-interest is pursued by individual actions alone. Culture derives from mutual reciprocity which gives the individual the resource of trust to overcome the prisoner's dilemma, the tragedy of the commons, the tyranny of small decisions, and all such formulations of the predicament of individual decisionmaking in a world without trust, culture and community. Culture is also the basis for collective identity or sense of self that derives from membership in a group, team, community, or nation. (p. 145)

This perspective incorporates both the individual and the organization, and expresses how one affects the other and the mutual interaction between the two. At a more psychoanalytical level, Schoenberger (1997) suggests that "culture is inherently and deeply implicated in what we do and under what social and historical circumstances, in how we think about or understand what we do, and how we think about ourselves in that context. It embraces material practices, social relations, and ways of thinking. Culture both produces these things and is a product of them in a complicated and highly contested historical process" (p. 120). Going to the heart of both the individual and the community, Schoenberger offers an encompassing view of culture:

> What culture does in this sense, is to create a way of life. One could also say that this is a world in which humans, through their actions and their special relationships, produce culture at the same time that culture produces them. This allows us to keep an eye on the practical, historically specific constraints to action and innovation at any particular juncture without eliminating the possibility that changes may, indeed, occur. (p. 121)

Why don't organizations change in the face of good information and obvious turmoil in the external environment? Best (1991) suggests that to understand the role of culture in change it is first necessary to acknowledge the specificity of an individual firm's culture, as manifested in a belief system that governs actions. This characterization implies unanimity of action. In contrast, Schoenberger (1997) is more forceful in suggesting that corporate culture is about power. To the extent that change must occur, the survival of the firm depends on it:

> Cultural change inevitably involves a struggle over power. By this I do not mean the rather circumscribed one envisioned in, say, organizational sociology over who runs this or that bit of the world, but a struggle over who has the power to construct and defend a specific social order and practices, behaviors, knowledge, and understandings appropriate to it. At the same time, cultural change involves a struggle over identity—over who and what the firm is. When these struggles are sufficiently acute, they amount to a kind of cultural crisis in which competing models of the social order, and the material and human resources and identities tied to them, are threatened with devaluation, and oblivion. (p. 122)

Considering the relationship between culture and power for an additional moment, this lens allows us to probe more precisely how a firm can actually disregard signals in the environment that foretell serious, even fatal, problems. Firms, especially those that become dominant in their industry, live long lives, become technological leaders, in effect set the social rules and define the norms of behavior among participants in the firm as well as across firms within the industry, operate with a large degree of impunity, and hence wield power over their environment. The extent of this reign, its longevity and pervasiveness, presents a firm with a sense of invincibility. Hubris sets in, inasmuch as a manager who receives continuous positive feedback for past behavior will not only begin to believe in his own powers of decision making, but will focus even more precisely on receiving and processing only those signals that reinforce his claim to success. Firms are creatures of habit. A large literature attempts to capture the means by which firms learn to change. Glasmeier and Fuellhart (1996) have utilized this definition offered by Levitt and March (1988, p. 320) elsewhere:

> Organizations are seen as learning by encoding inferences from history into routines that guide behavior. The generic term "routines" includes the forms, rules, procedures, conventions, strategies, and technologies around which organizations are constructed and through which they operate. It also includes the structure of beliefs, frameworks, paradigms, codes, cultures, and knowledge that buttress, elaborate, and contradict the formal routines. (Glasmeier and Fuellhart, 1996, p. 15)

Thus, paradoxically breaking out of the mold is most difficult for those individuals in the organization who are in charge of identifying the need for and executing change. One way of thinking about this paradox is to consider the fact that all corporate knowledge is situated knowledge. As Schoenberger (1997) suggests,

> Powerful as they are, the way firms interpret the world is structured by their position in it. This situatedness may produce what could be thought of as a predisposition to systematically read evidence about the state of the world in certain ways—or, alternatively, to systematically misread certain kinds of evidence altogether so that information, though available, becomes uninformative. (pp. 125–126)

This problem of myopia becomes all the more troubling as the source of change gravitates from one within an existing paradigm to that of one that is boundary-spanning and fully transcendent of the status quo. In other words, change can be detected within the existing context and hence is interpretable within existing norms and procedures. Change can be fundamental, challenging all previous rules and behaviors. Like technological change, gradual alterations occur proximately all the time and are often surmounted with little difficulty. Radical technological change may not only arise from completely outside of the existing system of social relations, but can arise in such a peculiar location as to be virtually undetectable until it is too late.

Regional Culture and Identity

The final point in this discussion has to do with change, which takes place in space and also originates in one place and affects another place—in many instances, a long distance away. There is a significant literature on regional change. In one vein of reasoning, regional change occurs as stimuli from outside create new and increased demand for the fruits of a region's labor. This alteration in current circumstances is in some very important way both facilitated and thwarted by conditions that arise within the local context. Why some places grow while others stagnate has something to do with initial resource endowments. However, as we know from the last 20 years, comparative advantage can be and is regularly created by the purposeful actions of people in places (Porter, 1990).

Aside from the vast body of understanding about what causes regional change, in this book we are interested in how situatedness affects the ability of economic change to occur. Chinitz (1960), writing about the ability of different types of regional economies to cope with change, suggests that the costs of production in a single location are the relevant unit of analysis when considering regional resiliency. Indeed, he was one of the first spatial economists to attempt to formally link attributes such as regional and industrial culture with long-run economic success. Markusen (1985) indirectly comments on the potential of industrial culture to generate spillovers within a regional economy that could thwart change. More precisely, Saxenian (1994) is one of many who argue that distinctions exist among industrial cultures apparently within the same industry that vary across space. In an even more encompassing treatment, Storper (1997) suggests that in fact regional culture and those attributes of place distinguish successful firms, thus reversing the direction of causality.

Of importance here is both the enabling and disabling influences of regional culture that condition and in many ways shapes over time the plausible responses of firms within an industrial system. In other words, an industrial culture takes shape in space. If the system is successful through time, then a gradual accumulation of norms, beliefs, expectations, social relations, and cultural identity take form and reinforce the existing system. This identity is both about the industry itself, the behavioral conventions (Schumpeterian vs. Chandlerian) and organizational qualities (permeable vs. impermeable; open vs. closed; tradition-bound vs. spontaneous) Chinitz (1960) discusses the situation in which the system of relationships unfolds. These two distinct but interrelated and overlapping contexts are both sources for positive and persistent change; however, they also can interact and facilitate myopia as the regional context becomes the vessel in which firms seek identity validation and through which weak signals are filtered out and often ignored.

Summary

In this chapter, sources of change have been separated from reactions to change that arise from embedded systems of industrialization. I have especially tried to distinguish those qualities of change that are all-encompassing from those that are peculiar to the situational context of individual countries. It has been suggested, too, that there are few pure forms of organization or institution and that we must understand how a specific firm in a specific locale took shape and why in any attempt to read success in coping with change. Vestiges of the past act as a bulwark against new threats, in particular because they provide a clarity with which to understand the current context and help interpret change. At the same time, however, when the stimulus of change is monumental enough, remnants from the past cannot be sewn together into a meaningful understanding of the current context, as both the shape and the terms of the environment may have changed. By setting forth these key stimuli of change and then providing some of the major structural factors that have shaped the reaction time and organizational capability and their regional context, I hope to suggest something about the contingent nature of development.

4 The Burden of Being First

The Foundations of Watchmaking and Timekeeping Technology and Britain's Ascent and Decline as the Dominant Watch Manufacturing Region

Volumes have been written on the history of timekeeping. Most notable is Landes's superb and richly documented historical account of the clock and watch industries, *Revolution in Time* (1983).[1] Landes's book provides the reader with a detailed depiction of the social and economic forces that surrounded the industrialization of timekeeping through the ages. This book begins where, in some important ways, Landes's leaves off, by exploring how and why the major producers of watches from the 18th century onward organized their industries; selected and then acted upon strategic initiatives; coped with major dislocating events such as war, economic crisis, and technological change; and ultimately competed among themselves for a stake in this worldwide industry.

Perhaps my most important deviation from past treatments of the subject is the attempt to situate the actions, reactions, and outcomes of different national industries within a larger economic and geographic context. The question I attempt to answer is this: Why did firms in the watch industries in specific countries choose the strategies they did? The answer may emerge after examining the impinging forces that structured their options and the then-dominant production systems available at different points in time. The consequences of the choices firms made in different countries are also addressed.

This chapter begins with a brief recitation of the history of timekeeping, drawing on both Landes's and Dohrn-van Rossum's (1996) books about time. With this introductory foray into history, the technological aspects of the book's focus—the personal timekeeper, the watch—are discussed. Having briefly described the essential technology, I then proceed to summarize the history of the early and in some sense most ill-fated of the early watchmaking systems—that of Great Britain. Britain was an early industrial leader and the first country to industrialize both in Gerschenkron's (1962) sense of being a lead country and in the specific case of watches. Its industrial supremacy was not based on technological superiority per se, but instead was the result of a knowledge system based on learning by doing. This system was effective as long as technology remained an accretion of improvements and innovations that combined known techniques (Landes, 1968, p. 283). This advantage was easily eroded, however, when other countries armed with superior technique developed systems of production that simply overwhelmed the British system.

Timekeeping in History

The recorded history surrounding the keeping of time and timekeeping technology dates back to at least the 11th century. Clockmaking in its various forms was undertaken by many cultures around the globe. Some researchers have wondered why the clock industry reached technological supremacy in Europe rather than in China, which was a much more advanced culture in the 12th century. Of the various arguments presented, one view suggests that Chinese culture had little use for accurate timekeeping (Landes, 1983). Timekeepers were the exclusive interest of the aristocracy and political leaders. Why didn't a significant industry evolve? Evidently the lack of need, combined with an insufficient system of intergenerational information transfer, kept the Chinese clock industry in a basic and imprecise state.

The development of a far more advanced timekeeping industry in Europe has been attributed to the evolution of an expressed need to mark time (Thompson, 1967), combined with strong mercantile interests, well-developed industrial antecedents, and a culture infatuated with new products. Industrialization created an insatiable demand for the keeping of proper time among textile barons, for example, who were vitally concerned with a punctual work force.

Clockmaking constitutes a separate and complete technological trajectory from watchmaking. Its industrial heritage is linked to basic metalworking trades such as blacksmithing, whereas the watch industry finds its roots in the manufacture of jewelry (there are important exceptions, as we shall see). Clocks were early urban status symbols. A major sign of localized development was the town's tower clock, which signaled to all travelers that they had arrived at a significant place. The nature of clock production and placement was such that localized, but itinerant, craftsmen emerged, not so much to manufacture clocks as to maintain them. Early clockmaking technology produced equipment that required constant adjustment and repair. Early clockmakers were a variant on the blacksmith whose skills were needed to keep the clock in tune and in repair.

As clocks became more persuasive, a strong demand developed for smaller timekeepers that could be placed in homes. Because of the basic underlying physical principles of clocks, only the wealthy had the domestic space for them. Thus, the initial spatial distribution of clock production was associated with the locus of demand: individual capitals, wealthy local merchants, and political leaders. This pattern was fluid in that clock manufacturers were a roving lot whose skills were up for sale to the highest bidder. The early history of the clock industry is filled with stories of rich merchants who "stole" or more likely bought the services of well-known clock producers whose patrons were residents of rival towns (Dohrn-van Rossum, 1996).

What's All the Fuss About?: Precision

One important quality has defined the reputations of the manufacturers of the world's premier timekeepers: precision. The early development of the watch as an instrument to tell time was heavily constrained by the ability of craftsmen to produce an accurate, durable, and reproducible product. In the early years of watchmaking, competition between countries focused on achieving technological superiority on a very small scale. Even so, the creation of what could be considered an accurate personal timekeeper took almost 200 years to achieve. A very long period of time in the early life of the watch industry was spent developing the ability to produce a device that yielded accurate time. It was not until the end of the 19th century that watches of calibrated precision were widely available. Once the basic technological problem of precision had been more or less solved, competition for market dominance revolved around pricing strategy, fashion, and distribution channels.

History of Watch Technology

Evolution of the Watch

Since its creation almost 300 years ago, the watch as a device has remained remarkably the same. Up until the 1960s, alterations in the watch occurred mostly to the exterior in response to fashion and consumer tastes. The internal mechanism remained stable. The advent of electronics led to the first significant change in the internal functions of the watch. The impact of this new technological development was profound. Centuries-old traditions in the manufacture of watches were revolutionized overnight. Prior to electronics, timekeeping accuracy was associated with the precision with which metal parts were cut, filed, and fitted together. The most accurate watches were made by a single craft worker who cut each part, polished each metal edge, and placed each tiny part in the watch movement. Tuning the watch to achieve accuracy required very careful attention to detail. A short technical description of the evolution of watches helps clarify the meaning of electronics to the industry. This section draws heavily from Knickerbocker's (1976) treatment of watch development.

The Standard Spring-Powered Watch

A standard mechanical watch consists of three groups of parts: "the ebauche," or movement blank; the regulating components; and other generic parts. The ebauche includes the framework (or backbone) of the watch, the gear train, and the winding and setting mechanism. Regulating components make the movement work at a correct rate. The miscellaneous parts include the case, crystal, and so on.

A mechanical watch is driven by a mainspring, which transfers stored energy (in the spring coil) to the gears that move the watch hands. The release of energy is controlled by the escapement mechanism. Although numerous escapement models were developed over time, the mechanism is constrained by the anchor fork to give up a precise amount of energy. The anchor fork rocks back and forth, allowing the escapement wheel to advance in tiny increments. These increments, converted by other gears, move the watch hands.

The anchor fork moves in conjunction with a balance wheel that moves back and forth. The balance wheel is motivated by a hairspring that coils and uncoils, keeping the balance wheel in motion. As the anchor fork disengages from the escapement wheel, it transmits enough power to the hairspring to coil it. As the hairspring uncoils, it rotates the balance wheel in the opposite direction. This motion rocks the anchor fork in the opposite direction, starting the next cycle of the regulating mechanism.

Not all watch movements are the same. Differences in movement quality relate to the technical composition of the parts used. Precision, ornamentation, and movement miniaturization differentiate a high- from a low-quality watch. Internal jeweling in watches does not reflect differences in quality. Jewels are used to reduce friction between touching metal parts. The majority of jewels used in the interior of a watch are made out of synthetic materials and do not add significant value to the watch. Within mechanical watches there is a qualitative difference between jeweled and pin-lever watches. The pin-lever watch (e.g., the Timex) contains a more simplified movement compared to a jeweled watch. A pin-lever watch has few moving parts and does not use jewels to reduce friction between metal parts.

The Electric Watch

In this century, the first major technological advance in watch movement manufacture was the introduction of the electric watch. The electric watch movement was made possible by World War II R&D developments in the miniaturization of motors and batteries. The electric watch was only a partial step away from the mechanical watch. The mainspring and many of the components of the escapement were eliminated and replaced by current from a battery that drove a tiny balance-wheel motor. The electric watch was introduced in 1957 and was available worldwide by the 1960s. Because electric watches were no more accurate than most mechanical watches, they did not make major inroads into the medium- and high-price watch markets.

The Tuning-Fork Watch

The second major innovation in watch technology, the tuning fork, did have a profound effect on the watch industry. The tuning fork is stimulated by an electric current from a battery. The current causes the tuning fork to vibrate at 360 cycles per second. A tiny strip of metal connected to the tuning fork transfers the vibration to a set of gears which, like a conventional watch, drives the watch hands. Because the tuning fork vibrates 31 million times a day, the mechanism is far more accurate than a mechanical watch. However, it also is quite fragile and does not take well to rough handling. Tuning-fork technology was invented in the early 1950s and became commercially available in the early 1960s. A women's version was eventually introduced in the early 1970s.

The Quartz Crystal Watch

The third, and most significant, innovation in watch manufacturing occurred in the late 1960s with the use of quartz crystals to regulate increments of time. When electric current is passed through quartz it vibrates at very high frequency. Microcircuitry subdivides the crystal's frequency into electric pulses that drive the watch. In some cases the quartz is used to power a stepping motor, which is connected to a gear train that moves the hands. Quartz technology also can be used to stimulate a tuning-fork device. In solid-state watches the pulses are fed into integrated circuits that convert the pulses into second and minute time increments. This last type of watch requires no moving parts. The face and hands of a solid-state watch are replaced with nonmechanical methods to display time.

Changes to the Watch Face

Prior to the introduction of the quartz crystal watch, changes in the timekeeping mechanism were completely internal to the watch. With the advent of the quartz crystal, time could be displayed by conventional means (analog) or by digital display. Two primary displays are important: light-emitting diodes (LEDs) and liquid crystal display (LC). LEDs are semiconductors that emit light (much like a light bulb). Originally used in calculators, LEDs became fashionable in watches in the early 1970s. Because LEDs required considerable power, they were not illuminated at all times. A push-button activated the display. The high-energy usage and the tedium of the push-button display relegated LEDs to little more than an intermediate technology on the way to a more effective display. In contrast, an LCD display consists of a glass sandwich with a thin coating of electrically sensitive chemical between the glass plates. When a current is passed through an LCD, the chemical changes its crystalline structure. The altered crystals reflect light coming from an outside source. While less power-hungry than an LED, LCDs' brightness and precision depend on the brightness of the external illumination.

The advent of quartz altered both the internal and the external features of the watch. The quartz watch had high levels of accuracy, and could also incorporate numerous functions. Within an incredibly small space, a quartz watch could include multiple timekeeping mechanisms, including alarms and other sophisticated functions. In combination, quartz technology revolutionized the watch industry.

Early Antecendents of the Watch Industry

It is hard to identify with any assurance the country that was the seminal producer of watches. The highly dispersed form of manufacturing prior to the Industrial Revolution resulted in myriad, scattered sites of production. Master craftsmen in France, Britain, Germany, and the Netherlands were all working at approximately the same time to perfect small, personal timekeepers.

In the early 18th century, producing a beautiful bauble took precedence over the manufacture of a highly reliable product. Much of early watch manufacturing was more akin to jewelry making than to instruments making, where precision was deemed important. In part because of their conspicuous presence—early watches were on display, either worn around the neck or on a long chain—and in part because of the technological potential that lay hidden within a watch case, watches were regarded as an important part of a nation's industrial prestige, commerce, and trade.

Early recognition of the national significance associated with being the locus of precision timekeeping meant that much of the industry's early development occurred behind closed doors, as journeymen horologists worked in secret to produce technical marvels. This competitive quest for precision reached its zenith in the development of the chronometer, a story recently and adroitly captured by Dava Sobel (1996) in her book *Longitude*. As Sobel reports, John Harrison spent his life developing a portable chronometer that could accompany captains of the British fleet on seafaring journeys around the world.

Excessive competition illustrated by the quest for the chronometer is also emblematic of the extraordinary lengths to which British watch manufacturers were willing to go to preserve an industrial culture even if it meant the long-term demise of the industry. Dogged persistence in maintaining a way of life symbolizes more than simply the unique competitive experience of the 18th- and 19th-century British watch industry. The next section of this chapter delves into the industrial antecedents of British industry with the purpose of exploring the role of the industry structure and geography in shaping both its rise and fall as the first large-scale watchmaking system.

The Burden of Being First

It seems appropriate to ask why Britain became an early leader in the manufacture of watches. Given that industry experimentation was occurring simultaneously in numerous locations on the Continent at approximately the same time, including, importantly, Switzerland, why were the British the first to achieve serious export capacity in the 18th century? The historical antecedents of the watch story are essential as we attempt to understand, situate, and explain the ultimate decline of the British industry by the middle of the 19th century, for it was the construction of culture and tradition that remains the lasting explanation of this particular industry's demise. To this we will add context which suggests that while culture and tradition were important, location and industry structure, combined with external pressures, served to limit the options the British industry faced in the middle of the last century.

As I discussed in the previous chapter, factors other than the dynamics of production systems have played major roles in the relative position of one country's watch companies versus another's. Failure to adopt the most modern production technique was only one of many problems that plagued industry leaders. Equally debilitating were exogenous events such as wars, exchange rates, and protectionist trade policies. Closer to home, problems included rigid distribution channels, institutional resistance to change, severe price cutting, and opportunistic behavior by industry competitors.

The Antecedents of the Industrial Revolution and Their Role in Shaping the Watch Industry

The question Why Britain and watches? can only be answered by situating the development of the industry in the context of the nation as an early leader in economic development. The British watch-producing industry was a superb technical complex and a major producer of watches for wide distribution in the early 18th century. Perhaps known best for its success in creating a durable and accurate chronometer, the British industry ruled the world of timekeeping at one point in history.

In addition, Britain was a pioneer in the world system of trade. The ability to be a pioneer rested on the early development of institutions, the implementation of agrarian reform, the generation of a surplus that could be reinvested in manufacturing activities, and the formation of a large and growing internal market. The state also strongly reinforced the gains from risk-taking and entrepreneurial endeavors. The lack of regulation embraced in the term "laissez faire" allowed unfettered capitalism to unfold. Thus, the formation of the British watch industry must be seen as an outgrowth of an expanding market and newfound wealth generated by a trade-based culture that both consumed domestically and sold watches internationally to the growing, yet still quite small, merchant class of the 17th century.

Although the location of the first manufacture of watches is attributed to Nuremburg, Germany, in the 1500s, Britain quickly rose to become the world's center for the technical development of the early industry. In its nascent state, the manufacture of watches focused on the inherent complexity of fabricating metal parts and then assembling them to produce a personal timekeeper that would actually keep time. As with many new inventions, a considerable period of time passed when the product was simply being fabricated for the first time, and no standard existed. Thus, while the invention came about sometime in the early 1500s, it would be another century before serious concentration of the industry would emerge.

The first organized sign of the British industry was the formation of the Clockmaker's Company in London in 1631. David Ramsey, clockmaker to the king, was elected the master of the company. The most important right bestowed upon the company at its incorporation was the ability to regulate the watch trade within 10 miles of London. This authority extended to the ability to examine all products considered of a kind, like a watch (Glasgow, 1885). Since the industry emerged at the time in which guilds were the primary form of industrial organization, the early British industry conformed to the model of a firm headed by a master craftsman and serviced by many apprentices.

The number of watches produced at any one time and by any one master remained small for almost 100 years. During this time, innovation was constant as master craftsmen solved some of the more vexing problems confronting the industry, most especially those related to the power system in a watch. A heightened level of innovation was fueled by national contests developed specifically to build the first easily used portable chronometer. John Harrison spent 20 years of his life achieving this feat. He was rewarded with the princely sum of £20,000 for his efforts (though the task wore him down and probably contributed to a shortened life).

Throughout the next century, until the early 1800s, watches remained an item of high quality and high cost, and hence of limited circulation. The British were masters of the high-end watch market, fashioning heavy, round globes of intricate parts into finished watches.

The British watch industry had two main geographic focal points: Lancashire and Coventry. The industry began, as it did in many countries, in a major city: London.[2] Its first activity was the repair of clocks and watches made in other countries. Once production of British timekeepers took hold, the industry moved quickly out of London to less expensive areas within the countryside. Of course, this did not occur before the industry had been organized as a craft guild, in which master watchmakers oversaw the work of many individual craftsmen.[3]

The shift in the location of the production of the watch movement took place at the end of the 18th century. While watch assembly and repair continued in the capital city, movement manufacture took place closer to complementary activities. Its early location near Liverpool can be attributed to a series of events linked to the development of inventions that made the regulation of time within a watch easier and more consistent. Because Liverpool had been an early site of inventions associated with the chronometer, many new ideas arose in and around the area. Being a center of trade also assisted the industry's development, as it was quite easy for producers of watches to access markets outside of Britain through the port of Liverpool.

The relocation of watch movement manufacturing to the countryside around Liverpool occurred during the late 18th and early 19th centuries when it became obvious that the more labor-intensive and hence more costly activities could be broken off from final assembly and shifted to lower cost locations nearby (Glasgow, 1885). As one early author noted,

> It soon became evident that it would be economizing to have the rougher part of the watch made by less skilled workmen than those employed to complete it, and that it would also effect a saving to have the work done in a village, where living was cheaper than in a large and populous town. It was thus that Prescott, being close to Liverpool, became the seat of the movement trade, as, besides having a small local watchmaking trade, it had a reputation for the production of the best files and other tools used in the manufacture, and although from various reasons chronometer watchmaking declined in Liverpool, movement making is so firmly established in Prescott and its vicinity that there is little likelihood of it now migrating to either London or Coventry, which are the present manufacturing centers. (Glasgow, 1885, p. 38)

Still, the number of watches produced was limited. Early records suggest that a very successful producer, Liverpool manufacturer Robert Roskell, in the late 18th century was capable of producing 30,000 watches in a year. These were shipped to the Americas. Roskell's success was emulated by others who clustered around Liverpool and shipped watches around the world (Glasgow, 1885).

Dual System of Production

The structure of manufacturing in Britain at the end of the 17th century consisted of a dual system of small-scale manufacturing for local markets and a factory system organized by merchants to serve a broader geography (Mantoux, 1928).[4] Watchmaking, though initially an urban industry, largely linked to watch repair and very limited manufacture of single pieces for wealthy patrons, was still a cottage industry as late as 1725, with the emergence of the metalworking region of Lancashire concentrated around the town of Prescott (Aikin, 1795); see Figure 4.1.

Metalworking in this region can be traced to the transformation of the primary metal production sector of the 17th century. Prior to the 1700s, iron forging was highly dispersed, since it relied upon charcoal made from trees. Both the small-scale and highly decentralized nature of

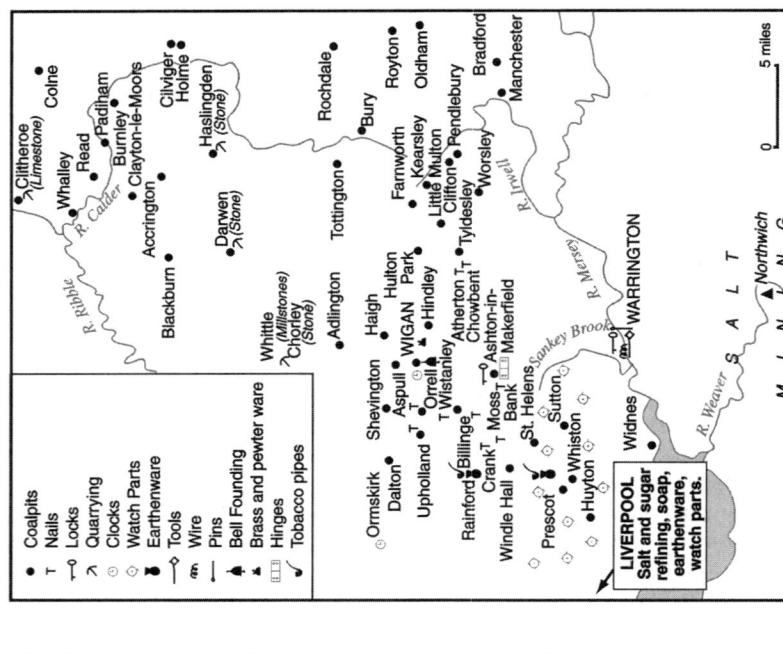

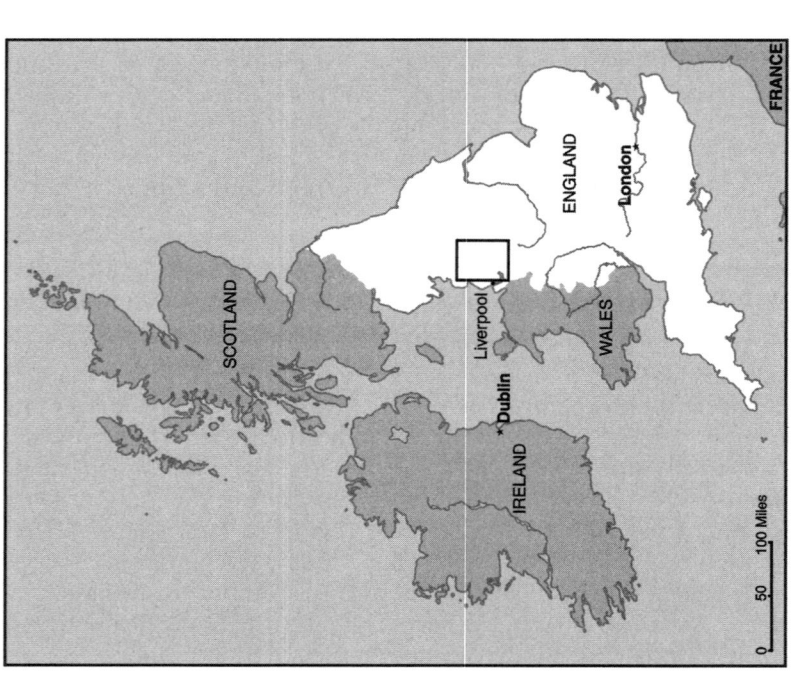

FIGURE 4.1. Mining and metal industries—about 1725 (Liverpool, England region). Cartography: Erin Heithoff. Sources: Map-Art; Bagley (1961), *A History of Lancashire*.

iron forging resulted in an inefficient and backward industry (Mantoux, 1928). While the inefficiency of the industry would seem to have been enough to force it off its standard, international imports from Sweden solved the problem of quality metal shortage and reduced the stimulus to improve the industry. Ironically, the fear that the entire country would be denuded of trees hastened the shift of the industry toward the region around Manchester, where coal deposits were abundant.

Alongside the metal-forging industry, a secondary metal-forming industry—more decentralized and specialized—existed (Wadsworth and Mann, 1931). Early on, small-scale metal-fabricating firms located near the coalfields, presumably because of the availability of a fuel resource needed to make molten metal. Its location also figured into a larger social and economic context since metalworking was an industry that allowed its participants to maintain an attachment to the land (Mantoux, 1928). This was important to agriculturalists who participated in manufacturing to supplement their farm income and to fill time during winter months. Thus, while the presence of coal is important in explaining why the watchmaking industry formed in Lancashire, it is by no means either the sole or the most important explanation for the formation of the cottage industry in the Lancashire region. A second major factor explaining the location of early watch manufacturing was the proximity of Lancashire to the port of Liverpool, which by the middle of the 17th century had become the second most important port after London for British mercantile shipping trade to markets throughout the world.[5] The metalworking industries, including watchmaking, and the textile industries grew up alongside one another, in the shadow of the burgeoning seafaring industry of trade.

The metalworking industry had humble beginnings. Perhaps the most pervasive product was nail manufacturing, which served both the growing domestic market and the shipbuilding market of Liverpool. The growth of the U.S. market for nails also was an important stimulus of demand, which further cemented the relationship between metalworking and trade (Aikin, 1795; Bagley, 1961; Barker and Harris, 1959).

The most important derivative industry leading up to the watch industry was the manufacture of metal tools and in particular very fine files, fundamental technologies needed for the production of small watch parts. Some of the first references to this complementary industry and the capability to make watch parts date to 1670. By the middle of the 18th century, watch parts were already being "put out" among small farms and cottages scattered across the southwest Lancashire plain. Parts makers, each of whom concentrated on the manufacture of one or two particular parts, lived within the vicinity of the village of St. Helens (Barker and Harris, 1959, p. 127).

Another factor contributing to the industry's evolution within the Lancashire region relates to the links between agriculture and early manufacturing practices. Up until the middle of the 18th century, cottage workers retained access to the land around their homes and farmed part-time. Hence, production costs were lower than in the more central areas, including Liverpool and greater Manchester (Mantoux, 1928). This was important since the wage bill was by far the most expensive item in the total cost of watch manufacturing. The cost of raw materials, other than the gold and silver used in the production of expensive cases, was a trivial sum. Indeed, the Lancashire putting-out system produced rough movements at such low prices that by the end of the 18th century, if not before, the Lancashire manufacturers were supplying most of the great watch firms in London, Coventry, and Liverpool, with both movements and the tools needed to finish them (Barker and Harris, 1959, p. 127). The elaborate system of production was sufficiently fragmented and decentralized to allow finishers to obscure the actual place of parts and movement manufacture. As Barker and Harris note,

> The man who performed the final operations of finishing and assembling the watch places his name and town on the dial. This concealed the fact that all the earlier precision work inside the watch had been undertaken by a number of domestic craftsmen in and about Prescott. The contribution of southwest Lancashire to national punctuality, in an age of industrialization of which the essence was accurate timekeeping, deserves more widespread recognition than it has been so far accorded. (1959, p. 127)

Distance was an important impediment in the early 18th century. Roads in general were terrible, thus making travel between Liverpool and London arduous and uncomfortable. So much production was moving out of the region toward the port that to maintain and ensure a steady flow of goods the road system had to be drastically improved (Bagley, 1961). By the early 1700s, Prescott, the early center of the watch parts industry, managed a toll road that was maintained in top condition to ensure that goods could move year-round to the port of Liverpool.

Thus, we can date the Lancashire center to the early to mid-1700s. By the late 1700s it was producing 150,000 to 200,000 watches per year (perhaps half for export). Labor in Lancashire was cheaper than in London, so specialty work on watch movements was done there in dispersed cottages and finished in the capital. According to Landes (1983), "From Prescott to Liverpool, eight miles as the crow flies, the countryside was dotted with the cottages of spring makers, wheel cutters, chain makers, case makers, dial makers—every specialty that went into the making of a watch" (p. 231). An even more elaborate description of the region in 1795 provides a flavor of the industry complex that had grown up outside of Liverpool in the town of Prescott:

> [Prescott is] the centre of the manufacture of watch tools and movements, of which we shall proceed to give an account. The watch tools made here have been excellent beyond the memory of the older watch-makers; and the manufacture has been extended by improvements in making new tools of all sorts and the invention for first cutting teeth in wheels of watches, and afterwards for finishing them with exactness and expedition. The drawing of pinion wire originated here, which is carried as far as to fifty drawings, and the wire is completely adapted for every size of pinion to drive the wheels of watches, admirable for truth and fitness for the purpose but left for the workmen to harden. . . . They make here small files, the best in the world, at a superior price, indeed, but well worth the money from the goodness of the steel, and exactness of cutting. . . . They make watch movements most excellent in kind, which is greatly owning to the superior quality of their files and tools. They likewise excel in what is called motion work, such as dial wheels, locking springs, hour, minute and second hands, etc. Main springs, chains for movements, and watch cases, were not part of the original manufacturer, but are now made here. . . . The tool and watch-movement workers are numerously scattered over the country from Prescott to Liverpool, occupying small farms in conjunction with their manufacturing business, in which circumstance they resemble the weavers about Manchester. All Europe is more or less supplied with the articles above-mentioned made in this neighborhood. (Aikin, 1795, cited in Smith, 1973, p. 12)

Statistics from the *1854 Prescott Directory* provide a basis for understanding the extent of the production system disaggregation:

Watch- and clockmakers	5
Watch balance makers	20
Watch escapement makers	10
Watch material makers	4
Watch movement makers	120
Watch pinion makers	31
Watch toolmakers	43
Watch wheel makers	28

As reported by Smith (1973), former general manger of the Lancashire watch company, very few watch- and clockmakers could be counted among the watch complex participants. This suggests the role played by the region as a sourcing site for the rest of the nation's industry, but not necessarily the site of new process and new product innovation. Still, the technological superiority of the industry cannot be underestimated, as the following description of the output of the complex in the same year demonstrates:

> It may be thought a bold assertion, that the best movements for the very best watches made up in London are frequently the efforts of ingenious artists of that Branch in the Town of Liverpool and its vicinity, after which they are put together, adjusted (or what is more properly the technical term finished) by Artists in the Metropolis—not that a watch always bears the name of either finisher or motioner, the Wholesale Dealers or principal Shopkeepers are the names generally which appear upon the time pieces. . . . In the year 1772, it is supposed there were made up in Liverpool and its neighborhoods not less a number than five thousand watches some of which were very valuable. (Smith, 1973, quoting John Holt Walton's 1795 description of watch manufacturing in Lancashire in the Holt and Gregson manuscripts at the Liverpool Public Library, p. 13)

But there was a side effect to this high level of disaggregation, as the managing director of the Lancashire Watch factory, T. P. Hewitt, noted on the eve of launching the only attempt at factory production of any consequence in 1893:

> Our system of manufacture made it impossible to compete with makers of cheap watches, except by submitting to reduction of wages and profits, which when attempted were destructive, as it drove out capital and labour which found better renumeration elsewhere. We did not fail for lack of skill on the part of our workmen, as the very men we failed to find employment for in England, went abroad, became the most formidable competitors with those of us who stayed behind, and began to send watches into this country. . . . The fact is that it was the absurd *method of manufacture* that lay at the root of our decay. Just let me for the moment attempt to describe how I was educated in the way of "getting up watches" as it was called. I commenced work in my father's movement-making shop in which were a few hands for assembling the parts as they came from outside workers. These workers were scattered over an area of some six or seven miles, without any means of public conveyance. We had no recognized gauges in the trade, each movement maker having his own arbitrary standards. Therefore, after the frames were made, we had to carry them about from workshop to workshop in order to have all the other parts made to each individual frame. On going to Liverpool to finish my apprentice-ship, I found this same system in vogue among watch manufacturers; indeed the wastefulness was intensified as, instead of carrying about dozens, as I did in Prescott, we carried them one by one. Now, gentlemen, bad as this system was, it demanded some virtues on the part of the men who worked it—notably, a stout pair of legs, and a good temper under disappointment. If you followed a methodical man, all went well, but if it was your misfortune to follow a not too industrious or too so-

ber man, well, you can imagine the situation that would arise. . . . Well, gentlemen, this system was somewhat improved when we made standard frames in Prescott, and we to some extent adopted the factory system, but it did not go far enough, as first it was requisite to get the whole of the branches of the watch concentrated under one roof, and employ all the machinery we could, so that all parts of the watch could be produced simultaneously, and require only a minimum of fitting. . . . Although we all know the impossibility of making watches under this old system, we had to recognize the fact that, like Sinbad the Sailor, we had the system on our backs, and although we were more and more exhausted the longer we carried it, the problem still remained how and when we could get rid of it. (quoted in Smith, 1973, p. 11)

For unrecorded reasons, the link between watchmaking and the formation of a pool of skilled engineers never materialized. The surrounding metal industries were dependent upon the knowledge of industrial processes drawn from firms outside the region. Although not stated as such, this may be one explanation why the Lancashire watch industry never really progressed beyond putting-out parts manufacturing organized around a dense system of production for assembly in other locations.

As happened with its antecedent industry, nail manufacturing, external demand for more volume, combined with the discovery of new methods of mass manufacturing (defined on the basis of the day) signaled the slow and inexorable decline of the world's pioneering watch region. Many distinct, yet overlapping, forces precipitated the end of Lancashire's hold on the world watch industry.

Unlike other industries, which succumbed to at least some mass-scale production, watches retained their status of a product for the well-to-do and small well-heeled foreign markets. The British held tight to both the prevailing product and production styles far longer than they should have. Because manufacturing strategies depended on a monopsonistic relationship with finishers in London and Coventry, independent producers of components had no incentive to produce a uniform product.[6] This persistent dependence is captured in the following passage: "The only competitors of firms in London were Coventry men who were interested 'a little in one class of works and movements.' Coventry was more a center for the assembly of finished watches, as were London, Birmingham, Liverpool, and 'a few outstanding places in the country' " (Glasgow, 1885, p. 30).

Thus, domestic distribution channels sent the wrong signals just at the time the market for watches was changing fundamentally. Rather than change either strategy or product, the British succeeded in retaining domestic market share (at least temporarily) by raising tariffs on the superior, cheaper Swiss product. One noted authority encapsulated this problem near the end of the 19th century:

> But hitherto the attention of English watchmakers has been directed to the improvement in the quality of watches, rather than to their cheapness of production, and it is greatly to be questioned whether the introduction of the factory system, and the wholesale adoption of machinery in manufacture, would at all benefit the trade in this country. English watches have always commanded their price in the markets of the world, and the good name they have ever borne has led to their imitation by foreign producers, both the forgery of English makers' names and of the English hall-marks in the cases, and by sending unfinished cases here and getting them marked at the Goldsmith's and the provincial halls through the agency of case-makers and others. And what would become of their esteem if, instead of continuing to maintain our high position, we were to compete with the slop trade of Switzerland or America? There are many people who will tell us that some wretched thing in the form of a watch "goes well enough for them," but is such an argument sufficient to lower our standard of excellence? (Glasgow, 1885, p. 34)

The belief in the supremacy of the British industry was undermined by British producers themselves, some of whom engaged in illegal importation of Swiss watches that they then marked as British products. The seeming unanimity of the complaints about Swiss producers only served to disguise the behavior of British watch producers. Further contributing to the narrowness of Britain's product scope, merchants failed to send the required signal to British producers about the need for a change in product and strategy. Indeed, although the evidence is sketchy, it appears that merchants, especially those in London, were satisfied to keep British production as is, thus assuring a continuing supply of luxury watches. They did not need the British watchmakers to move into new markets for lower cost, more abundant watches, as they had ready access to Swiss producers and distribution links to markets other than those in Britain. By maintaining separate markets, merchants could play one producer off against the other; this was particularly effective when the British industry began slipping into decline. There is some anecdotal evidence that the merchant function itself was highly segmented, with upscale distributors owning shops in key markets such as London, while roving merchants distributed supplies of less expensive watches manufactured in other locations, such as Switzerland and later the United States.

Accusing the Swiss of unfair labor practices and of making cheaper products bought time in Britain, but it did nothing to help encourage exports. Between 1800 and 1875, when world watch production skyrocketed from 350,000–400,000 to about 2,500,000 units, British production decreased from 200,000 to about 170,000 units.

Attempts at Change

The Coventry center made an attempt at factory production. Watchmaking in Coventry started relatively late (in the mid-1800s), with one or two large companies dominating the trade. As in Lancashire, the production strategy of firms in Conventry was first centered in cottages. The formation of the first factory was highly simplified: in 1843, the Rotherham Watch Company consolidated its workers in one building.

Many components continued to be produced in Lancashire and sent to Coventry for assembly, however. Despite factory production, watches produced in Coventry were never intended for a mass market but instead for the merchants of the masses. Thus, Coventry's industry died, supplanted by explosive growth in the motor vehicle industry.

Another failing that served to weaken the industry's ability and willingness to change was the lack of technical skill development. Early records of the effect of the byzantine apprenticeship system reveal a gross underdevelopment of competencies needed to move from one generation of technology to another:

> The want of technical education has kept the trade back more than anything else. Under the old apprenticeship system, the pupil was usually kept running errands and making himself generally useful for the first year, or perhaps more, after which he was promoted to the bench, when, after learning perhaps to make pivots, a little filing, etc., he was often left to his own resources to pick up whatever he could. It was with the empirical knowledge thus obtained that most of the past generation of finishers and others were formed into workmen, and it is not astonishing that the watch manufacture of the country has not taken a higher rank amongst the sciences, when it is considered in whose hands it has hitherto been left; indeed, the surprise should be at the progress made by men who have all along been working by what may be termed the rule of thumb. (Glasgow, 1885, p. 35)

Could it be that the British watch workers, having been brought up in a system characterized by hierarchy and ritualistic exchange of know-how, resisted the application of machines to their work because they were well aware of their own incapacity to perform? It seems reasonable to suspect that a system based on rituals and learning by doing rather than rules and more codified knowledge could engender a knowledge base that was simply insufficient to move from one technology to another. Workers could have resisted change because their previous history had resided in a knowledge metered out in an often highly inequitable fashion. Given that a worker's wage was based at least in part on what he or she could accomplish technically, why should anyone have agreed to the implementation of new capabilities when there was no guarantee that returns would be had from such a gesture?

Perhaps an equally debilitating force reducing the flexibility and influence of the British watch industry was the imposition of tariffs by the United States beginning in the 1860s, which reduced demand for metalworking products in general. As Barker and Harris (1959) report,

> These protective duties did great harm to the domestic watch making industry of the district as well. A 30 percent tariff was imposed on watches entering the United States and a considerable encouragement thereby given to American producers. The output of American-made watches grew eightfold between 1862 and 1872, in which year three times as many watches were being made in the United States as in Great Britain. (p. 87)

In the literature on Great Britain's watch industry, one word recurs: conservative. Most sources describe the British craftsman as a stodgy man who, knowing the right way to make a watch, saw no reason to change (Hoke, 1987). Unfortunately, this conservatism was replicated at every level of the production system. Rather than seeking constructive ways to change the industry and make it more competitive, watch industry leaders formed a powerful association and lobbied Parliament for protective import restrictions, particularly against Swiss products. Additionally, geographic dispersion and cottage production made it difficult to concentrate capital investments in the industry. Most watchmakers owned their own equipment and could not afford to invest in the expensive machines that would have allowed them to compete with the Swiss. As Barker and Harris (1959) report, "Meanwhile the Swiss were developing their own industry and many of their products were competing with the British on the home market: imports of watches from Switzerland rose from 42,000 in 1853 to 90,000 in 1855 and 160,000 in the early 1860s" (p. 370).

As with other British industries, reports also suggest an insufficient culture of entrepreneurialism capable of recognizing the value of and need for making a potentially large capital investment. This cultural proclivity is often blamed for the outright failure of many British industries to make the transition from being "industrial leader" to "industrial competitor." No doubt this claim has some truth. However, as Gerschenkron (1962), Sandberg (1974), and others report, switching technologies and whole systems of production was not simply a matter of insufficient motivation. To change production systems would have required wholly new labor relations, new distribution channels, and a substantial abandonment of sunk investments in existing technologically laggard, and yet still productive, enterprises.

This reluctance to change and the impact of cheaper Swiss imports were not the sole culprits in the British watch industry's decline. More convulsive was the introduction of the "American system" of mass production. The U.S. armaments industry pioneered a form of parts manufacturing that was applied, albeit incompletely, to watch manufacturing. The system of mass production reduced the role of skilled labor in watch manufacturing, expanded production volume exponentially, and drove down the price of watches to within the grasp of many working people. More importantly, it allowed the organization of production on a scale and at a price that simply could not be achieved using old methods of cost reduction, including reducing the wage bill. As Barker and Harris (1959) report,

> The men who owned their own workshops sometimes chose their own masters. They sometimes worked for more than one at the same time. This free choice of employer was to the domestic worker's advantage when his handiwork was in great demand but when trade deteriorated during the 1860s, the choice became severely restricted and the men considered themselves lucky to find anyone who would give them work, even if they had to accept part of their payment in kind. (p. 92)

Even as the British attempted to introduce this new technique for organizing production, institutional and cultural myopia limited its transfer. British industry's earlier success with high-precision chronographs and other special precision timepieces "reinforced its rigid structure" (Davies, 1993, p. 44). The high-precision timepieces precipitated a highly specialized and extensive division of labor. Watchmaking was labor-intensive and low paying, while the end product sold for great sums. The extreme division of labor and extensive emphasis on exterior watch case artwork yielded a highly complex device that was financially out of most people's reach. British watches were far more complicated than they needed to be for a mass audience. They were easily underpriced by Swiss and U.S. watches.

Why did the British refuse to produce cheaper and more accessible watches by adapting a system of manufacturing similar to that pioneered in the United States? Doing so was not outside the technical ability of the nation's industry. In 1855, the British sent evaluators to assess the new U.S. system and to ascertain the ability of British industry to follow suit (Rosenberg, 1969). By the 1860s the British armaments industry had successfully adapted American organizational techniques to weapons production (Davies, 1993). Why couldn't the same be done for watches?

First, skilled labor resisted the notion of making a cheap watch, believing it to be an insult to the existing industrial culture to produce anything less than a beautiful timepiece. The British style included making "complicated" watches based on a very costly design using fragile components easily warped during use. British watch design also made the British watch very large for its day, far bulkier than the cheaper and relatively accurate Swiss and U.S. pocket watches.

But the workers' response was more than just an effort to protect the value of skilled labor. The history of the prefactory system of production organized by manufacturers and entrepreneurs can be summed up in one word: brutal. The watch parts makers of the Lancashire district were witness to more than 100 years of debilitating exploitation of workers in the textile industry. The memory of children literally sold by their villages to work in the early textile factories, where the means of control was the whip, left an indelible mark on the minds of working men and women of the 19th century. The watch industry avoided this mass exploitation, in part because watches were still considered luxury goods produced in small lots and thus not subject to the gains associated with factory production. The lack of standardization further removed watches from the ability of entrepreneurs to reap huge gains from factory assembly of former cottage producers (Mantoux, 1928). It took skill and patience and some great personal and physical sacrifice to spend the required countless hours filing minute metal parts to a form that allowed them to work in conjunction with others.

Third, undertaking and perfecting interchangeability required developing complex and highly accurate machine tools capable of working at very fine levels of precision and accuracy. The British armaments industry had developed machine tools to produce a rough system of interchangeable parts manufacturing. But these machines were far too coarse and inexact for watch production. And, as I reported earlier, the necessary engineering skills needed to perfect machines capable of holding the required tolerances for the mass production of watch parts did not reside within the watch region (Barker and Harris, 1959).

Alongside the development of the American system came restricted access to this new and burgeoning market. The windfall of the Civil War, which created tremendous demand for domestically made watches at a time of restricted market access, was forcefully followed afterward with arguments about infant industry protectionism and further market constrictions, diminishing any hope that the British would have access to the U.S. market (Aldcroft, 1968; Sandberg, 1974).[7]

Finally, and perhaps more decisively, the British Horological Institute that was created to protect the interests of the skilled watch manufacturers could not cajole them into accepting the new technology. The union actually restricted apprentices' learning of the new technology, and the institute retreated from taking a position on the need for change by arguing that it was a "benevolent, educational, and social and not a commercial manufacturing organization" (Davies, 1993, p. 45). Unlike the armaments industry, which was totally reorganized by order of the national government, watch manufacturing faced no similar absolute authority or imperative. The institution critical to the industry's perpetuity, the Horological Institute, participated instead in its slow demise.

Strategic assessments made by industry leaders also came into play in the decision not to pursue the American system. The British industry's claim to fame was precision and beauty. British watchmakers looked to the Swiss as the real enemy because the Swiss were capable of making precision products of exceptional beauty. The system of manufacture employed by the Swiss was very much akin to the British model of an extreme division of labor implemented within a cottage-based putting-out system. Thus from the perspective of keeping an eye on one's competitors, the British drew more inspiration and felt more directly competitive with the Swiss, even as the U.S. market emerged as the most important to the British. This is a fine example of how emulation is based in time and space. Had the British looked to the U.S. model as the one to beat, perhaps the long-run outcome would have been very different, since the U.S. system proved capable of unseating both the British and the Swiss in short order.

The Long Road Downhill

By the middle of the 19th century, any effort to reverse the trend of British industrial decline was futile. Many remaining employers continued to reduce wages as a means of trying to preserve their businesses. Others strove for efficiency by applying some measure of mechanization, albeit on a small scale. Many formerly independent contractors moved into workshops owned by their employers; by 1860, more than half of the watch workers were working outside of their homes (Barker and Harris, 1959). There were even efforts to economize on the use of labor and materials by focusing on a few standard sizes. By 1885, the historic system that had once provided watches for kings, "disintegrated" and in 1885 "the confined and decayed little factories" were reported to be almost deserted (Barker and Harris, 1959, p. 373).

By the end of the 19th century, much of the industry was erased from the landscape as other more lucrative endeavors took over the watch industry factory shells. An attempt at standardized factory production in 1895 with the creation of the Lancashire Watch Company lived only long enough to produce a few watches before even its valiant efforts failed, in large part due to poor marketing and the resistance of former watchmakers to yield to the yoke of contemporary watchmaking techniques.

By 1914, many of the companies with famous names and established brands had disappeared. Most of the failure can be attributed to an unwillingness by manufacturers to pursue new modes of production. Yet some firms in both Lancashire and Liverpool pursued the use of machine-tooled movements, if only half-heartedly. A union of paternalism and poor business sense, combined with bad marketing, led to the demise of other firms. While there were isolated examples of successful firms, this was no counterweight to the industry's otherwise stodgy performance.

For Coventry, the demise of watches occurred in lockstep with the expansion of other industries, particularly bicycles and automobiles. Lancashire, a notable textile and linen center, also recovered, but only momentarily, from the loss of watches. Interestingly, in the later history of Lancashire's linen industry, many of the same problems associated with resistance to change hobbled it, too.

Summary

No single factor can be held accountable for the demise of the British industry. Certainly, the myopia and rigidity of production system leaders hobbled the industry. Moreover, had not national policy restricted access to British markets, the British industry may have been forced to change. Distribution channels hampered attempts to introduce more modern products by essentially threatening not to sell British products that deviated from past practice. Institutions created to perpetuate change, such as the Horological Institute, lacked the institutional legitimacy and authority to either lead or coerce participants to make needed change in the industry. And the Swiss, opportunistic on the one hand and unencumbered by monopsonistic distribution channels or rigid industry patterns of behavior on the other hand, quite easily found a niche in the illegal selling of watch parts to renegade British watchmakers, who in turn placed their name on the dials and encased the watches, so that they became British watches in name only.

What can we conclude? A highly fragmented system is vulnerable to external shocks that reorder the economic environment in which firms must operate. Alone, any one of the factors that confronted the British system could have been successfully managed. Combined, and coming from several directions and in rapid succession, no single response would do. The technological discontinuity of the U.S. system of mass production challenged the fundamental basis of the industry. As in other examples of the time noted by Usher (1920), who wrote of the decline of British industry at the end of last century, the rapidity with which change was needed and the relentless onslaught of the several forces calling for change proved to be insurmountable. One can only speculate that the singular focus of the British on the Swiss, as the country's major industrial competitor, could have distracted British firms from truly comprehending the meaning of the new U.S. industrial system. Locked in a competitive battle, from which there appeared to be no escape, the British may have been blind-sided by the far more ominous threat embodied in the new system of manufacturing pioneered by U.S. firms. For it was the U.S. system that finally sounded the death knell for the British industry.

Thus this chapter tells a story of the sweeping effect of technological and broader environmental change on one nation's industry. The specific moment of the emergence of the British watch industry, occurring in tandem with the Industrial Revolution, might simply suggest that unique factors associated with being first ordained the outcome. And yet, as the ensuing chapters suggest, history repeats itself. As technological progress is made, the prevailing system is eventually unseated. It is not simply a change in technology that revolutinizes the system, but rather it is the sum of all ancillary adjustments that occur in the existing system in response to the original stimulus for change. In the next two chapters, the histories of the Swiss and the American industries are recounted; they, too, were forced aside by new technologies and new and unforeseen economic circumstances that proved quite intractable and difficult to surmount.

5 Why Switzerland?
The Rise of the Jura System of Watch Manufacturing

Why Switzerland? (1970), an aptly titled book by Jonathan Steinberg, a lecturer at Cambridge University, recounts the origins and development of this tiny country squeezed between two behemoths: France and Germany. My book, too, asks, Why Switzerland? Why did this tiny, multicultural, multilingual country that straddles the European Alps become the center of the world watch industry in the 19th century, at a time when Britain and the United States had already established dominant paths within it? The Britain of the 18th century clearly was the economic powerhouse of the industrialized world, capable of producing watches for her own population and for populations around the world (Baillie, 1928). The United States, though an upstart compared with Britian, by the middle of the 19th century pioneered a new system of manufacturing that within 25 years overthrew Britain's dominant position in the volume watch industry. Accounts of the decline of the British industry usually reference the unmistakable influence of the U.S. industry, but it was in fact the Swiss who rose to prominence by the middle of the 19th century and took control of much of the watch market on the Continent.

To put to rest the first question, Why Switzerland?, one needs to examine the economic prosperity of this tiny country. As early as the 16th century, Switzerland was "one of the richest, most populous, and militarily most formidable European states" (Steinberg, 1979, p. 20). Her military supremacy over her neighbors was brought to an end by a series of military losses to the French. While it may be hard to fathom how military losses could have benefited a country's development, these losses laid the basis for the country's enduring stance of neutrality, which in turn promoted prosperity.

Neutrality explains a great deal about how Switzerland rose to such prominence amidst much larger and more resource-rich neighbors. By maintaining neutrality from the 16th century onward, the Swiss benefited from their militaristic neighbors' losses. Being neutral meant that as other countries fought and destroyed, or were destroyed by, their rivals, Switzerland stood on the sidelines ready to provide the food, arms, and other equipment essential for the waging of wars (Meier, 1970; Steinberg, 1976, p. 23).

Although a neutral country, Switzerland could not escape its own internal religious wars. In fact, throughout the 18th century, this source of constant cleavage actually served to strengthen the resolve of the Swiss toward neutrality. In searching for an explanation of why Switzerland persevered and thrived while other nations suffered, Steinberg suggests that

> what differentiates Swiss history from the European pattern is the outcome. Swiss communities built from the bottom up, growing out of free peasant or urban associations, and are in a curious sense bottom-heavy, rather like those dolls which bounce up no matter how often the child pushes them over. The weight is at the base. The communities have deep equilibrium to which, as the point of rest, the social and political order tends to return. (p. 27)

The "bottom-heaviness" acknowledged by Steinberg refers to the consensually based system of decision making that has come to define sociopolitical relations within Switzerland. The highly decentralized, canton-based structure of political representation ensures lengthy deliberation around common issues that make up daily life. This deliberative system impacted other institutions, including economic practices.

This deep-seated drive to return to a position of stability says much about how the Swiss emerged as the world's center for the watch industry in the 19th century. For more than 150 years the Swiss watch industry bounced back to a position of preeminence in the face of many trials and tribulations, including protectionism, war, and technological discontinuities. This seeming invincibility was not to endure forever, however, as the more recent history of the Swiss watch industry illustrates. Faced with a single apparently insurmountable obstacle or technological rupture—for example, the emergence of the American system—the Swiss endured and in some cases thrived. But in the face of many obstacles that converged in space and time, the Swiss watch industry was eventually unable to persevere. The bottom-heaviness of the Swiss culture of deliberation also meant slowness to change (Steinberg, 1976).[1]

In this chapter I recount the early history of the Swiss industry—an industry that vied for control of the world watch industry from the 18th century onward. In its earliest days, the Swiss industry looked, at least on the surface, much like its most formidable rival, the British industry: it was highly disaggregated and spatially dispersed. But, as with every story, the devil is in the details. For while the Swiss may not have significantly challenged the British system until the beginning of the 19th century from a technological or a commercial point of view, nonetheless, within less than 50 years Swiss watchmakers were not only close on the heels of the British watchmakers, but were far ahead of them in important ways. This capability did not emerge out of thin air. In fact, significant evidence suggests that by the early 19th century the Swiss were quickly approaching technological parity with the British; by 1830, they were capable of making a thorough assault on the British position of dominance. British industry also must share the blame for its own demise. An inability—some report an unwillingness—to act in the face of significant information about the new competition caused the British industry to fall further behind the new industry leaders. Some authors suggest that Britain was blinded by the international recognition it earned from being the first to manufacture a chronometer, while others argue that the competition to create the first truly operational device to measure longitude actually stunted the British watch industry of the 18th century because the best minds were drawn off in search of chronometry.[2]

The Swiss watch industry's rise to preeminence alongside but not in direct confrontation with the British watch industry was due to a different attitude and a distinct organizational structure. A review of Swiss watch manufacture starting in the 17th century reveals that at the time of the early development of watches, while the British dominated production for their own market, external trade was far more fragmented. An insatiable worldwide demand for watches hid the rising capability and power of the early Swiss industry. In the 18th century, the world was still quite geographically fluid; thus, market penetration was not the exclusive domain of any one country.

To explain the rise of the Swiss industry, it is essential to first recount the role played by the French industry in the early continental manufacture of watches. Paris was always a center of watch innovation, but never a center for watch production. French competition with Great Britain to produce an accurate timepiece began in the mid-1700s, when the two countries vied for dominance of the oceans. To become dominant, both needed accurate timepieces that would allow them to calculate longitude (Landes, 1984). Gaining this technology would give one country a significant military advantage. An early "arms race" to create the chronograph ensued; its consequences were far more practical than the weapons buildup of the 20th century. While competition between the British and the French took up valuable resources, the Swiss benefited mightily from the talent and ingenuity of French watchmakers who traversed Switzerland's trade routes with considerable ease. In this instance geographic proximity paid off. What can be said of the French industry is that its preeminence as a producer of brilliant and extravagant products destined for kings and queens could never withstand the onslaught that mass-producing watches for a larger audience would cause. Well into the late 18th century, France could still claim to be home to the most famous names in the watch industry, such as Breguet (perhaps the most famous); nonetheless, this technological mastery did not translate into successful larger scale manufacture. Although several early innovators lived in Paris, they all had to go either to Great Britain or to Switzerland to see their ideas put into production. By the end of the 18th century, France was importing watch parts and whole assemblies from its southern neighbor, Switzerland, which by that time had risen to world-class stature (Baillie, 1929).

Several authors assert that France's expulsion of the Huguenots in the 16th century permanently hobbled French watch production. Protestant values made this group more efficient, dedicated, and compliant producers. Many emigrated to Switzerland, settling in areas that have since become top watch-producing centers of the world (Brearley, 1919). Hence, France deserves credit for its watchmaking innovators, but also blame for forcing out a group who would go on to produce watches in Switzerland.

The French industry did not entirely die out.[3] It retained a role in the production of fashion watches and was known for the manufacture of midpriced mechanical watches. Several attempts were made to revive the industry from the early 1800s onward.[4] This decline was punctuated by important bright spots, which included the machine production of parts for ebauches dating to the late 1700s. The French clearly did not lack creativity or competency, but they were incapable of matching the marketing and business know-how of both the British and the Swiss.

The Emergent Systems: Switzerland and the United States

Geographic space, explosive growth in the global market for manufactured goods, and unique intervening events allowed the development of two separate systems of watch manufacturing in the mid-19th century. Earlier, we saw that the emergence of both Swiss and U.S. centers of watch manufacturing combined to bring about the decline of the British industry. The early development of both complexes each constitutes a story in itself, and represent necessary journeys in the story of the world watch industry. Here we depart from conventional wisdom by first asserting that the difference in context explains the emergence of these formidable competitors, each in its own right and on its own terms capable of unseating Britain as the center of this industry. Second, we suggest that the Swiss system of production deviated organizationally from that of the pioneering British in that the Swiss industry grew up based around much higher ratios of watchmakers to parts manufacturers. Hence, there was greater variety in terms of both product and price. What distinguishes the Swiss system from that of the British is not the geographical organization of the industry, as the historical record indicates that the Swiss employed a putting-out system similar that created by the British. Rather the difference is found in the early technological know-how imported from the adjacent region of France, combined with an organizational logic that laid the foundation for skilled manufacturing to define the nature of production. The Swiss overtook the British by first becoming excellent and far-reaching traders, and, second, by becoming the watch parts supplier for the world. Low costs, high precision, and a period of technological indeterminacy combined to derail a competitor that had become self-satisfied with existing patterns of trade.

Early History of the Swiss Watch Industry: One Nation's Loss Is Another One's Gain

Geneva did not have a significant watch industry prior to 1550 (Jaquet and Chapuis, 1970). According to numerous accounts, however, in the mid-1500s, the expulsion of the Huguenots from France became the single most important early factor in the development of the Swiss watch industry. The relocation to Geneva of some of France's most skilled and talented watchmakers created a ready-made technological capability that launched the nation into the world of watch manufacturing (Baillie, 1929).

Geneva was particularly welcoming to the new residents. Although the largest city and the capital of Switzerland in the 1500s, Geneva was declining as a mercantile center. The city needed industry and lacked either watch- or clockmakers (Landes, 1983). Thus, the Huguenots were accorded special privileges and given significant assistance in establishing the watchmaking industry. Not surprisingly, Geneva became a preeminent center of watch manufacturing—the city had been an early center of jewelry manufacturing, which in a critical sense provided the earlier basis for differentiating the Swiss from the British. Importantly, the city also was a center of international trade. As one early observer noted in the year 1750, "There is scarcely a part of the globe, however distant, to which European nations send their ships, where one does not find the Genevese" (quoted in Jaquet and Chapuis, 1970, p. 113). Perhaps this quality, more than any other, made it possible for a valuable watch industry to flourish, within 100 years, supported by an established gold and silver industry (Enright, 1995). While the French Huguenots were critical in injecting remarkable skill into the Swiss industry, prior to their arrival, Switzerland's gold- and silversmiths had already sowed the seeds of a craft-based watch industry that produced very high-value jeweled watches (Jaquet and Chapuis, 1970).

By the late 1600s, clock- and watchmakers' guilds began to thrive in Geneva. Although the city had only about 25,000 people by the mid-1700s, many were highly skilled craftsmen and held important keys to production. Early on, the Swiss system developed a social division of labor in which the production of the watch movement was separated from watch finishing. This early specialization distinguished the Swiss industry from others in that many more watchmakers were spawned than in the mode of production pioneered by the British. Emphasis was placed on the production of a wide variety of watches, with many different qualities and prices. With this division also evolved an early spatial division of labor that kept the high-skill occupations in the city and sent the more repetitive (and often dangerous, toxic materials-based) tasks to the mountains of France and Savoy.

The industry developed far more rapidly in Geneva than in any other country. As Baillie (1929) reports, "In 1686 there were 100 masters; in 1716, 165; and in 1766, 800 masters with some 3,000 workers" (p. 251). The industry's growth was not coupled with extraordinary innovation; Britain and France were the innovation centers. Up until the 17th century, the Swiss excelled at emulating the competition rather than overtaking it. Low wages, out-labor based on women and children, and high skill set the Swiss apart from their competitors, a separation that was to become more profound in the ensuing 100 years.

An important aspect of Geneva's success was its ability to determine and meet the needs of overseas markets through middlemen and the nascent merchant banking trade in Savoy and Turin. The Geneva product usually undersold its British counterparts—from which it was often copied—by 60 to 75 percent of the British price. Trade with the Middle East, Asia, and India also was quite well developed early in the industry's history. The Swiss provided the world's elite with beautiful jewel-encrusted baubles, often sold in pairs. In 1785, Geneva's production peaked at 85,000 units per year.

But it was not simply opportune location that afforded the Swiss the opportunity to become the world's watchmakers. As Landes (1979) reports, by the last quarter of the 18th century the desire for large watches, obvious symbols of wealth, was passing. Status seekers were buying French-made watches, for the French were manufacturing a thinner and more refined watch. Here again proximity to France worked in favor of the Swiss, who copied the more refined, thinner French watch. Meanwhile the British ignored the new trend. One author has suggested that thinner watches were easier to smuggle, and this served to fuel the pressure for thinner watches (Jaquet and Chapuis, 1970). The drive toward thinness is more likely the result of broader fashion trends, which included a shift to form-fitting, lightweight clothing. According to Baillie (1929), the Swiss carried "the thinning process too far and the English not far enough" (p. 261), but this innovation in watch design changed the course of events and allowed the Swiss to become leaders in watch manufacturing.

In reading the history of the development of the world watch industry one is reminded of the degree of instability and variety that accompanies the propagation of a new technology. French masters continued to influence the direction of the industry, though their role in later years was almost exclusively related to the development of complicated and expensive watches and important machine tools (Ballie, 1929). The French were not incapable of imagining new organizational forms of the industry, such as machine or factory manufacture; indeed, it was a French watchmaker who created the first machine system for the production of ebauches. But the French were unable to take advantage of their own innovations. The British, too, continued to pursue specialized watch manufacturing. My intention here is not to recount the blow-by-blow development of the leading edge of the industry, but rather to demonstrate how different systems of manufacturing emerged and then ultimately dominated the industry. Readers interested in the remarkable and fascinating details of watchmaking are encouraged to examine Landes's (1983) work above all others.

With the expulsion of the Huguenots after the revocation of the Edict of Nantes,[5] a new watchmaking region developed in mountainous Switzerland. On the heels of the edict, the French creation of a "master craftsman" designation also led to the emigration of many French watchmakers whose skills were not sufficient to earn them the title of "master" (Jaquet and Chapuis, 1970). This early exclusion of nonguild members served initially to preserve Geneva's advantage. Over time, however, the differences in cost, coupled with the emergence of some very talented watchmakers living in the mountains, undercut Geneva's industrial dominance.

In the late 1700s, the Neuchâtel region emerged as a competitor with Geneva (Figure 5.1). Initially it had participated in the industry as a parts production location for watch manufacturers in Geneva. Its production strategy mirrored Geneva's in battling British industry: cheap labor, longer work days, and more women and children workers.[6] This was cou-

Why Switzerland? ■ 95

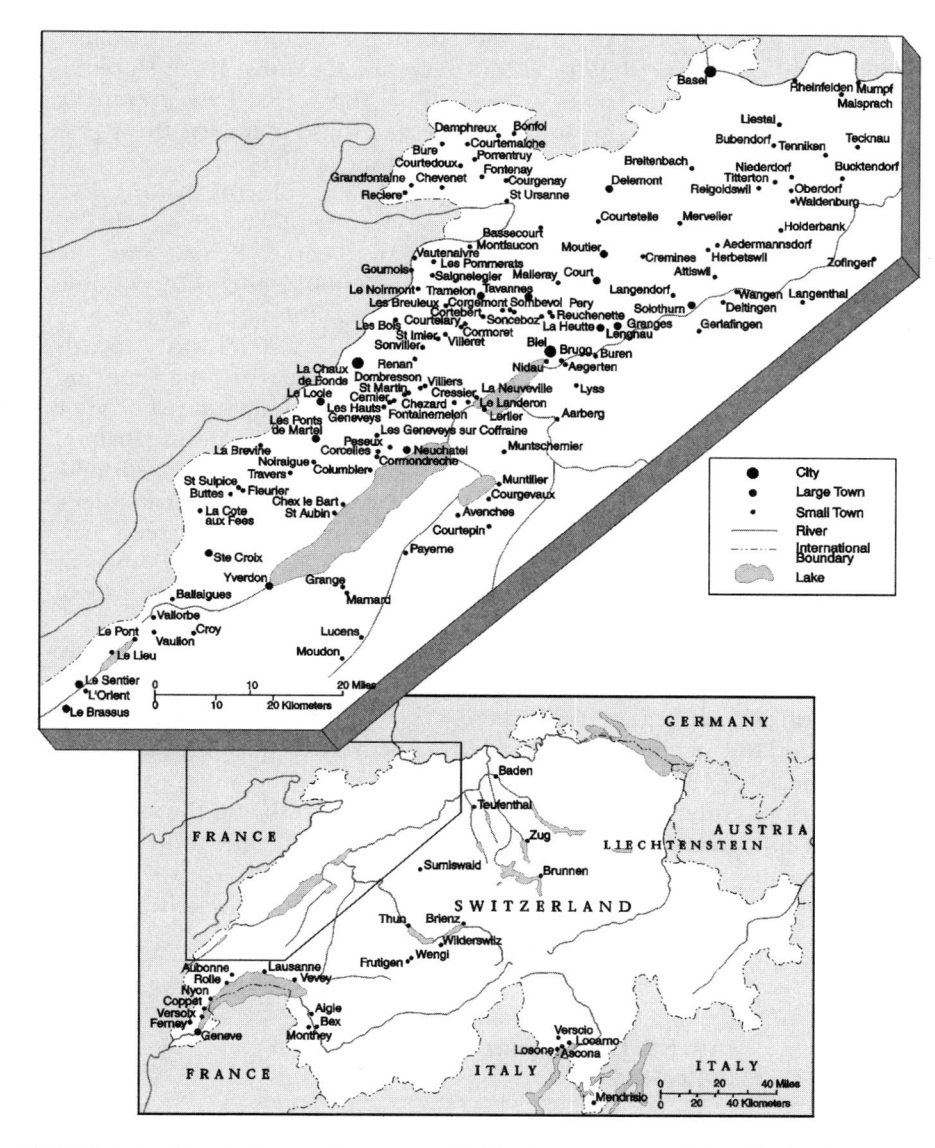

FIGURE 5.1. The Swiss watch region, 1920. Cartography: Erin Heithoff. Sources: Map-Art, National Map of Switzerland.

pled with clever engineering and the creation of standardized machine tools to replicate parts (Baillie, 1929; Jaquet and Chapuis, 1970; Harrold, 1984). Thus, the manufacturers located in the Swiss Jura eventually achieved a position superior to that of Geneva guild members through a combination of hard work and mechanical innovation.

Watchmaking in the mountain region has a colorful history. Evidence suggests that the industry was started by one of its natives, a blacksmith named Daniel Jean Richard, who began to copy a British watch he had acquired. Le Locle, a small city of about 13,000 residents, now home to the Tissot Watch Company, claims Richard as the father of the Jura watch industry. However, a careful reading of Swiss watch history reveals a core of men who were actively engaged in the industry at the time and were coincidentally establishing new firms in small mountain towns after serving apprenticeships in any one of the world's watch regions (Britain, France, Germany, or Switzerland). Towns became known for the individual operations owned and operated by key merchants, major assemblers, and company owners. The industry spread widely, filling the steep, narrow, Jura mountain valleys. Between 1752 and 1788, the number of watchmakers in the Neuchâtel region grew more than sevenfold, from 464 to 3,634 (Table 5.1; Landes, 1983, p. 267).

The location of a second watch region in Switzerland was no geographical accident. The mountainous area of the Jura is geographically situated on an important trade route between Germany, France, Italy, and Switzerland (trade was strong both north and south and east and west [Steinberg, 1976]). Merchants passed through the valleys on their way to various commercial centers. The climate is harsh and the land suitable only for seasonal farming and cattle raising. A light industry was needed to supplement the area's subsistence farming (Pinot, 1979). The long winters provided ample time for the manufacture of watches, which functioned as both a cottage and a family industry.

TABLE 5.1. Watchmakers in Neuchâtel, 1752–1812

Year	Number of watchmakers
1752	464
1764	1,032
1778	2,087
1788	3,634
1808	4,316
1812	3,220

Source: Jaquet and Chapuis (1970).

Context Creates Culture

The origins of watch manufacturing in Neuchâtel were quickly followed by the formation of a system of specialized workshops. A 1750 census of Le Locle revealed 41 watchmakers, 15 case makers and goldsmiths, 5 enamelers, 6 spring makers, 1 maker of fusee chains, 9 engravers, and 4 makers of watchmakers' tools. At the same date, a census of La Chaux de Fonds listed 61 watchmakers, 68 clockmakers, 2 makers of repeating work, 8 case makers and goldsmiths, 10 engravers, 8 toolmakers, 1 spring maker, 2 makers of fusee chains, 1 enameler, and 3 gilders. The key to the industry's profound technical capabilities came in part from its early subdivision of production tasks into two components: the manufacture of the watch's internal assembly and the final assemblers. This specialization of tasks was perpetuated by class and cultural distinctions between land owners/merchants and laborers, and served to ensure a highly fragmented system of production (Pinot, 1979). Thus, an early description of the region's culture provides an important basis for understanding how the industry developed such a profound level of technical and social specialization.

Villages in the Jura were tightly settled and organized around family-based systems of land tenure. Land ownership was passed down through families, which served to bond individuals to particular areas within the region. The poor quality and limited availability of land, however, made it impossible for most families to survive solely on agriculture. Since land was passed on to the first son, there was a need to provide a means of subsistence for other members of mountain village families.

The long winters and poor quality of soils restricted agriculture. This increased the need for supplementary income while at the same time preserving farmers' hands for precision machining (Landes, 1983; Pinot, 1979). The early form of cottage production was a putting-out system: a contractor would bring the materials to the farmer, who would work on the parts production over the winter. Upon receiving payment for the parts, a farmer could reinvest the money he was paid in more agricultural land or livestock. Watch manufacturing translated into increased property acquisition.

The costructure of agriculture and watch manufacturing perpetuated the region's independent-minded culture. Landowning farmers could work at their own pace in manufacturing watches. The intense, focused handwork also refined skills in the population, as individuals capable of producing a complete watch on their own could move on to producing a brand for themselves by contracting for parts manufacturing from others. Manufacturers were the first to see that the system of outworking reduced their control and increased their risk. The merchants were effectively squeezed out of the new system as the manufacturer also became a merchant. In some cases the manufacturer provided the materials to the workshops, much as the merchants did under the old system.

This system was soon deemed inefficient due to the difficulty in coordinating the production of a myriad of parts across the mountainous Jura. To increase control over production and to reduce transportation costs, contractors offered higher salaries to watch parts manufacturers who lived in villages. Like the farmer, this worker usually owned his own home and tools and therefore remained an independent worker.

Mechanization and Toolmaking

The early development of machines that standardized parts manufacturing (Jaquet and Chapuis, 1970) contributed significantly to the evolution of the region's remarkable hold on precision machining of parts at relatively low cost. The Swiss ability to underprice the British was not simply due to the production of inferior watches at lower costs. Instead, the Swiss ability to manufacture precision parts resulted in an accurate, less costly product. Machines were used in the manufacture of Swiss watches relatively early on—certainly before the industry changed from a purely craft-based to a more factory-based system (Harrold, 1984). By the late 1700s the Swiss had made significant progress in the development of machine tools that could assist in the manufacture of watch parts (Jaquet and Chapuis, 1970).

As the industry evolved from a cottage to a protoindustrial system of production, further innovations were made to machines to reproduce metal parts in large amounts. The Jura emerged as a center of toolmaking equipment. This allowed the production of standardized equipment—something that the British never accomplished and which sped their downfall.

At the end of the 18th century, the Jura industry had grown in size and scale. Production was outstripping the ability for the scattered ebauche (rough parts) makers to keep up with industry demand. A convergence of interests led to the development of the first machine- based ebauche factory just over the border in France. Frederick Japy, a prominent French Swiss watch manufacturer, had a longtime interest in and desire to establish a more mechanized system for the production of ebauches. With the existing system of decentralized production taxed to its limits, the factory in France quickly established an important market presence, supplying more than 40,000 ebauches for the industry in its first year. Swiss watchmakers, in particular four from La Chaux de Fond, were quick to conclude that their dependence on the factory in France could have serious consequences for the Swiss industry. Moreover, they recognized that the industry was booming and that money was to be made in ebauches. As a consequence, in the same year the four watchmakers established a factory in Fontainemelon, near La Chaux de Fond. By the 1790s these two factories were producing hundreds of thousands of ebauches. Following the ascendance of the Neuchâtel center, Geneva turned to high-end manufacturing in this market.

Since both Geneva and Neuchâtel could undersell Great Britain, Switzerland soon dominated British markets (and markets in other parts of the world). Switzerland's ability to undersell was based at first almost exclusively on lower wages, but later technological improvements allowed the Swiss to make a cheaper watch in terms of materials as well (Davies, 1993). The cheapness and quality of the Swiss watch was such that British manufacturers increasingly turned to imports, in some instances hiding Swiss watch movements in cases signed by a British watch master.

Creating a Competitive and Cooperative Environment

The gradual progression of Swiss watchmaking competency from the ability to fabricate inexpensive, relatively accurate watches to the creation of watch innovations came as a surprise to the British. Up until the mid-18th century, the majority of watch innovations came either from Britain or France. The Swiss followed a strategy that combined rapid introduction of new products with a cultural system that fostered competition and emulation. The British remained wedded to the belief that only the most sophisticated watches with a specific movement design could be considered a chronograph, or device capable of "precise measurement of time," while the Swiss were willing to award the title of "chronograph" to versions of technology that were less expensive to manufacture but still capable of maintaining precise time (Landes, 1979, p. 23). This increased the number and type of watches considered high quality and dramatically expanded the market for Swiss watches at every price level.

More important to the development of Swiss skills levels was the cultural organization of the Jura industry, which was both competitive and emulative. The British looked down on all but the highest priced watches. As Landes (1984) remarks, "It never occurred to them that so diligent and enterprising an industry, highly emulative and competitive within its own ranks, would generate a flow of innovations and improvements that would enable it one day to surpass its predecessors within its own ranks in quality as well as quantity" (p. 22). Supported by their government, Swiss watchmakers organized competitions and standardized tests to encourage the development of innovations. And because of the dense spatial fabric of the industry, ample numbers of skilled workers demonstrating significant creativity emerged from inside the industry (Jaquet and Chapuis, 1970).

In addition to the sponsorship of tests and competitions, which facilitated the region's advance toward technological supremacy, the Swiss watchmakers created other institutions to perpetuate skill. The first horological school was established in Geneva, followed soon by the creation of another school in Neuchâtel. It was not long before additional institutions were founded in the Jura to further strengthen the region's skills base. Few restrictions were placed on individuals wishing to learn the watch trades; indeed, young girls were among those trained in the region's watch schools to carry out precision assembly tasks.

The emphasis on education, training, and structured innovation in the 18th century was impressive by any standard. While the British were content to allocate responsibility for perpetuating skill and innovation to the old master watchmaker under an outmoded guild system, the Swiss were doggedly determined to push the limits of learning and knowledge creation. The first scientific society in Geneva was established in 1776 by a watchmaker. Over the next three years this society sponsored competitions for the creation of the best steel and brass, measurement tools, and gearing. Recognizing the link between the study of mechanics and watch manufacturing, an apprenticeship program and educational curriculum were established. The first school dedicated to watchmaking was founded in 1834. Students were trained in the most complex and sophisticated aspects of watch design and manufacturing, using the most modern techniques and designs.

In Neuchâtel, a society for the advancement of the industry was also created. Known as the Patriotic Emulation Society, the organization's goals were heroic in proportion and guided by a simple aim to attain "everything that may contribute to the public good." Like the society in Geneva, prizes were offered for solutions to vexing industry problems. And although the emphasis was predominantly on resolving technical problems, nonetheless, in 1826 a competition was conducted for the creation of a plain watch that could be easily manufactured. Self-study was a particular predilection of the Emulation Society, which commissioned numerous research reports investigating needs and trends in the industry. Libraries, exhibits, and museums were created to promote the watch industry to the common person.

These societies had relatively short lives. The revolutionary period of the 1840s drew the public's attention away from commerce toward politics. In later years, much of the innovation efforts fostered in a very public forum was taken up by the watch industry itself and its various organizations and associations. (These will be discussed in later chapters.)

The Swiss not only created their own watch technology but were quite facile in absorbing innovations previously rejected by other watch manufacturers such as the British, who shunned new developments that appeared to differ too much from past practice (Brearley, 1919). A case in point is the development of the stem-wound watch. Until the early 1800s, watches were wound with a key inserted in the face of the watch. This rather cumbersome procedure worked well with the traditional British round, thick, and heavy fusee-driven watch. It was incompatible, however, with lighter, thinner watches. While a British watch craftsman was the first to develop the side-winding stem watch, the Swiss really brought the innovation to market (Harrold, 1984). British failure to capitalize on the design stemmed from their perception that the side-stem watch would appeal only to handicapped people incapable of using both hands to insert the key in the watch face (Brearley, 1919). This lack of insight into the changing nature of markets and customer preferences was ultimately self-destructive for British industry.

From Craft Production to Protofactories

Mechanization allowed Swiss industry to continue expanding rapidly through the first half of the 1800s. Larger shops sprang up in the 1850s; by 1870, 10 percent of the labor force worked in these "protofactories" (Landes, 1983, p. 295).

The transformation of work and the movement to a factory system was not without stress and resistance. The factory system differed substantially from previous industrial reorganizations. While the contracting system subordinated the manufacturing worker to the contractor, the worker still owned his own tools (Pinot, 1979). If sufficiently skilled, he could work overtime and make his own watches, which could then be sold to independent merchants. The degree of control exerted by the urban worker over his craft allowed many artisans to develop high levels of skill.

Modernization and mechanization led to the development of factory work. This drastically reduced the freedom and flexibility of former cottage and urban craft workers. But the early independence of watch parts manufacturers remained ingrained in the culture. Every worker sought to control his or her work life. This independence was an important part of the region's culture, which in later years led to serious labor strife and stressful relations between factory owners and workers. Thus, as with many industries on the Continent, watchmaking began as a putting-out system that built skills in the region. This was followed by the development of protofactories, and finally by the mechanization of the industry.

Early attempts at factory-based manufacturing did not actually result in a streamlined system of mechanized production. In 1883, in the Jura, there were approximately 14,000 watch workers. Women were some of the first to work within the factory system (Table 5.2). Until the end of the 19th century men were still predominantly working as part of an urban collective industry (Pinot, 1979).

Like other industry leaders, the Swiss stumbled in the face of the coming trend toward factory production. A rather difficult but brilliant watchmaker named Pierre Frédéric Ingold proposed the concept of factory-mechanized production in the early 1830s. He could not find anyone in Switzerland who was interested in his ideas. Moving on to France, he found backers for a watch factory, but these plans too came to nothing as the artisan guild balked at the idea. Forced to venture further afield, Ingold eventually went to London and suffered the same reception as in France, and for the same reasons: guild resistance. In this case persistence did not pay off. In 1845, Ingold traveled to the United States, received a warm introduction, but was eventually forced to make a hasty retreat back to Switzerland. The failure of his ideas to take root in any of the watch industries of the day is an instructive example of the power of paradigms and the resistance to paradigm shifts. The incremental learning process necessary as a basis for change is often long and difficult. Resistance to change takes on many forms (as we will see later), and is made all the more powerful as a force that locks in the existing means of operation, one that is strengthened when the external context supports continuance of the status quo.

It seemed inevitable that watch manufacturing would become both factory-based and more highly mechanized. Outside influences were important stimulants of change. After extremely good years from 1864 to 1873, the Swiss industry fell on hard times (Table 5.3). Almost without warning the U.S. market was closed to Swiss products. While wartime constraints on foreign exchange certainly limited imports of all sorts of

TABLE 5.2. Gender Breakdown, Watch Industry of 1870

Place	Male	Female
Geneva	2,095	1,139
Neuchâtel	9,412	5,360
Berne	9,050	4,639
Soleure	512	295
Vaud	2,498	1,135
Fribourg	318	103

Source: Jaquet and Chapuis (1970, p. 169).

TABLE 5.3. Watch Movements, 1864–1887

Year	Swiss exports to the U.S.	Waltham Watch Company production
1864	169,000	38,103
1865	226,000	44,632
1866	262,000	52,168
1867	207,000	72831
1868	209,000	64,482
1869	206,000	42,089
1870	330,000	55,042
1871	342,000	66,655
1872	366,000	74,530
1873	204,000	79,346
1874	187,000	64,847
1875	134,000	49,243
1876	75,000	84,737

Source: Moore (1945, p. 65).

goods, including watches, the Civil War also gave a tremendous boost to the expansion of the American watch industry. Ten years later (1876) the Philadelphia Centennial Exposition rubbed further salt in the wounds of the Swiss when U.S. watch firms displayed not just fine watch specimens but a whole new way of manufacturing watches (Baillie, 1929; Harrold, 1984; Jequier, 1991). Unable to continue ignoring the success of U.S. factories about this time (exports to the United States had all but ended), the Swiss hastily sent investigators overseas to report on the upstart's discoveries. The Swiss reacted with cool heads to American success. They assimilated new production techniques—through greater concentration of production—but never reached the scale of U.S. production. They retained their hold on the market for custom and novelty watches, and continued to produce for their main markets in Germany and the United Kingdom. As Landes (1979) notes, "The Swiss were quick to note that the multiplicity of their enterprises and the competition and emulation that prevailed within their industry were the best assurance that it would not overlook the opportunities offered by the new technique. At the same time, the publication of a constant stream of data on the performance of high-quality watches of varying construction reassured them that, except for the best American watches, the machine-made watch was still not up to the finest Swiss work" (p. 30). The Swiss strategy, then, was to move toward interchangeable parts while retaining an interest in special finishing, complicated pieces, and high fashion. Interchangeabilty had been a technology long on the back burner; the Swiss had been familiar with it conceptually, if not practically, since the early decades of the 19th century, and they had the machine tool capacity to outfit new factories and modify others.

There has been considerable debate about the reasons for the Swiss loss of the U.S. market at the turn of the century. The U.S. government raised tariffs in support of the domestic industry and restricted access to markets, allowing U.S. industry to effectively dominate market channels formerly controlled by the Swiss. Perhaps the Swiss were too interested in quantity and not interested enough in quality, shipping many cheap medium-grade watches to the United States. Domestic watchmakers, seeing the gaping market niche, produced a watch through partial interchangeablity that was attractive to the U.S. consumer (Baillie, 1929). The introduction of the railroad watch standard (discussed in Chapter 7), which specified the use of a U.S. machine-made watch, also contributed to the Swiss market share contraction (Harrold, 1984). While the loss of a large share of the U.S. market was momentarily devastating, in time the Swiss recovered both a major portion of it and developed other markets, thereby retaining their position as world leader of the watch industry.

The Swiss response to U.S. technological challenges was decisive. Over a period of 20 years, 1885–1905, they proved more than capable of making needed technical progress (Dosi, 1984). Over the course of two decades, the Swiss adopted aspects of the American system that were cost-effective. The Swiss shifted from their almost sole reliance on small-scale cottage production and limited factory manufacturing to an intermediate form of production that combined mechanization and partial vertical integration. Standard parts were mechanically manufactured on a large scale in centralized factories, while flexibility was maintained through dispersed design and assembly activities. Even the more complicated parts were eventually mechanized using "versatile machines which were susceptible of all manner of adjustment, and hence required some skill to operate" (Landes, 1983, p. 40). In a reverse of the usual technology flow between the United States and Switzerland, by the turn of the century U.S. watch firms were sending key employees to Switzerland to buy Swiss machine tools because they were deemed far more accurate in cutting metal parts at small scale than were the tools made by U.S. manufacturers. Within the existing mechanical technology system, the Swiss industry achieved new levels of profitability and international renown.

The shift to a new system of reorganization was by no means seamless. The initial burden of the period of instability brought on by the closure of the U.S. market was borne by the workers. Piece rates were cut, higher levels of demand were required, and longer hours resulted. Throughout the previous 150 years of industrial development, the Swiss managed to forestall the development of unions.[7] Swiss merchants and then manufacturers were benevolent and paternalistic, a side effect no doubt of the high degree of education in the population and the steadfast social cohesion of Swiss mountain communities. In addition, merchants and manufacturers lived side by side with their workers in mountain villages; thus, one's coworkers were also one's neighbors. When the economic crisis came, however, capitalists initially knew little else to do except to drop wages. Given the geographic concentration of the industry and the dependency of workers on this only source of proximate employment, the conditions at the end of the 19th century changed the labor relations in the industry, a transformation that had major implications for the industry in later years.

The lack of factory-provided social welfare services led many local associations to furnish the workers' needs (Barrelet and Ramseyer, 1990; Rappard, 1914). Societies supplying insurance, health care, and retirement benefits were created to protect watchworkers and their families. The culture of the region developed a level of independence from the employer that became integral to the factory system and the culture of innovation. Workers and skilled craftsmen innovated and then protected their innovations from appropriation by the factory owner. This was to become a problem later when the watch industry once again faced a need for radical change.

There is no doubt that by 1910 Swiss mechanical watches dominated the world watch industry (Knickerbocker, 1974). The Swiss controlled the micromechanical export industry thanks to cost-competitiveness, superior manufacturing competency, high levels of precision, and extraordinary attention to detail and style. The vertically integrated parts manufacturers achieved economies of scale through volume production. This benefit was passed on to assemblers in the form of low-cost movements. In the most labor-intensive aspects of the industry, the vertically disintegrated system of assembly and case manufacture kept overhead charges low.

Summary

How did a remote rural region of the Swiss mountains become a center for horological innovation? While there are many competing theories, the weight of empirical evidence suggests that culture, institutions, and geographic concentration combined to yield an industrial structure that demonstrated great flexibility, the internal capacity to learn through interaction with external competitors, and a heightened sense of the importance of competition to preserve systemic nimbleness. Manufacturers recognized that the original, highly fragmented organization was vulnerable to cost competition associated with the vertical integration of competitors. Firms in the region adopted the best features of the American system while adapting it to local conditions.

Still, there was a gap in time between the onset of competitive crisis and Swiss reaction to it. Although the antecedents of machine-based manufacturing were clearly and comfortably within the grasp of Swiss watchmakers, nonetheless it took a serious and long-lived competitor to shake the Swiss out of their complacency. The ability to react in this instance was linked to the complementarity of the skills required to move to large-scale machine-assisted manufacturing, which already resided within the Swiss industrial culture.

Of critical importance was the prior experience with machine-based manufacturing, which while not initially widespread in adoption, still functioned as a touchstone manufacturers could reach back to in order to progress forward. In this instance, movement along a technological trajectory fit within an existing paradigm and hence had high potential for successful assimilation. As later chapters suggest, both the magnitude of required change and its coupling with other, often exogenous, events bears directly on the ability of a system and a culture of production to respond successfully to demands brought about by technological change.

6 The American System of Watch Manufacturing

Like the watch manufacturing industries in Britain and Switzerland, the industry in the United States began small and scattered, with several watchmakers, who had begun in repairs, experimenting with the manufacture of high-quality handmade watches. The earliest American watchmakers did not start from scratch; instead, they built upon the experiences of the British and to a lesser extent the Swiss watch industries. The emergence of the U.S. industry represents the convergence of capabilities in the clock, machine tool, jewelry, and distribution trades. Although the U.S. industry has roots dating to the early 1800s, it would be another 40 years before a serious watch industry emerged. It would take still another 20 years to solidify the gains of the machine-based production system that came to characterize the U.S. system of watch manufacturing. In the case of the U.S. watch industry, this convergence brought together watch manufacturing based on know-how from Britain with a robust and growing machine tool industry and a concept of manufacturing known as the "American system of mass production." The result was the American watch industry (Hoke, 1990; Hounshell, 1984).[1]

I will begin my history of the American watch industry by exploring a set of key factors that led to a high degree of standardization in the industry. My second goal is to review the history of the two leading American watch companies and demonstrate how the early structure of the industry resulted in a degree of instability and uncertainty that became very problematic as the industry aged and approached the end of the 19th century.

Early History

The U.S. watch industry was established by a small number of key individuals. Two pioneers, Aron Dennison and Edward Howard, were responsible for establishing numerous factories, many which were formed from the failed attempts of earlier investors. Harrold's (1984) chronological depiction of the development of U.S. watch manufacturing firms, shown in Figure 6.1, shows the outgrowth of and cross-fertilization that characterized the U.S. industry

From its earliest days, U.S. watch manufacturing progressed in fits and starts, with a few watchmakers starting and refining the production of high-quality watch products. By the 1850s, Dennison's and Howard's factories, both operating out of Boston, were producing watches of recognized quality. The success of these two operations, notwithstanding repeated reorganizations early in their life, attracted the attention of other knowledgeable manufacturers and businessmen interested in investing in this new growth industry.

Between 1840 and 1860 at least 18 watch factories were established. The success of these new enterprises rested on the entrepreneurial spirit of their founders and the devotion of their skilled workers. Many of the factories established in midcentury were playing off a limited knowledge pool, based around British watch models. As a consequence, there was a great deal of uniformity in early watch design and business organization.[2]

The geography of the early U.S. watch industry reflected the close link between clocks and watches. Thus the earliest factories were founded in Connecticut and Massachusetts, the country's two centers of clock manufacturing. By 1840, manufacturing of watches began to fan out geographically, but still remained within close proximity of the early core. The geographical closeness of the industry in its early stages was reinforced by self-selection and the restricted mobility of early machine manufacturers and watch workers. Early mobility was circumscribed by the availability of investors and the need for proximity to established population centers to tap private capital pools in order to establish factories. It would be another 20 years before watchmakers began to roam the country, venturing out to the edge of the frontier in anticipation of new adventures and the promise of a more prosperous life. By the 1860s, watchmakers thought little about the personal and financial risks associated with moving from a factory located in Boston to one located somewhere else along the Eastern Seaboard. In fact, mobility provided some degree of financial security in a dynamic industry. Since watchmakers' skills were in short supply, if a job in one place went sour, watchmakers could be assured a good job in a factory somewhere else.

The very uncertain and unstable technology was costly and time-consuming to build. Private capital was needed to finance the long periods between factory establishment and market-grade watch production. The venture capital of the time was modestly receptive to watch enterprises.

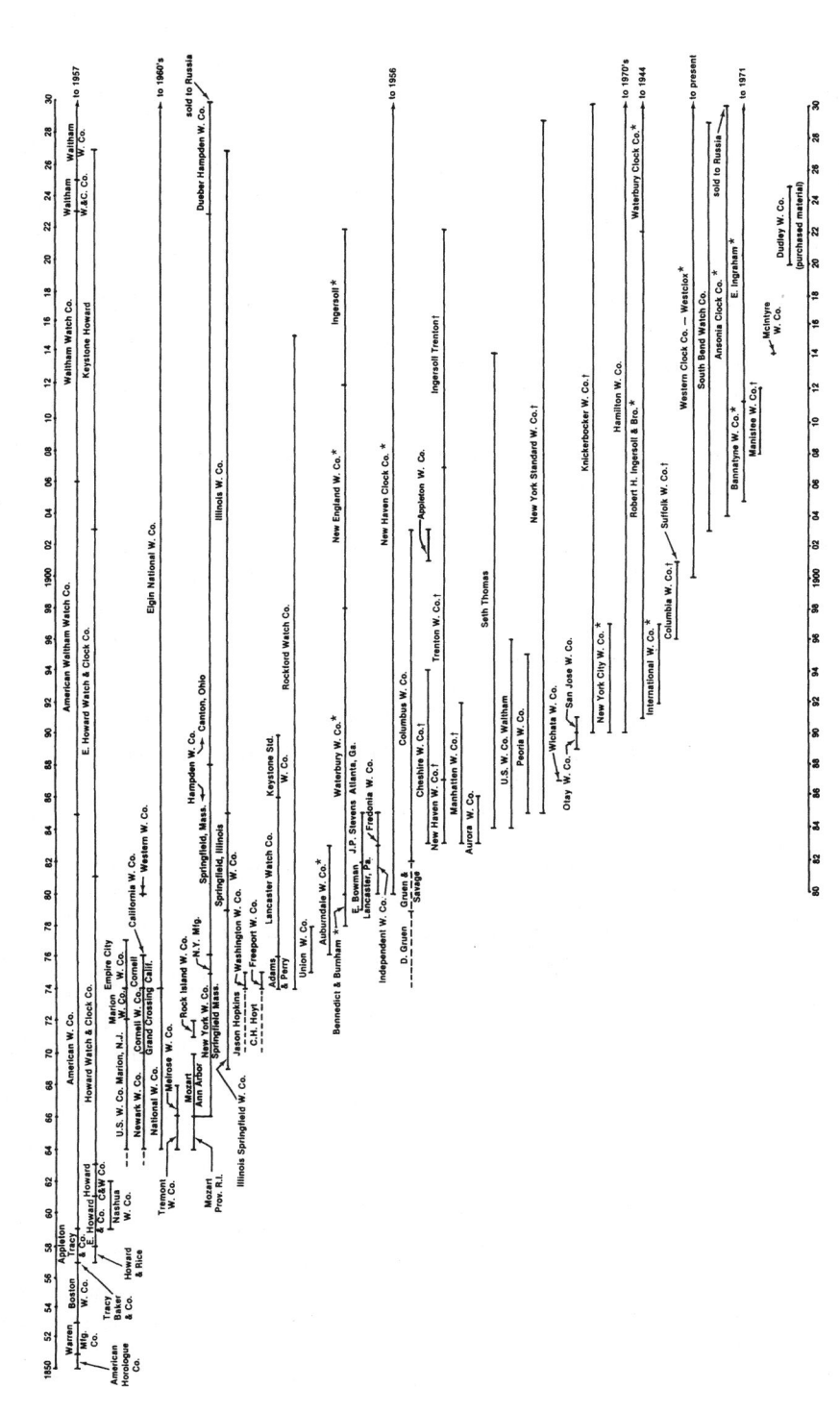

FIGURE 6.1. American watch factories. *Dollar watchmakers, †Inexpensive jeweled watchmakers. Copyright 1981 by M. C. Harrold.

This was particularly true given that the other "high-tech" industry of the day, armaments, was being financed in part by the federal government. At the same time, proximity was important because the high level of capitalization needed to operate a factory involved great risk; hence, early backers wanted these investments located close to home so they could intervene in the management of a company should business falter. Thus, between 1840 and 1860, watch manufacturing reflected a pattern of concentrated dispersal.

Although linked geographically, the American watch industry was only modestly associated with the much older clock industry. This limited closeness meant that geographic proximity and past practice did not act as a constraint on the early configuration of the industry. The openness of American watchmakers to new ideas was in part the result of a lack of skilled labor. Shortages of skilled labor were managed at first by recruiting labor from Europe, and later by sending employees of a watch company off to England and in some cases Switzerland to observe watchmaking techniques. This intermingling of labor forces facilitated the transfer of technology between countries and also made the theft of industry secrets a common practice. Although it was recognized that the Swiss and the British systems of production organization would not satisfy the volume and scale envisioned for the U.S. industry, nonetheless, there was plenty to learn from the likes of the British and the Swiss about materials, machining processes, and finishing. The early development of the U.S. industry is a good example of industrial emulation with subsequent modification.

The machine tool industry grew up alongside the watch industry and helped propel the development of the watch industry forward. The United States was the first country to use standardized machined watch parts in large volume. However, the practice of parts manufacturing using dedicated machines was not an American invention. Both British and Swiss watch firms not only experimented with, but actually used machine tools and procedures necessary for machine-manufactured parts as early as the 18th century. Despite this early lead, two major influences hampered the widespread introduction of this capability in either European country.

The first well-developed application of machine manufacture of key parts took place in the Geneva factory of the Swiss watch company Vacheron and Constantin. Unfortunately, this technical innovation was never transferred from this firm to the whole of the Swiss watch industry. Even with a free flow of ideas, however, the Swiss home-based and later protofactory-based system would have resisted the introduction of machine manufacturing because it challenged the status quo and was believed to be incapable of producing a high-quality watch. The American system might not have emerged almost unchallenged in the 1860s had the Swiss industry adopted this capability earlier and wholeheartedly.

The peripatetic pace of new firm formation in the U.S. watch industry reflected the early stage of the product cycle. Firms were started up and rushed into the prototyping stage in order to secure financial backing. Many companies began with little more than a handful of watches to leverage as assets, and many of these watches were made from parts that were simply purchased from an existing firm. The lack of evidence required by financial backers of these start-ups almost guaranteed high rates of business failure.

More destructive was the continuous turnover of labor as workers were enticed to join start-ups with higher wages and better benefits. Because of the dense cluster of firms and the overall shortage of skilled workers, machinists could hop with relative safety from firm to firm in search of better working conditions. The continuous turnover of labor thrust established firms into a perpetual state of chaos, as the loss of skilled machinists could shut a firm down with little warning.

Failure rates were also high because of the tremendous capital investment required to establish a new watch factory. Machines were very expensive and had to be tailor-made for individual factories and their watch models. In addition, commissioning a machine did not mean that it was delivered on time or in working order. Indeed, there was no guarantee that a commissioned machine could produce the proposed part at all. The long lead times required to produce the basic equipment needed to establish a factory meant it could take several years to completely outfit a factory for mass production. Added to this headache and risk were the ongoing fears of labor abandonment and the inevitable emergence of new designs and new competitors to make vulnerable a firm's sunk costs.

The creation of new businesses, watch firms among them, was encouraged by the rapid pace of settlement in the United States. Like traveling salesmen, business promoters were literally selling factories to communities and selling communities to firms. Communities looking to put themselves on the map were the easy targets of business promoters who first traveled the Eastern Seaboard and then later the Upper Midwest in search of willing local boosters ready to subsidize the development of a new firm. Watches were the high-tech industry of the 1850s (Harrold, 1984).

The establishment of new factories in the face of very high-risk financial uncertainty did not happen accidentally. The rapid pace of new business openings and closings in the formative years of the industry (1840–1869) was fueled by the availability of surplus equipment, a potentially mobile labor force, and demand associated with soldiers' needs during the Civil War. Secondhand equipment could be bought at auction, sometimes at very low cost, which encouraged the establishment of seriously undercapitalized firms. Buying surplus equipment, however, had a downside: used equipment was usually designed to produce an outmoded model of watch. At a certain point, so much excess surplus equipment was available that entrepreneurs with virtually no experience in the watchmaking business could buy pieces of equipment and sell themselves to a community, with no idea of the difficulty of producing watch parts, much less assembling the parts into finished products.[3]

The Waltham Watch Company

America's first watch company was started in a suburb of Boston in the 1850s. At its zenith, the Waltham Watch Company epitomized U.S. manufacturing ingenuity and craft. One of the first factories unrelated to the armaments industry to take up what became known as "the American system of manufacture," the Waltham Watch Company utilized a highly rigid system of industrial organization, borrowed from the armories, and employed a remarkable group of machinists and engineers capable of building precision machinery to manufacture tiny parts for watches (Hoke, 1990). As America's first watch company of any consequence, Waltham holds a special place in the history of the industry. It was a pioneer in process and product technology, and in its early life accounted for two out of every three watches sold in the country. It also was a trailblazer in marketing strategy, being one of the first companies to develop a line of products appealing to consumers across the spectrum.

Built in the womb of America's first industrial region, the Waltham Watch Company was established in 1850 by Aaron Dennison, the father of the U.S. watch industry. Dennison began his professional career in watchmaking as an apprentice with a firm in Britain. He subsequently returned to the United States determined to start a watch company. Along with Edward Howard, a clockmaker and manufacturer, Dennison set up shop in Roxbury, a suburb of Boston, and commenced to make watches. With little local capacity to draw upon, Dennison spent the first eight months of the company's existence traveling in Europe and acquiring skilled labor, know-how, and parts for the manufacture of watches. During that time, a room was set aside in the Roxbury clock factory to start the development of necessary machinery and parts. Most of the first employees were from England, having come over to the United States to work in the watch repair trade.

Although Dennison was a brilliant inventor, he proved to be less successful as a businessman. From the very beginning of the watch company's life, Dennison oversold its capabilities in order to generate the capital needed to fund the factory. Investors' expectations of a quick buck came back to haunt the firm, as it would take three years before any truly usable product was produced. In the meantime, the company's creditors grew increasingly impatient.

Early in its life, the company moved out of Roxbury and relocated to Waltham, about 9 miles away. One explanation for the firm's relocation so early in its life was that Dennison had underestimated the complexity of the process and physical requirements of machine- and factory-made watches; a much larger plant was needed simply to house the equipment. Another explanation was that the air in Roxbury was polluted by the area's surrounding industry. The manufacture of watches, products of small size and of high precision, required very clean air and very bright natural light. Still another explanation for the relocation was the capital shortage the company faced, and therefore the need to find other investors to bankroll the company early in its life.

The site of the new facility was a former farm along the Charles River. The plant was built by a newly formed company, the Waltham Land Company, which was created with the assistance of the savings of local residents. Given the cost of the machinery to produce watches, Dennison lacked the capital to build the necessary physical plant to house this machinery. His original conception of the firm was as a community of watchmakers, with employees living close to the facility on land owned by the company, in houses built by the private sector. Expecting private interests to construct housing and other necessary facilities, Dennison was soon disappointed and found that to be assured of a stable work force the company would have to build its own housing. It thus fell into the unenviable position of being landlord as well as corporate owner.

By 1854, the company was making 30 watches a week with the help of 90 employees. The company's internal structure was very simple: Dennison managed the plant and Howard managed the finances. While the company had little problem moving its product, selling all the watches it could make through a few retailers, the profits from production were simply not enough to finance new capital equipment and operating expenses. Dennison's passion for invention and not for factory management did not help the situation much.

After a series of bad years and multiple draws on new sources of capital, the company went bankrupt. Its property and equipment were sold at auction. The factory was turned over to a number of men who had surreptitiously acquired the firm. The new owners were unable to get along and the business fell into mismanagement. Dennison was called back from England, where he had been scouting for new technology and skilled workers to start up a new venture. Thus, although he lost ownership of the company, he regained his job, this time unburdened by the responsibility of ownership. Meanwhile, the factory changed hands once again, with controlling interest now falling to Royal Robbins, a cunning and sophisticated businessman who would run the company for the ensuing 26 years, in what turned out to be the best years in its life.

As part of the firm's continuing reorganization, it established an exclusive distribution contract with a sales agent, Robbins and Appleton and Company. An agent was important to Waltham's early life because markets were small and specialized and sales required considerable hands-on assistance and service. The company also lacked the funds needed to support its own sales staff. This exclusive contract assured Waltham of consistent prices, but also tied it to a rigid distribution channel over which Waltham Watch had little control. Over time, this arrangement became an impediment, prompting the firm to develop its own sales force.

In no more than two years, the company once again was threatened with disaster. The panic of 1857 caused markets to dwindle and demand to dry up. With no relief in sight, Robbins faced a difficult choice: he could shut the plant and risk losing all his workers, or he could propose that they remain with him but accept a wage cut. Given the nationwide depression, the workers accepted a wage cut of 50 percent. It is important to note that wages were not a worker's entire compensation package. Many of the skilled laborers lived in company-subsidized housing and would not have wanted to give up this valuable benefit at a time when finding another job was difficult, if not impossible, to do (Gitleman, 1965).

The Waltham Watch Company successfully persisted through the end of the 19th century, producing watches of distinguished quality at a high volume. The company built a production system capable of generating large volumes of precision products, which were easily sold through retail channels around the country. Even the Swiss were surprised at the precision of Waltham watches, given their relatively low price. Aided by a national trade policy that regulated the importation of Swiss watches, Waltham achieved a preeminent position by the 1870s. As one chronicle of the time noted, "The best piece of good fortune that befell Robbins was the rapid increase in the use of the machinery that began with the Civil War. At this date the fundamental principles of machine design had been developed, the technique of mass production had been established, and the means of accelerating and extending transportation had been worked out" (Moore, 1945, p. 41).

Thus, after 10 years of hard work, risk, and sacrifice, on the eve of the Civil War, Waltham had established itself as America's premier watch company, a position the firm would retain for the next 20 years. While its position as industry leader was unassailable, its position of relative security was not to last for long. The demand stimulus unleashed by the Civil War and the restriction on international trade required by the war effort would entice other entrepreneurs into the industry. Meanwhile, as the first mover, Waltham had to contend with high and ongoing capital costs, huge sunk costs, and a labor force whose members were continually lured away by newly formed competitors.

The 1860s–1880s: New Stimuli and New Entrants

The Civil War was an important impetus for new firm formation. By the middle of the last century warfare was increasingly executed on the basis of timed engagements. Every infantryman therefore needed a watch. Nonetheless, watches were still very costly items to own (costing one-tenth of the annual wage of a worker of the day—$1/day vs. $40 for a watch). Consequently, while watch demand did not swell as it might have had there been a truly cheap watch to buy, nonetheless the expansion of new markets enticed new producers to enter the competitive fray.

The 1860s–1880s represent the golden age of U.S. watch manufacturing firm expansion (Harrold, 1984). The dramatic growth in the industry was fueled by changes in American tastes in manufactured goods and a significant increase in the nation's population. The American system of mass production had taken hold in many industries. Large shares of the nation's work force could now count on a regular factory job with a wage large enough to generate savings. Between 1860 and 1880, per-capita income rose by 70 percent and the population expanded by almost 12 million new people. As more people entered the industrial work force, they needed watches to get to work on time. Thus, like many other consumer durable industries, the watch industry expanded to satisfy this growing nation's demand for new goods. But, even though Waltham had succeeded in producing watches in volume, the product itself was still priced outside the reach of most of the nation's citizens. It would take the emergence of new competitors to force the price of a watch down to a level affordable by most people. While the production of a cheap watch was inevitable, it did not necessarily follow that the industry leader would be the one to bring such a product to market, even though Waltham had the insight to recognize this trend and the know-how to accomplish the task.

The need for an inexpensive timepiece was widely recognized by everyone in the industry. The Swiss had already produced an inexpensive watch, the Roskopf, but it never really caught on. Several U.S. designs were made, but none received the required backing for manufacture. The

Civil War was important in reinforcing the perceived need for an acceptable inexpensive timekeeper. However, far more important than stimulating demand for cheap watches, the Civil War led to restricted foreign access to the U.S. market. Thus, the budding American firms had virtual unimpeded control over the domestic market for affordable watches (Harrold, 1984, p. 21).

One explanation for the retarded introduction of cheap watches was the firms' need to demonstrate an ability to produce a high-quality watch in order to retain financial backing and maintain wholesaler privileges. Notwithstanding latent demand for affordable watches, the distribution channel of American watches was tightly controlled by wholesalers who had exclusive rights to sell to retailers. Wholesalers are notoriously conservative, given their intermediary role. Payment comes at intermittent times of the year when inventory turnover occurs. Thus distributors dictated to firms the nature of access to markets and controlled when they got paid for goods produced. To gain entry into the nonmilitary market, companies had to manufacture products that could be sold to well-established clientele. This seemingly innocuous institutional peculiarity probably stymied innovation by structuring (some might say dictating) demand. It certainly controlled market access and more or less defined what the market would wear.

New Competitors and the Emergence of a Midwestern Watch Region

In the period 1860–1880, as many as 25 new companies were established, some out of the ashes of others. Even when new firms were created, many owed their existence to the availability of used equipment. While in some cases, such as the Elgin Watch Corporation, entirely new machines were purchased to start the firm, nonetheless the same people set up and brought to reality new watchmaking capacity. The rate of true product innovation was low, although the development of process technologies continued into the next century.

From a pattern of highly localized spatial expansion, encompassing Massachusetts, Connecticut, New York, and New Jersey, watch factories began to fan out to areas west. Between 1860 and 1880, new factories were established in Pennsylvania, Illinois, and Ohio. At the same time, factories in the center continued to proliferate; for example, seven new plants were added in Connecticut.

One key to this westward expansion was the mobility of labor. Another key was the tightening linkage between the watch industry and the jewelry industry. Although the most far-reaching edges of watch manufacturing and product technology had not yet been achieved (for true interchangeability still eluded the industry),[4] the basic factory technology had begun to stabilize. In the absence of true interchangeability, which might have acted finally as an effective barrier to entry, new entrepreneurs continued to enter the industry. As the industry matured, in place of new start-ups operated by entrepreneurs with an intimate knowledge of basic watch technology, by the end of the 1860s new firm formations were comprised of producers who were businessmen and profiteers first and watchmakers second (at best). Moving to the foreground of importance in starting a watch firm were traits such as business acumen and the stomach for taking risks in the face of unreliable rewards.

The Elgin Watch Company: The Newcomer Takes the Lead

By far the most enduring firm of this period was the Elgin Watch Corporation. The firm began in Chicago, when it was founded by John C. Adams and Benjamin Raymond, former mayor of Chicago (Alft, 1984). Why Chicago? The location of this trade center, at the intersection of the growing American railroad system, and parts west, provided a fortuitous home. The mentality of the day, "There was money to be had," pervaded the city region and every ripe venture was examined carefully for the possibility of profits.

Elgin, Illinois, in 1864 was quite a distance from Chicago, and hardly a speck on the map. However, it had three important things going for it: it was on the Galena and Chicago rail line; Mr. Raymond was a native son; and Elgin was a community on the make, having recently landed a major business enterprise—Borden's Dairy Company.

Perhaps because Elgin was *not* set up by a watchmaker, the factory was not bound by industry tradition. The plant was organized in a highly efficient and systematic fashion using the most skilled labor that could be hired away from (of course) the industry leader, Waltham. To woo the most proficient factory skills of the day, Elgin cleverly offered three things in great demand: higher wages, land to settle and homestead, and the opportunity to be part of a new venture unencumbered by the practices of the industry leader (Harrold, 1984, p. 28). Important people with very established watch lineages left en masse from Waltham. While it took much longer to start production than originally planned, the burgeoning railroad market provided sufficient security to Elgin investors to see the factory through to production, starting in 1867.

From the beginning, Elgin and Waltham were peculiarly linked. Elgin's factory owners recruited their first skilled workers from Waltham, offering them significant wages and signing bonuses. This gesture allowed Elgin to quickly establish a production system that mimicked and yet improved upon the Waltham model and thus supplied the firm with the credibility to raise needed operating capital. Credibility was sorely needed, too; before the first Elgin watch was produced, the company had to raise $250,000 to finance machinery and inventory (Alft, 1984, p. 55). Even this level of funding proved insufficient because the company almost went under in its first year. Wealthy local families came to the rescue, and the construction of a permanent factory building finally occurred in 1865.

Almost immediately Elgin successfully competed with Waltham. Within five years, the company was producing tens of thousands of movements. The economic crisis of the 1870s forced the company to reluctantly cut its prices, but these price cuts were a blessing in disguise that yielded greater market share (Harrold, 1984).

To make its watches affordable, Elgin sold movements that were then cased by jewelers. This strategy permitted customers to select a watch case they could afford, giving them the best of both worlds: a high precision watch movement with high reliability and a case for it that fit an individual's pocketbook. It also meant that Elgin did not have to invest capital in the expensive machines needed to fabricate cases. Given that the material costs of making a case, compared with material costs for a watch movement, were high, Elgin's strategy reduced the company's financial exposure, allowing it to spend funds on ongoing technology development of a limited line of high-quality watch movements.

Price competition has always played a key role in the U.S. watch industry. Elgin sold watches through wholesalers, a strategy that made considerable sense given Elgin's initial servicing of the growing Western market and its channel strategy, which emphasized the manufacture of uncased watches. By using wholesalers, the company did not have to pay a sales force.[5] In 1885, volatile prices due in part to very different market strategies led to the formulation of a "watch trust" that resulted in a uniform price throughout the supply chain. This lasted for six years. Elgin eventually left the arrangement when Illinois laws ruled out trusts. However, the effects of the trust remained intact as the two key firms retained identical pricing (Alft, 1984).

By the late 1890s, Elgin had moved close to becoming the largest producer of watches in the United States. With more than 50 percent of the domestic market, Elgin's plant employed 3,000 workers and manufactured 1,800 watch movements daily. To retain market share, Elgin provided rebates to wholesalers that allowed significant underpricing of Swiss watches. This same strategy proved deleterious to Waltham and contributed to the firm's continuing instability.

The seemingly endless growth of market share came to a halt in 1893 when the long-lived economic recession smothered consumer spending. At first the company felt it was invulnerable to the national economic decline. But a complete about-face occurred nearly simultaneously as the company announced quite unexpectedly on August first that it would reduce its work force by 50 percent. By the end of August, the factory had shed half its employees; it continued to cut payroll through the following year.

The rapid era of new firm formation came abruptly to an end by the beginning of the 1880s. Of the approximately 25 watch companies started between 1860 and 1880 only a few survived beyond the 1880s. It was far easier to hang out a shingle that read "watch company" than it was to fill orders and endure the long lead time and huge expense between incorporation and the production of volumes of commercially available watches. The costs were so high and the uncertainty of achieving decent manufactured volumes in a timely fashion so great, that many factories reportedly imported Swiss parts to have something to sell and to meet investor expectations.

Sowing the Seeds of Their Own Demise

The American firms, particularly Waltham and Elgin, transformed the world watch industry. At the beginning of the 1880s, the American system was producing in excess of two million watches per year. Waltham and Elgin were responsible for 80 percent of the industry's total output (Table 6.1).

By this time, however, the golden years of the U.S. industry had already passed. A series of factors, including a high rate of new firm formation, rampant price competition, economic recession, and a renewed Swiss industry, contributed to the U.S. industry's steady decline over the next 50 years. The two major watchmaking companies set upon a destructive course of price competition that left Waltham in almost complete shambles. Meanwhile, Elgin gained price points that were dependent upon volumes that could not be guaranteed during an era of macroeconomic instability and reinvigorated international competition.

TABLE 6.1. Total Output, Waltham and Elgin Corporations, 1880

Company	Time pieces produced
Waltham	1,500,000
Elgin	700,000
Total	2,200,000

Source: Harrold (1984, p. 35).

The very different styles of operation exhibited by Elgin and Waltham were a factor as well. Waltham's factory, as the older and more established of the two, was originally organized as a giant disintegrated system of production. While the firm manufactured and assembled standardized parts, the watch assembly process was highly fragmented and carried out by hand. A description of the Waltham factory in the mid-1860s is reminiscent of the Swiss cottage industry in which each watch model was manufactured in its own area within the factory. The highly sectionalized system reflected the slow evolution of a firm that started out with a workshop configuration employing 150–200 people, and then grew into a huge factory employing several thousand workers (Harrold, 1989; Moore, 1945).

From its beginning, Elgin developed an entirely different system of highly rationalized production. Unlike Waltham, which produced many models of watches, Elgin concentrated on far fewer models and worked to perfect each one. Lot sizes at Elgin were substantially larger than at Waltham. Elgin grouped types of watches efficiently. But the organization of production, while different, was only partially the cause of Elgin's success.

The other reason was Elgin's pioneering application of time–motion studies. Under the influence of Thomas Avery, a highly pragmatic manager who had grown rich from selling lumber to reconstruct Chicago after its Great Fire and then used his wealth to gain a controlling interest in Elgin, the firm discovered how it could drastically cut prices while still maintaining quality. Avery established his now famous program, "Popular Pricing Policy," cutting watch prices by 50 percent in 1878. Over the course of the next 10 years, Avery cut prices in half and then in half again (Harrold, 1989; Moore, 1945). While the short-run effect of this gesture momentarily boosted demand while transferring volume leadership to Elgin from Waltham, it also threw the watch industry into a free fall, forcing Waltham into a major reorganization (Moore, 1945). Unlike Elgin, which enjoyed concentrated ownership and therefore uniform direction, Waltham was owned by hundreds of small investors, including many former employees, which made coordinating a consistent business strategy almost impossible. The inability to completely control the direction of the firm only served to cripple it further as its once smaller rival pushed ahead in the competition. The effects of this ruinous price competition afflicted U.S. industry from that point forward, weakening all producers and reducing their range of options by the end of the 19th century.

As the 1880s ended, the American watch industry had entered an era in which profits were slim or nonexistent; external competition, especially from the Swiss, was growing; and massive reorganization was just around the corner. But price competition was not the sole culprit weakening the U.S. industry. Three major developments—the creation of the dollar watch, the establishment of railroad watch standards that locked out foreign competition but locked in a standard of design that made watch capability virtually indistinguishable among firms, and the innovation of the wristwatch—proved too much for the domestic industry. By the early part of this century, it was a shadow of its former self.

The creation of the dollar watch was a profound diversion, which forced established watch firms in a new direction. This diversion drew away the attention of America's jeweled watchmakers, a regrettable mistake inasmuch as it allowed the Swiss to reassert themselves by the end of the 1880s.

Making Watches for the Masses

For almost 100 years, watchmakers chased after the seemingly unattainable goal of making an affordable watch that kept accurate time. America's premier watch companies manufactured precision timepieces based on a system of jeweled interiors. While significantly standardized in production organization and execution, the jeweled watch still required considerable hand labor in assembling its minute parts (Hoke, 1990). With the application of machine tools to watch parts manufacture, the price of a watch had dropped from $40 in the 1860s to around $10 in the 1880s. But this still placed the watch out of reach of the working man or woman. Given that wages at the time were on the order of a dollar a day, jeweled watches proved unaffordable for most people.

Entrepreneurs on both sides of the ocean recognized the huge unmet demand for affordable watches. Interest in creating a cheap watch dates back to the middle of the previous century. Numerous companies in Switzerland and France tried different strategies and technologies, but failed to produce a consistently inexpensive watch. The creation of a cheap, affordable pin-lever watch must be attributed to George Roskopf, a German watchmaker living in Switzerland. The inexpensive watch industry in the United States evolved primarily in response to this new development (Harrold, 1984, p. 54).

To appreciate the enormous impact a cheap watch would have on the industry, we need to consider the impediments to its development and commercialization. As I mentioned previously, various facets of the know-how needed to produce a truly affordable watch had been known for quite some time. Indeed, the problem of manufacturing a cheap watch was not chiefly the result of a technological bottleneck, but instead reflected cultural rigidity that defined a watch only in terms of its precision and technical complexity. A good watch could not be "cheap" because cheap was associated with poor quality. For an inexpensive watch to break through this barrier, a complete reconceptualization of the watch as both a technology and as a consumer product was required.

It took Roskopf, a clever, altruistic Swiss watch designer and assembler, motivated by societal interests, to break out of the existing industrial model and manufacture and distribute the first dollar watch. In the middle of the 19th century, the Swiss economy was subject to tremendous swings in economic growth. The large mass of workers in the Swiss watch industry earned too little to be able to purchase one of the watches they manufactured themselves for world markets. The new watch was based on a set of principles that, when converted to manufacturing principles, revolutionized the industry. As Roskopf noted:

> Moral principles: 1. To provide the working classes with an accurate watch at a reasonable price. 2. Consequently, to suppress all luxurious and unnecessary work in the external presentation of the movement and the case, and to apply the economies thus made to the advantage of the working parts of the watch. 3. To utilize only those materials which were of excellent quality, and to remunerate the workers properly in order to maintain a high standard in the quality of the workmanship. Technical principles: 1. Reduction of the number of working parts to the fewest possible. 2. Simplification of the escapement. 3. Improvement of the system of winding. 4. Provision of the greatest possible motive force. 5. Simplicity and solidity of the case. (cited in Jaquet and Chapuis, 1970, p. 167)

Roskopf's remarkable vision challenged the fundamental principles of the Swiss watch industry. Swiss watches were known for both their ornamentation and their high technical quality. In fact, much of the industry's innovativeness rested on improved materials in the watch's presentation. This ostentatiousness extended to the design itself, which included parts deemed superficial in function; these were eventually abandoned, however, in the general manufacture of watches. The radical nature of the Roskopf watch "message" held it back from early manufacture. If the highly fragmented Swiss system had embraced the very inexpensive machine-made Roskopf, U.S. industry might never have gained control of the low end of the watch market.

Since he was a designer and an assembler rather than a manufacturer, Roskopf had to convince the Swiss of his design's merits. This proved to be very difficult. The existing system of home work resisted the introduction of a watch based on cheap parts manufactured by machine. As Jaquet and Chapuis (1970), looking back to the era of the first truly affordable watch, note:

> Today [1953], we may smile at the idea that it would be difficult to produce a good metal watch for 20 francs. But, in order to realize the extent of these

difficulties, it is necessary to go back in the imagination to the year 1860, when watch making, in the Neuchatel mountains, was largely a home industry. The workman had his accredited employers, and he remained faithful to them. It was by no means easy to persuade him to change either his type of work, or his methods, or the quality of the work he did, and now here was a project to introduce at La-Chaux-de-Fonds the manufacture of a metal-cased watch with a plain movement, of a design which conflicted with every notion of manufacturing accepted at the time. (p. 168)

This digression into the evolution of the Roskopf inexpensive watch allows us to proceed with the American effort to produce cheap watches for the masses. First, though, I must note that, like the British before them, the Swiss industry overlooked the manufacture of inexpensive watches—a major market niche that would seriously affect the industry at large. The lack of activity called the culture and system of manufacturing into question and signaled an important weakness in the disintegrated, labor-intensive, highly individualized system of production. But the problems presented by the Roskopf watch to the existing Swiss production organization only 20 years later gravely challenged the U.S. system of machine manufacturing as well. For while the American industry was to eventually come out with the cheap watch, like the Swiss, established American watch firms were not the ones to pioneer this product. Indeed, like the Swiss, it would take another organization, in this case a clock company, to develop and distribute this product.

Borrowing (and in some cases, perhaps stealing [Harrold, 1984, p. 59]) from the Roskopf design, many American firms tried to manufacture inexpensive watches. Interestingly, most of these firms were in the clock manufacturing industry where the technology of working with sheet metals was highly developed. One of the well-known producers and distributors of dollar watches was the Waterbury Clock Company, later the Waterbury Watch Company, which mass-distributed the first truly affordable watch. While its initial idea was sound, its significance was eroded by linkage with a quirky distribution gimmick based around the sale of clothes. The watch was defined by its free distribution with the purchase of a man's suit. This widespread distribution strategy eroded the market value of the watch. Eventually declining demand forced the company into bankruptcy (Brearly, 1919).

Like many failed enterprises, at its core the Waterbury Watch Company was a solid business. From the ashes of the Waterbury Watch Company came the most famous cheap timepiece, the Ingersoll dollar watch. Created by the Ingersoll brothers in 1892, this mass-distributed punched-metal watch was sold for a dollar (wholesale price). Like Roskopf, the Ingersolls adopted the strategy of designing the watch and the case as an integrated system. There were several advantages to this, in particular the creation of an aesthetically pleasing watch case that could hide a multitude of sins found in an inexpensive movement.

Perhaps the Ingersoll brothers' most important innovation was the introduction of new distribution channels for watches. Heretofore, watches were distributed through wholesalers who sold to retailers, who were for the most part jewelers. The intimate links in the distribution channel functioned to rigidly restrict access to these watches by the common citizen. The Ingersoll brothers originally sold their dollar watch through a mail order catalog. Although mail order demand was encouraging, it did not reach the level necessary to produce the dollar watch at a profit. First shunned by traditional distribution channels, the dollar watch eventually found its way to market via a variety of channels, both conventional and nonconventional (Brearley, 1919). The Ingersoll Company also adopted an ingenious repair policy to assuage customers' concerns about the reliability of a cheap watch: the company promised that if their watch broke after an unreasonably short period of time, the company would effectively replace it, free of charge. Of utmost importance to sales success of the Ingersoll watch was the discovery that sales through small retail outlets were more profitable than sales through department stores and jewelers. Hence, a mass distribution channel completely outside the confines of the traditional method of distribution became the hallmark of the dollar watch and its successor, Timex.

In response to the development of the dollar watch, jeweled watch producers forced prices down for their least complicated seven-jeweled watch. Numerous low-cost jeweled watch producers responded to the Ingersoll challenge by cutting costs. Soon, for slightly more than twice the price of a dollar watch, the consumer could buy a quality, inexpensive jeweled watch ($3.50 vs. $7.00) (Harrold, 1984).

The establishment of a market for inexpensive watches enticed new producers into the watch industry and precipitated a domino effect in price competition. Newly established firms increased competitive pressure on the established dollar watch firms and led to further cutthroat price competition from which only a few relatively large firms emerged. The same competitive pressures befell the cheap jeweled watch companies, further driving down the price of watches. This competition also had an impact on the larger, higher quality jeweled watch producers such as Waltham and Elgin, which responded aggressively and virtually drove the low-priced jeweled watch producers out of business. A few statistics tell it all. From 1880 to 1930, the ratio of jeweled watches to dollar watches produced in the United States completely inverted (Table 6.2). As Harrold (1984) insightfully notes, "It was as if the watch industry had been invented all over again" (p. 64).

TABLE 6.2. Production of Watches, Jeweled and Dollar, United States, 1880–1930 (in millions)

	1880	1890	1900	1910	1920	1930
Jeweled watches	3	11	23	41	63	80
Dollar watches	1	5	18	58	118	186

Source: Harrold (1984, p. 64).

New Standards Create Momentary Breathing Space

The creation of the dollar watch challenged the fundamental economic basis of the American watch industry. Firms like Waltham and Elgin had little choice but to pursue the low-price market, even if it meant manufacturing and selling watches at a loss. Oldtimers at Waltham remembered that Dennison had foretold the future when he spoke of creating a watch that working people could afford, even though firm management refused to allow him to develop such a watch (Moore, 1945, p. 46). In response to cutthroat competition, Watham continued to introduce new models even though this drove up costs and caused inventories to swell. In the face of high costs and declining demand, Waltham sought a merger with another watch company. At one point, Waltham even considered merging with Elgin. But terms could not be reached and the parties separated without executing a satisfactory deal.

Just when the situation could not look more grim, the economy recovered and the Waltham company's balance sheet moved strongly into the black. Of major importance in this short-run turnaround was the emergence of the railroad standard, which created a captive market for U.S. watch firms. There can be no denying that this development staved off disaster for the industry in the waning years of the last century. However, this new source of demand stimulus came back to haunt the industry, for it eventually stifled creativity and new product development, just as new styles and consumer preferences demanded something other than a pocket watch. A critical influence on the design and structure of the watch industry at the turn of the century was the establishment of railroad regulations that stipulated performance standards for watches used by railroad workers. The standard setting effectively created a secure market for American firms while locking out foreign competitors, particularly the Swiss.

Railroad standards for watch reliability and repairability emerged with the recognition that the safety of passengers and freight could only be guaranteed by the accuracy of the railroad system. Begun in 1872, the initial regulations were incremental and voluntary. A serious crash outside Chicago impressed upon the rail industry the importance of accurate timekeeping. Two trains crashed head-on on a single-track stretch of railway; it was posited that one of the engineer's watches was a few minutes slow, causing a slower moving postal train to be on the same tracks as an oncoming freight train (Harrold, 1984; Sauer, 1992).

The first watches designated as "railroad quality" were high-quality, but common watch models already in production. But as the 1880s progressed, watch companies began to design watches specifically for railroad usage. Nonetheless, numerous watch types were still in circulation, some designated "railroad," others not. Many were foreign watches produced through varying means, in particular through hand manufacture. After the Chicago train crash, railroad officials found it imperative to tighten the regulations. Working with a jeweler, they established technical standards for railroad watches. The specific rules were designed to regulate watches in use, and served to homogenize watch design and demand. Of key importance to the railroads was the stipulation that a railroad watch, any railroad watch, had to be easy to read, easy to repair, and easy to keep on time. As Harrold (1984) notes, "The standards specified companies and models so that trainmen knew what was allowed and jewelers knew what parts to stock, which was probably the reasons for requiring American watches" (p. 75). This standard setting effectively ruled out British and Swiss watches, which were handcrafted and thus required specialized parts.

With the emergence of railroad regulations, new competitors entered the industry. Now, the industry included four watch segments: expensive, moderately priced lines of jeweled watches, railroaders, and dollar watches. As with other industries, product segmentation commonly occurs when maturity arrives. With a product like watches, segmentation presents significant problems for market leaders, as they are forced to maintain a position in all segments to retain their lead. In the case of watches, this presented serious problems in the management of inventory, materials, design, and distribution channels. As I suggest next, latecomers can challenge the status quo because they are unencumbered by sunk investments and relatively outdated production technology. A reasonable strategic response by dominant players is to push toward standardization across segments, thereby reducing the amount of capital tied up in multiple production systems needed to produce specific models. But this often requires a significant write-off of sunk capital, a policy painfully and—in the case of watches—very reluctantly and belatedly pursued. The inability to instantaneously shift to a new production configuration, as implied by neoclassical economics, often results in high short-run costs, rising inefficiencies, and severe turmoil.

The Hamilton Watch Company of Lancaster, Pennsylvania

The Hamilton Watch Company of Lancaster, Pennsylvania, was a relative latecomer to the U.S. watch industry. It was built from the remains of a number of failed companies that had started locally in the fertile plain of southeast Pennsylvania. Spurred by a belief in the need for more accurate timekeepers than were currently available, local businessmen in the rural community bought the physical capital of the Lancaster and Keystone Club Watch Companies and in 1892 reorganized them under the name Hamilton Watch Company.

The company's targeted market was the railroads. The company's early advertisements boasted of the product's reliability, and the company made every effort to have its product known as, tested as, and used as the national railroad pocket watch.

For the first 16 years of the company's life, its sole goal was the production of the reliable Locomotive "940." This watch was big and bulky, but its precision set it far apart from competitors. It was not until 1908 that the company introduced a special woman's watch, but when it finally did market the Lady Hamilton, it tapped into a limitless market.

The demand, combined with the development of railroad watch standards, allowed firms to substantially upgrade and standardize production. This also raised the level of risk and indebtedness, however. The railroad watch model was so uniform that little distinguished different companies' products. As Harrold (1984) notes, "For the most part, however, these watches had no readily identifiable features, being more distinguishable by performance than appearance" (p. 51). A negative side effect of the creation of standards was the "technological lock-in" that occurred around the watch's production. Major investments to upgrade watch products and produce the needed volume of surplus parts served to restrict the design and production visions of major watch producers. Moreover, because they were externally uniform, major watch producers balked when the growing market for wristwatches, which was built around fashion, demanded smaller movements and highly externally differentiated products. More critical still, conversion would have required writing off huge sums of money tied up in accumulated spare parts inventory. For some companies such a write-down would have seriously challenged their economic basis. As Harrold (1984) perceptively notes, "By the late 1890s, existing watch companies had watches in circulation that spanned almost 50 years of watch manufacture. Since owners cherished their watches, and buying a decent watch was still a significant investment, the companies were required to maintain parts inventories for thousands of models. This inventory had to be maintained, further increasing the industry's expensive overhead" (p. 51).

A long recession that began in the 1870s and continued through the 1890s produced fierce price competition in virtually every domain of the watch industry. Weakened by market instability, key industry players adopted an inward orientation that slowed their response time to new developments in the field.

By the end of the 1880s, a radical incremental technological change—the wristwatch—was in the offing. Internal contradictions within U.S. industry, including profound price competition among the major watch firms that drove prices through the floor and the arrival of new domestic competitors producing a very low-priced product built upon technology far simpler and less costly to manufacture than that of jeweled watches, added to the period's instability.

Perhaps more important in explaining what inhibited the ability of U.S. firms to switch quickly to the production of wristwatches was the institutionalization of a uniform regulatory regime for the production of railroad watches using machine-manufactured parts. This effectively closed off a major segment of the U.S. market to hand-manufactured watches. Standardization of production around a market segment (railroad watches) stabilized and homogenized technical product development, thereby eliminating the stimulus for new investments to maintain parity with the emerging new technology. This series of fortuitous events, or missteps, by the U.S. industry allowed the Swiss industry to reposition itself and serve an even broader market. Was the Swiss production system uniquely poised to take advantage of the insulated product design behavior that came to characterize the U.S. industry? To the extent that fashion was a major influence over the emerging wristwatch market, the production of individual brands by many small firms provided the variety desired by consumers. Differentiation was demanded and the Swiss were able to respond.

Summary

By the mid-1860s, the American watch industry was fully formed and functioning vibrantly. Some 20 companies, parts manufacturers, and machine fabricators comprised the industry (Abbott, 1881). It was a time of great experimentation and rapid incremental change as firms sought to control the precision technology of watch manufacturing and overcome the need to import some of the finer parts from Europe, such as hairsprings (Harrold, 1984). New developments in machine technology allowed for continued movement toward true interchangeability (a quest not achieved until the mid-1930s).

The U.S. watch industry had its coming out in conjunction with the

Philadelphia Centennial Exposition of 1876. In line with the U.S. image of the rugged frontiersman, the Waltham Watch Company "impressively demonstrated a number of machines and some of their finest watches and it became clear that American factories could saturate most segments of the market with well-made timekeepers" (Harrold, 1984, p. 3). By the mid-1870s, the U.S. watch industry was producing a watch that was more consistent and accurate than the Swiss handmade watch. U.S. industry aggressively attacked the Swiss where it had attacked the British, in the lower priced market segment. As Harrold (1984) notes, "As if to demonstrate the point, the industry unleashed their unjeweled watch in 1878 for $3.50, tearing away the low-priced market that had been the last Swiss stronghold" (p. 3). A close examination of the major U.S. watch companies of the 1870s suggests the depth of the differences between the two systems. It took 20 years and several fortuitous circumstances before the Swiss system of fragmented production helped Switzerland to retrieve its lost position as the world's timekeeper.

The 1860s through the 1890s were the golden years of the U.S. watch industry (Moore, 1945). Industry pioneers saw the opportunity to provide accurate, precise timekeepers to the masses through the application of machine manufacturing techniques. To undersell the Swiss, it was necessary to minimize variable costs. This could only be achieved by simplifying the watch, standardizing its parts, improving the watch's reliability, and distributing it widely.

The U.S. machine-based system, while momentarily enjoying the gains from standardization and low variable costs, faced the twin problems of needing very high volume and large variety. A respite provided by the establishment of the railroad standard in 1892 gave American firms a lock on the domestic market, and hence access to a large uncontested demand. However, this initial blessing of a market monopoly became a curse as the uniformity encouraged by the railroad standard clashed with future demand for variety, which was ushered in by the penetration of wristwatch technology.

7 More Than One Way to Win a War

We have now considered the origins and evolution of the watch industry from the 17th century to the early part of the 20th century. Three historic moments in the industry's evolution were addressed: the unsuccessful transition of the British industry from a craft-based to a more mass-scale system of manufacturing; the gradual emergence of a vertically specialized system of horizontal fragmentation organized around a highly skilled and malleable labor force (the Swiss system); and the equally gradual emergence of the "American system" of machine-based standardized parts manufacturing, capable of production in large volume. In the case of the American system, its success was linked to associated trends, including the benefits of a growing domestic market and the convergence of industrial technologies that led to the rise of a complementary machine manufacturing capability. At this point the watch industry moved toward producing a more mass-scale product, but a persistent engineering problem the need for high volume and high variety at low cost—continued to dictate industry behavior.

Why did the U.S. industry rise so quickly to prominence and almost as quickly decline? How was it that the Swiss industry was able to reenter the U.S. market as a formidable competitor? To what extent were the problems that plagued the U.S. watch industry internal to the industry itself (i.e., its institutional context), and to what extent were external factors (i.e., the macroeconomy, geopolitical developments) controlling the options U.S. firms faced? Similarly, what was it about the times that allowed the Swiss industry to reemerge as a forceful competitor? To what extent was its elevated position a result of changes taking place in the macroeconomy, versus those from within the institutional context of the industry itself?

In the first part of this chapter I examine the emergence of wristwatches and investigate how an early industry leader, Waltham, coped (or did not cope) with change. Here two main factors are considered: macroeconomic instability and managerial transformation. Rampant trade protection and a fragmenting industry, coupled with managerial upheaval associated with the shift from proprietary to managerial capitalism, destabilized conditions and left many firms (in many industries) with inadequate resources and clouded vision. In the second part of the chapter I delve into the problems that plagued the Swiss industry upon its initial confrontation with the Americans, and then examine how the Swiss industrial system responded to this challenge. Primary emphasis is placed upon the extent to which shifts in industrial structures are malleable in the face of technological and institutional change. As we will see, the price of converting to a model embodying aspects of the U.S. system gave the Swiss only a momentary respite from instability.

The Swiss Watch Industry Reemerges as a Major Industrial Competitor

At the end of the last chapter, I noted that the negative effects of 20 years of cutthroat competition in the U.S. watch industry were masked by the introduction of a new watch standard, the railroad watch standard, that effectively locked out foreign competition. Was the financial disaster of the 1870s–1890s the necessary macroeconomic condition that enabled the superior Swiss system to reenter the U.S. market after a period of trade protection, or was this experience the jolt needed to force the Swiss out of their midcentury complacency? There is no question that the discovery of a competent competitor located across the ocean called into question the entire Swiss system of manufacture. Up until this time, economic instability was managed by "battening down the hatches" (see Lazonick, 1991), reining in costs, and squeezing productivity out of existing capacity. When such efforts failed, wages were lowered, male adult workers were replaced with women and children, and, finally, and as a last resort, the work force was reduced.

But the Swiss response to U.S. competition resulted in more than just another round of cost reductions (though this did occur in the initial stages of the crisis, starting in the 1870s). First, the Swiss carefully studied the problem of new competition. Major institutional change followed: key firms consolidated, while others simply got out of the business entirely, making way for a new system.

The transformation was abrupt. Losing two-thirds of the U.S. market directly or indirectly affected every watchmaker in the Jura. According to Landes (1979), Swiss exports to the United States peaked at a value of 18 million Swiss francs in 1872, plummeted to 13 million Swiss francs the following year, and sunk close to 3.5 million Swiss francs by 1877 (Table 7.1; Landes, 1979). The recession lasted a long time. However, because of past investments made to capture new markets, many Swiss watch firms had grown large enoughto become essential to the national economy.[1] Moreover, the Jura region was almost totally dependent on watch manufacturing (Table 7.2; Landes, 1979, p. 33). For all these reasons, the freefall decline of economic conditions in the Jura was halted and a much needed large-scale change occurred.

The Jura Industry's Past First Haunts, Then Saves It

Machine manufacturing of watch parts was not new to Swiss industry. The first machine-manufactured watch parts were produced in Switzerland in the 1840s (Jaquet and Chapuis, 1953; Jequier, 1991; Moore, 1945). Such forays (those sufficiently radical to raise eyebrows) could

TABLE 7.1. Swiss Watch Exports to the United States, 1864–1882 (in Swiss francs)

Year	Francs
1864	8,477,192
1865	11,301,954
1866	13,093,408
1867	10,362,418
1868	10,469,728
1869	13,222,578
1870	16,512,162
1871	17,105,752
1872	18,312,511
1873	13,054,147
1874	12,119,941
1875	8,499,501
1876	4,809,822
1877	3,569,048
1878	3,995717
1879	5,292,098
1880	10,143,813
1881	11,809,122
1882	13,238,489

Source: Landes (1979, p. 29).

TABLE 7.2. Watch Industry Employment in the Swiss Jura, 1870–1920

	1870	1888	1900	1910	1920
Neuchâtel	13,689	14,629	18,024	16,322	19,016
Bern	14,772	19,157	22,359	21,832	26,085
Solothurn	806	2,395	3,965	6,111	6,329
Geneva	3,234	2,416	2,202	2,435	3,376
Vaud	3,633	2,827	3,136	2,814	3,333
Basel-land	94	367	648	1,106	1,500
Six cantons total	36,228	41,791	50,334	50,620	59,639
Proportion of industry	92.1%	94.7%	95.4%	95.1%	94.9%

Source: Landes (1979, p. 33).

be and were ignored. Roskopf and his cheap pin-lever design also were a thorn in the side of the Swiss industry. But the invention of the inexpensive watch was initially ignored by the Swiss for numerous reasons. The pin-lever watch, with its fewer parts and plain case, was not considered a "real" watch by conventional Swiss standards. Roskopf also chose a distribution channel considerably outside the norm. In addition, at that time, watch manufacturing was still a home-based industry and therefore Roskopf could not exert control over the highly fragmented system.

The crisis of the 1870s could not be ignored, however (Katzenstein, 1984). The Swiss no longer had other markets to escape to, as they did in the past. Except for Britain (Gourevitch, 1986), Switzerland's European trading partners, one by one, enacted trade restrictions that made the situation unstable. In addition, since the U.S. Civil War, tariffs had been used increasingly to raise revenues for the U.S. federal treasury. The Swiss watch industry was supposedly protected under a treaty signed into law in 1850 between the United States and Switzerland (President of the United States, 1851), and that carried with it favorable terms of trade, such as reciprocal treatment with regard to tariff levels and market access (akin to today's most favored nation status); still the turmoil unleashed in this period of industry-led protection was felt across the Swiss economy (Shannon, 1967). As the 1870s progressed, the Swiss simply had to adopt a new model of development or slide further out of control. The magnitude of external events called for immediate action. Swiss hubris was brushed aside and machine manufacture was introduced and took hold.

Let's review the important events. The Swiss dominated the world market until the Civil War, and then trade restrictions and heightened demand provided sufficient opportunity for the U.S. watch industry to expand. A new system, the consequence of a scarcity of skilled labor, emerged, and with it came measurable benefits of economies of scale, vertical integration, and machine manufacture. The Swiss, lost a key market momentarily to a more adept competitor. Meanwhile the success of U.S. watch manufacturers began to attract new entrants. U.S. domestic competition increased as the locus of control of the U.S. system of watch manufacturing passed from craftsmen to entrepreneurs. The 1870s crisis occurred, and finally the Swiss reacted by reformulating their system, introducing machine manufacture, but retaining the differentiated labor-intensive elements of assembly to provide system flexibility. Thus, the Swiss moved toward a system with significant economies of scale in the production of standardized parts while preserving labor flexibility to ensure innovation and maneuverability in the face of unexpected change.

In 1870, the Swiss Jura watch industry employed 35,000 people, most of whom worked in their homes. Thirty-five years later, the majority of all Jura workers were in factories. By 1905, Swiss watchmaking was predominantly a factory system with larger factories and high levels of mechanization. By 1893, when the Chicago World's Fair opened, the Swiss industry's worldwide preeminence had been restored. American firms were stunned, but faced such significant domestic problems that they didn't have the resources to analyze the Swiss recovery. The Waltham Watch Company was in an advanced state of turmoil and had much at stake (Figure 7.1; Harrold, 1984, p. 48). The U.S. industry had literally split in two (data from Harrold, 1984, p. 50). By 1890, jeweled watches as a share of total watches produced was declining annually; by 1900, there were more dollar watch producers compared with jeweled watch producers (Figure 7.2; Harrold, 1984, p. 50). A vocal and uncontrollable fringe of firms emerged; a significant contingent of watch assemblers came into existence and survived by buying watch parts on the world market, including from the Swiss, and assembling products for final sale (Figure 7.2).

The transformation of the Swiss industry required the abandonment of a centuries-old tradition based on the craft producer and the home manufacturer. In order to compete, the Swiss had to beat and exceed the United States on its own ground. The prevailing system based on market coordination gave way to a mixture of scale and control plus fragmentation. In part because of the specialization that continued to characterize the industry, small assemblers could make a living finishing watches even as the industry changed. But a period of long global economic instability challenged this new system even as it was being born.

More Than One Way to Win a War ■ *135*

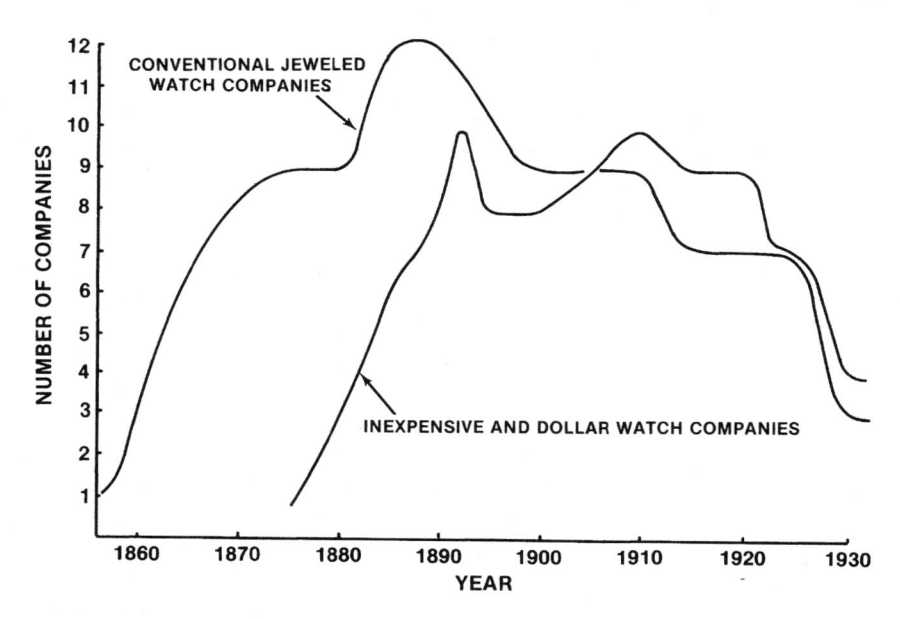

FIGURE 7.1. Watch company population, 1860–1930. Copyright 1981 by M. C. Harrold.

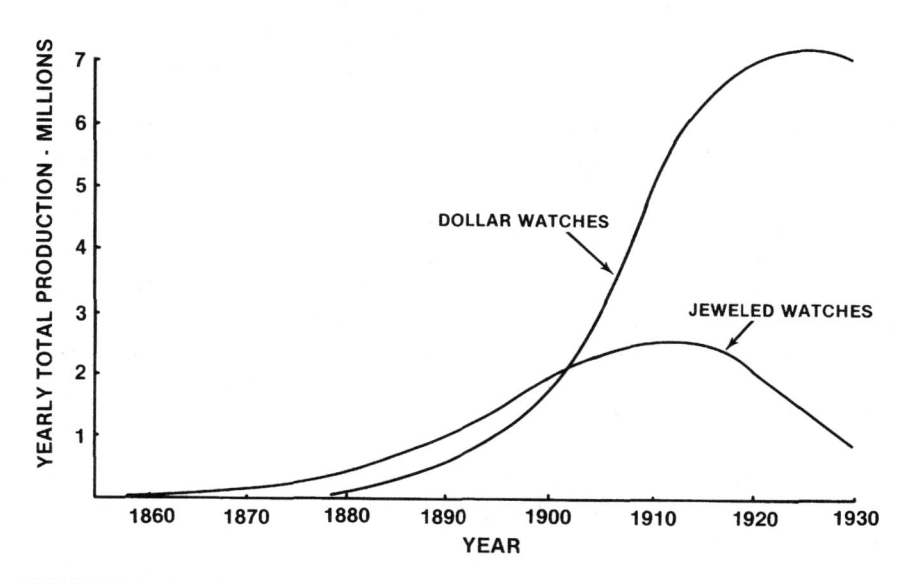

FIGURE 7.2. Yearly watch production, jeweled versus unjeweled, 1860–1930. Copyright 1981 by M. C. Harrold.

Costs of Redirection

The great strength of the vertically specialized, horizontally fragmented system of production that characterized the Swiss industry at the end of the 19th century was its flexibility relative to the U.S. model. But flexibility was only one systemic attribute. More important from the standpoint of technical change was the intense rivalry between the watchmaking firms of different Swiss villages and small towns and their employees (Pinot, 1979). In addition, the watchmakers were extremely independent: they owned their own tools and, while dependent on contractors, could set their own hours and effectively regulate their own productivity and rate of profit.

The rural collective industry functioned as long as demand increased modestly and the problems of spatial fragmentation did not become so onerous as to be unworkable. The first stage of economic consolidation, intended to restrict competition and regulate the quality of production, involved the replacement of the rural home industry with urban factory manufacturing. Geographic concentration in factories in cities allowed contractors to exert greater control over the work process and the final product. Moreover, the new system reduced transportation costs, to the considerable advantage of contractors. Retaining the structure of home work but agglomerating producers for better control allowed spatial concentration of the industry to take hold.

A delicate balance always existed between the captive watchmaker and the merchant. As profits increased, as they inevitably did with the expansion of markets, the watchmaker could himself become an organizer of the labor process, effectively creating another competitive outlet for watchmaking. Moreover, once the watchmaker himself became a contractor, he divided his time between making profits for himself and making products for his contractor. The absence of any effective industry regulation meant that the contractor was subject to a high degree of uncertainty, which spurred the formation of the first factories. Under this mode of regulation, the worker's skill was reduced and his ability to fabricate complicated parts declined. Pinot (1979) describes this transition:

> Under this type of production, the worker faces a new kind of life. He loses his quality of specialist and becomes a simple jobber. He works on machines sometimes up to 12 hours a day. He is under the supervision of a boss and must abide by quotas of production. He is paid a fixed weekly salary. He loses his independence, the possession of his house and tools; most of the time he does not even belong to the Bourgeoisie anymore. (p. 176)

This radical institutional transformation affected all aspects of Jurassian life. Because of the unstable nature of demand and the low wages in the industry, women, even mothers, left home to enter the industry, too. Fathers, who previously garnered respect as craftsmen in control of their destiny, now were reduced to the status of "workers," as interchangeable as the machined parts of the watches they assembled.

The decline of the homework system provoked a wide range of responses, most particularly in the structure of social relations between workers and employers. The urbanization of the industry produced overcrowding in cities. Health problems and the shortage of conveniences were pervasive. Employers no longer acted paternally toward their workers, which was a particular problem since no institutions provided for workers' health, insurance, and housing needs.

There are relatively few historical treatments of this period of economic, social, and cultural change (see Pinot, 1979; Rappard, 1914; and Jequier, 1991, for exceptions). Early worker-organizing efforts were in part a response to the economic and social problems of the day. The Swiss worker, despite his or her attachment to the labor process, had not attempted to organize over the previous century. This lack of collective action may be attributed to a number of factors. The home-work model of production brought with it two countervailing influences. First, the spatial fragmentation of the industry produced relative worker isolation. To the extent that there was collective action, it occurred at the level of the village or community, as one group of watchmakers competed with another group in the neighboring town (Jequier, 1991). Second, because the workers often owned their own tools and appeared to be masters of their craft, organizing was not seen as necessary or beneficial. Third, the Swiss population historically distrusted organizations that did not reflect the will and mood of the populace. Indeed, the Swiss tended toward more anarchism than toward collective action (Rappard, 1914).

But the institutional crisis of the late 19th century changed all this, bringing into being the conditions necessary for a widespread worker response. The rapid increase in factory size in the 1880s and 1890s presented its own problems. Jequier (1991) describes the situation:

> At Le Coultre and Compagnie, for example, the number of workers grew rapidly from 406 in 1889 to 482 in 1890. This distinct expansion of the work force itself created problems of order and discipline, which were exacerbated by the fact that at the slightest slowdown in business, the employers laid off workers to cut their variable costs. The gulf between employers and workers widened relentlessly, and the absence of information made any form of dialogue impossible. The fear of the employers was matched by the

resentment of the workers. The balance of power favored employers, who made skillful use of home workers, the "reserve army" so well described by Marx. A new social hierarchy came into being in the sphere of employment. The machine could do without skilled workers, and women, who were considered more docile, were willing to work for very low wages. (p. 327)

Thus, the benefits associated with the system of vertical specialization and horizontal fragmentation were an impediment when the system of industrial organization was reorganized. As this example suggests, the transformation to machine manufacture was neither instantaneous nor complete. In this case, the purpose of retaining vestiges of the past was clearly understood by employers who saw the need for some stability in order to buy time before the transition to new institutional relations could take hold.

From trust to mistrust, from shared identity to alienation, the necessary transformation of the industry met with labor resistance. At first, employers retained the upper hand and hardened themselves to the pleas and demands of workers. Secret unions formed in response. Not all employers were reckless and willing to replace their familial associations with the increasingly deskilled worker. Because the industry began in a region that had family-based social relations, many employers were distantly related to their workers. Thus, it was difficult to be completely dispassionate in response to workers' legitimate claims for stable work and decent wages.

The trade union movement grew in importance in the early years of the 20th century. Unionization did not affect conditions in the watch industry demonstrably until a merger occurred between the two major machine-based unions—the watchmakers and the machine tool makers. From that point forward, employers had to bargain in good faith or face the prospect of strikes. In addition to its effect on the watch industry, the power of the union movement overall made itself felt at the end of World War I, when unions joined with war-weary citizens in calling for a general strike to protest blockades enforced by the United States and Britain that had reduced the shipments of material goods including food into Switzerland. The effectiveness of the blockades had unleashed a sense of permanent insecurity in both citizens and industrialists. Food supplies were increasingly used as bargaining chips against the Swiss to induce them to reduce shipments of strategic manufactures to the Germans. Swiss industrial firms that had been supplying both sides in the war now found themselves locked out of formerly lucrative markets and blacklisted in others. Having depleted much of their reserves to buy scarce materials, and now facing the closure of markets, the employers' response was to reduce wages and then to cut employment levels when orders were shut off. In the past, the decentralized industry made organizing around collective concerns difficult, hence strengthening the bargaining power of employers (Rappard, 1914). But investments during the decades leading up to the war in response to the American system of watch production and the turn toward wristwatches had increased economies of scale and concentrated activities in fewer locations. By the end of World War I, the partly fragmented but increasingly mechanized system was more efficient than it had been at the end of the last century, but this efficiency was bought at the price of instability in the rank and file and growing vulnerability to labor struggles (Meier, 1970).

Innovation and the Absence of Sunk Costs Permit Repositioning

Despite great upheaval and simmering hostilities with labor, the Swiss response to U.S. technological challenges was resolute. Over a period of 20 years (1885–1905), the Swiss proved more than capable of meeting the technical challenge. With the advent of the U.S. industry, the Swiss lost considerable market share in the United States. However, due to the collapse of the British industry, and the inattention of U.S. producers to export, the Swiss manufacturers did not yield control of global markets. This proved to be critical: foreign trade allowed the Swiss industry to regain its financial momentum and ultimately to amass the resources needed to pursue innovation.

Over the course of the next two decades, then, the Swiss system adopted those features of the American system that were cost-effective. The system shifted from its strict reliance on small-scale cottage production to an intermediate form of organization that combined mechanization and partial vertical integration. Standard parts were mechanically manufactured on a large scale in centralized factories, while overall flexibility was maintained through dispersed design and assembly activities. Even the more complicated parts were eventually mechanized. Thus, within the existing technological system, the Swiss achieved new levels of profitability and international renown.

Why were the Swiss watchmakers capable of embracing an alien form of production technology that had been rejected previously by the majority of the industry's firms? The answer is that the Swiss watch industry had the ability to retrofit aspects of machine manufacture because Swiss and French watchmakers had in fact invented the first machine tools for production automation and parts manufacturing in the early 1800s (Bruton, 1979; Chapuis and Jaquet, 1956). Stimulated by the success of the first French factory that mass-manufactured movement parts, Swiss watchmakers made an initial foray into factory production of watch movement parts in Fontainmelon in the Jura (Bruton, 1979). At the time, the machine-based technology was viewed with suspicion by workers and was largely restricted to the production of movement blanks. After their initial rejection, machines were slowly incorporated into the industry over time. Thus, movement toward machine manufacturing may not have been as great a leap as some have suggested (Landes, 1984). In this case, the explanation can be found in the extraordinary competition within the watch industry, which continued to spawn innovations, coupled with a social agenda and cultural identity steeped in watch manufacturing.

The emergence of the American system was like a rude slap in the face that forced members of the Swiss industry out of their complacency. But complacency alone fails to explain initial faltering by the Swiss. The original highly fragmented system suffered problems similar to those that plagued the British in the middle of the 19th century. So why were the Swiss able to respond in a way the British had not?[2] Unlike the British, the Swiss industry re-created itself by calling on its deeply embedded ingenuity and recasting itself in the face of a new reality.

Market signals also were instrumental in the industry's turnaround. As consummate traders located on the Continent, virtually alone with no significant trade-based competition, the Swiss watch industry was allowed to observe new fashion trends and to fill new market niches as a totally new product evolved—the wristwatch.

The lapsed time before the wide adoption of machine manufacture was due to fear in the highly disaggregated craft- and family-based system that machines would displace the traditional industry. Only when a powerful crisis befell the industry did firms within it finally adopt the innovations that had been pioneered almost 100 years earlier.

By 1910, Swiss mechanical watches once again dominated the world watch industry (Knickerbocker, 1976). The Swiss controlled the micromechanical export industry through cost-competitiveness, superior manufacturing competency, high levels of precision, and extraordinary attention to detail and style.[3] The vertically integrated parts manufacturers achieved economies of scale through volume production of standardized parts. This benefit was passed on to assemblers in the form of a supply of low-cost movements. In the most labor-intensive segments of the industry, the vertically disintegrated system of assembly and case manufacturing kept overhead charges low.

The movement to wristwatches was taken up seriously by flagship Swiss firms such as Rolex. Rolex (founded in 1908) came of age in the era of wristwatches and soon was creating a product of remarkable accuracy, precision, and quality. It might be argued that relatively new upstarts such as Rolex actually propelled the rest of the industry forward by challenging the status quo from within (Dowling and Hess, 1996). Because the high-grade watch manufacturers embodied the full complement of skills needed to move to a new product type, they could approach the wristwatch with significant resources and resolve. Others were forced to follow suit.

Time on Your Hands

The Swiss watch industry was saved by more than its ability to adapt machine manufacture to the existing industry. Technological change in the form of the wristwatch provided the Swiss with a new product that was much in demand, one that U.S. firms at first chose to ignore. It was fortuitous that at the beginning of the 20th century, just when the Swiss were compelled to adopt new capital equipment and move toward the larger scale of standardized parts manufacturing, simultaneously a switchover to a new product occurred. Because of the Swiss industry's long resistance to the machine manufacture of watches, its capital investments at the time were relatively low and highly dispersed. As a result, few if any manufacturing family fortunes were lost with the advent of machines. All firms choosing to compete in the industry had to make some level of investment in the new technology. There also was a technological challenge in working at the more miniature scale required of wristwatches compared with that of pocket watches. The Swiss system of emulation honed local skills to a very high level of precision. Thus the know-how and technique to further miniaturize the watch were within the reach of the Swiss industry. This option and competency was not firmly grounded in the U.S. industry, where the bulky railroad watch stymied movement toward smaller watch sizes. This description of the seemingly effortless innovation of the wristwatch should not be taken too far. It took almost 40 years for the wristwatch to be fully accepted both by the buying public and the Swiss watch complex itself.

Wristwatch design began a slow ascent toward a standard in the years from 1880 to 1910. In 1880, the first small series of wristwatches was ordered from a Swiss company by the German navy to be worn to battle (Kahlert et al., 1986). Not until the Boer War of 1899–1902 did the wristwatch become an identifiable military tool. By World War I, the wristwatch was viewed as a vital item in battle. The laggard nature of the U.S. watch industry and its failure to seriously implement wristwatch production left the market almost completely to the Swiss industry, which was more than happy to supply timekeepers to both sides. Perhaps because the war was not fought on U.S. soil, and because the United States entered the war reluctantly and late, the needed stimulus that propelled the Swiss forward in wristwatch manufacturing was missing as a driving source of change in the U.S. industry.

While the military adopted the wristwatch quickly, civilians were much slower to change from pocket watches to wristwatches. There were several reasons for consumer resistance. The most important, from the standpoint of accessing the huge American market, was the initial concern that a watch small enough to fit on someone's wrist was too effeminate for men. American men were used to their hunting pocket watches, and initially resisted wristwatches. As Gohl (1977) notes, even by 1900, a trial shipment of wristwatches to a Swiss distributor in New York was eventually returned to Switzerland as *"unsalable* in the States" (Gohl, 1977, p. 587). Some evidence suggests that American watch distributors looked askance at wristwatches and therefore failed to provide the market stimulus needed to ensure their early and widespread acceptance (Sauer, 1992).

Luckily, two countervailing trends swamped initial reservations regarding the wristwatch. First, wristwatches were ideal for women, particularly those entering the work force. Early watch models designed for women were usually just smaller versions of men's pocket watches. The wristwatch, on the other hand, afforded women a greater choice of styles and exterior qualities. Second, the wristwatch suited those interested in sports and bicycling, which both called for a timekeeper that did not have to be pulled out of a pocket, opened, and read. Third, fashion trends were moving in the right direction. Clothing was becoming more form-fitting and French haute couture embraced the small, elegant wristwatch.

Although World War I marks the true beginning of widespread wristwatch acceptance, U.S. firms still hung back and only reluctantly and half-heartedly distributed watches (see Table 7.3 for figures on the Waltham Watch Company). Gruen and Company, a U.S. firm that distributed Swiss-made watches, was the first company based in the United States to market a number of different wristwatch models. It was not until the 1920s that American firms finally accepted the supremacy of wristwatches; by that time, however, they had lost tremendous market share to the Swiss. Although a few of the larger American firms retooled for the

TABLE 7.3. Production of Movements, Waltham Watch Company, 1926–1935

Year	Pocket watches	Wrist watches	Electric clocks	8-day clocks
1926	271,919	189,215	—	91,837
1927	165,628	301,426	1,214	124,301
1928	130,140	276,774	498	12,151
1929	172,931	263,991	952	70,823
1930	99,111	314,081	504	23,025
1931	33,666	82,868	5,234	24,546
1932	34,040	130,346	10,471	12,412
1933	36,159	121,044	15,300	6,245
1934	29,430	231,708	22,202	21,415
1935	26,363	332,374	33,515	6,800

production of wristwatches at the expense of "Uncle Sam" for the U.S. war effort in World War I, this action was not pervasive enough to save the industry. By the end of the war, the Swiss had literally regained the market share lost during the previous three decades and were clearly accepted as the leader in the distribution of wristwatches. The 1920s, with its emphasis on frivolity and design, served to cement the wristwatch's preeminence over the pocket watch and Swiss control over its distribution.

Why did U.S. firms fail to embrace the new product? One reason was technological lock-in (an illness that had initially plagued the Swiss and led to the late introduction of machined parts), another was the intense price competition that had evolved during the 1880s and 1890s. U.S. firms were weakened, fearful, and directionless in the face of radical incremental change.

Technological lock-in and path dependency are concepts frequently invoked in situations like that confronting the U.S. watch industry in the 1910s and 1920s. Unfortunately, most applications of these ideas fail to adequately enumerate the attendant qualifying factors that accompany such developments. What precipitates the condition of lock-in is rarely technological change by itself; instead, it is usually combined with other extenuating circumstances. In the case of the U.S. watch industry, key coincidental influences—extensive and unrelenting price competition, rigid distribution channels, the emergence of new lower priced competitive products, a long and sustained recession, isolated markets, global economic instability, and the material requirements of World War I—produced circumstances that were highly unstable and largely uncontrollable. If any one of the rigidities, particularly those external to the industry, had not manifested itself, the industry could possibly have changed. To take just one example, if the recession of the 1870–1890 period had not persisted and had consumer markets continued to grow robustly as in the previous decade, either an industry leader or some young upstart watchmaking firm might have pushed the U.S. industry off its mark and forced it to confront the challenges of wristwatches much earlier.

In addition, the rampant period of price competition drained the bank accounts of America's watch companies, leaving few resources for embarking upon entirely new models of watches. Wristwatches meant the abandonment of a significant stock of sunk capital. Thus, as Gordon Clark (1994) has pointed out, a "wait and see attitude" on the part of U.S. firms was not only unavoidable, but was rational given the overall condition of the industry. At the same time, the U.S. industry enjoyed a buffer against the forces of change because of trade protection. If the U.S. industry had not been so domestically oriented and protected from international market competition, the outcome might have been radically different. The lack of context in discussions of concepts such as path dependency and technological lock-in render these ideas so general and circular in reasoning as to make them pat, but not particularly helpful. As we will see later, in the Japanese case, lock-in is not a stable concept and is in fact avoidable, given certain exogenous influences and events.

The U.S. Industry after the Turn of the Century

From the 1910s through the 1940s, America's watch companies passed through very difficult times. Although a Waltham watch, with its machine-based mode of production, was highly accurate, it lacked design and style. The company was slow to move toward wristwatches, which gave the Swiss a decided advantage (particularly during World War I). The same could be said of Elgin, which had never invested in the level of product differentiation that Waltham had, and hence was lacking the range of products that increasingly came to define the market for watches. While Waltham and Elgin were granted a brief reprieve from Swiss competition during World War I (through numerous trade policy interventions, which once again forced the Swiss momentarily to the sidelines of the U.S. market), the Swiss began to encroach upon the U.S. market a few years later. Trade regulations designed to restrict imports of Swiss watches proved to be relatively ineffective in stemming Swiss watch movement imports. Illegal smuggling of Swiss movements into the United States, where they were to be cased by U.S. companies, combined with clever alterations of Swiss watch movement designs incorporating features (e.g., extra jewels) that were unregulated by trade policy, allowed the Swiss to infiltrate the U.S. market (Alft, 1984).

The renewed dominance of the U.S. market by the Swiss occurred even though the cost index between production and distribution was tilted decidedly in favor of U.S. manufacturers. The distribution system adopted by the U.S. watch industry was excessively dependent upon wholesalers, a costly choice of distribution channel (see Moore, 1945, Table 25). Even though U.S. watch production costs steadily declined, the U.S. watch was not extraordinarily cheaper than the Swiss product. This was particularly true after the Swiss fully incorporated the best elements of the American system (Moore, 1945, p. 214). High distribution costs combined with a very different product mix produced a product and a price point that were no longer accepted unquestioningly by the U.S. consumer. The convergence of the two watch production systems led one observer to note that as foreign producers adapted U.S. machine methods of production and gradually substituted the factory system for the practice of putting-out work to independent craftsmen, the preference for American watches became less marked.

Hamilton after the Turn of the Century

The Hamilton Company's obsession with precision was a double-edged sword. For the first 25 years of Hamilton's life, the company entered contests and competed in exhibitions in which it could repeatedly prove its product's superiority. What it lacked was knowledge of market trends. While other U.S. companies such as Elgin and Waltham moved into the wristwatch field in the 1910s, it took Hamilton until the late 1920s to introduce a wristwatch (Sauers, 1992). By then, the other U.S. companies had moved on to produce even more variety, manufacturing pocket watches of various sizes and qualities and a growing variety of wristwatches. However, the bigger threat came from the Swiss. When they saw that U.S. producers could manufacture a cheap and accurate timepiece, they pioneered the fashion watch, which would form a new market appealing to women.

By the end of the 1920s, Hamilton had begun to produce its "Art Deco" watch series. This foray into design-oriented watches was a start, but could not stave off a devastating decline in market share, hastened by the onset of the Great Depression.

Not until the eve of the Great Depression did the demand for wristwatches achieve parity with demand for the pocket watch. As one American expert noted, "Whatever it was—economics or demand—the year 1927 was the year of greatest change for the wristwatch market. That was the year in which the production of pocket watches was outstripped by wristwatches" (Gohl, 1977, p. 592). A similar view of the Swiss watch industry can be seen in export statistics reported in a comprehensive treatment of the wristwatch: "According to Swiss export statistics, the point of change occurred in 1930, when 6.2 million pocket watches balanced 6.2 million wristwatches" (Kahlert et al., 1986, p. 45). By these dates, the U.S. watch industry no longer boasted of supremacy over the Swiss.

World War I and Its Aftermath

World War I wreaked havoc on the watch industry on both sides of the Atlantic. U.S. and Swiss industries were forced into production for the war effort. Waltham, the U.S. industry leader, was given few options: it had to produce willingly for the war effort or be taken over by the federal government (Moore, 1945, p. 107). While other firms were approached by the federal government, only Waltham was initially willing, able, and interested in diversifying; it did so by beginning to manufacture fuses. This development proved less than successful because fuses were costly and hard to make, and the resources required to pursue this line of work drew capital away from watch manufacturing operations, just as the market was recovering due to increased wartime demand for watches and the need to pursue wristwatch technology (Moore, 1945, p. 107).

World War I presented considerably more complex problems for the Swiss. Switzerland had been neutral since the 16th century. As a neutral, the country does not take sides in international war actions. To assert a neutral position, the Swiss maintained separate economic and political relationships with combattants without directly participating in the war itself.

Thus trade occurred throughout the war with the United States, Britain, and Germany. Economic relationships with the belligerents predating World War I meant that Switzerland was caught between the warring parties both geographically and in terms of trade of manufactured goods. Dependence on trade with the Germans for raw materials, especially coal and metals, meant that to keep its industries going, Switzerland had to produce goods for the German war effort. This dependence placed Switzerland directly in opposition to the Allies who wanted all trade with Germany shut down to starve the German war machine.

The desire to starve the German side was at the heart of a growing awareness of Swiss trade practices with the Germans. At the outbreak of the War, Switzerland was not self-sufficient in growing food and had to import increasing amounts of grain and cereals to feed its citizenry. The United States was the logical provider of these goods. But as the war dragged on, the Swiss were found to be importing amounts considerably in excess of their own needs and reexporting surpluses to the Germans (Meier, 1970). World opinion toward Switzerland grew increasingly disparaging, especially after the publication of a report about Swiss relations with Germany. As Meier (1970), citing an article published on August 9, 1917, in the *New York Times* by James Louis Moore of Harvard, suggested,

> Switzerland, like other neutrals, rather than keeping its food supplies for its own needs exported abnormal quantities of them to Germany, thus thwarting Allied efforts to end the war.... The gist of the ill-feelings toward Switzerland, was the conclusion that if the neutrals "exposed their own populations to the risk of starvation, relying upon the generosity of America to supply the deficit, we must, with unyielding firmness, inform them that, in time of war, American interests come first." (pp. 68, 69)

Food supplies were only part of what was provided by Switzerland to Germany. Like U.S. watch companies, Swiss watchmakers were encouraged and enticed to produced for the war effort. They converted watch factories into fuse and aircraft bearing production centers. Optics was also a major pursuit of watch companies during the war. Though neutral, Swiss citizens were exhorted to assist in the war effort; even the unions used their considerable influence to encourage output increases during the war. Although being neutral theoretically meant supplying the belligerents on both sides, in different circumstances it is doubtful the Swiss people would have taken up production against Germany, given that over half of all Swiss had Germanic backgrounds.

After having recovered at least some of the U.S. market between the late 19th century and the first 10 years of the 20th century, Swiss exports to the United States once again plunged to low figures in 1921 (Table 7.4; Moore, 1945). The trade blockade, advanced by the United States to contain possible Swiss reexports of strategic materials to Germany, meant that access to the U.S. market was circumscribed from 1917 to 1919. The blockade, while damaging watch exports to the United States, was more than made up for with exports to other countries in Europe. And there was the ever-present question of export to Germany. As Germany did not have a watch industry itself, suspicions ran rampant that Swiss watchmakers were producing for the powers on the German side (Meier, 1970, p. 67). The Swiss defense was that Switzerland "sold optical instruments and jeweled bearings to the British for navigational systems . . . sold fuses to the Germans, and in this way . . . remained neutral" (L. Tissot, personal communication, August 1989).

Whatever the truth in terms of the volume of exports of watches and watch parts, and industry growth associated with Swiss sales to the belligerents during the war, the war yielded further instability for the watch industry. The increasing disparity between the Swiss people who profited from the war and those who lived in fear of uncertain access to necessary foodstuffs and other basic goods only added fuel to union dis-

TABLE 7.4. Watches, Movements, and Materials Imported into the United States, 1912–1926 (Value in Dollars)

Year	Completed watches	Watch movements	Completed watches and watch movements	Cases and parts	Total
1912/13					2,615,744
1919	3,542,033	6,520,129		2,584,563	12,646,725
1920	3,938,980	8,147,876		3,601,377	15,688,233
1921	1,864,951	4,461,051		2,842,603	9,168,605
1922			7,122,783	2,204,576	9,327,359
1923			6,740,822	3,312,864	10,053,686
1924			4,761,678	5,977,233	10,738,911
1925			7,269,706	3,060,663	10,330,369
1926			10,416,582	1,414,656	11,831,238

Source: Adapted from Moore (1945, p. 214).

content. After the war, the once sacred Swiss reputation as the makers of the world's finest watches was tarnished by the role played by the industry in the supply of war materiel (Jequier, 1991). Even as trade routes were reestablished in 1918, the inevitable deflation associated with the war's conclusion hit the Swiss watch industry hard. Unemployment in the watch industry once again became a serious problem, with more than half of all watch workers unemployed in 1921. Factory sales dropped by one-half (from 13.7 million to 7.9 million units from 1920 to 1921). Inevitably, with the return of chaos, the hybrid system of partial vertical integration was challenged by the reemergence of small producers. Eschewing machine manufacture and exploiting the excess capacity for parts manufacture in the industry, small assemblers began producing inferior products, which further eroded the Swiss watch reputation. Uttinger and Papera (1965) reported:

> The crisis occurred in 1921, and represented the worst setback the industry had ever suffered. Exports declined by 48 percent and the number of unemployed rose to 3,000. It was during the crisis of the early 1920s that the development of the "Chablonnage" (the export of all individual parts necessary for assembly in another country) occurred. This marked the beginning of the "export of the industry" to other industrial nations. Vicious price wars on finished watches as well as parts plagued the watch industry in those years. (p. 203)

The Interwar Years: Optimism Gives Way to Decline

As the early decades of the 20th century wore on, the world's watch producers confronted considerable economic uncertainty. In the United States, both domestic and foreign companies faced unstable markets and were jolted by exogenous shocks that significantly impinged on the orderliness of daily affairs. As uncertainty grew, each country's industry sought stability through market regulation. When that failed, they tried brute-force price-cutting competition. Unfortunately, market regulation served to support production organizations that proved rigid and difficult to transform as even greater forms of instability unfolded.

The Swiss regained considerable market share in the United States in the 1910s and early 1920s. The Swiss complex was only partially transformed by the war. Reconversion was relatively swift. Although importation of Swiss products was restricted during the war, the U.S. market reopened quickly after the Armistice and Swiss watch products flooded in. This was a serious problem for U.S. firms. The federal government responded by increasing tariff rates 30 percent, effectively choking off access to the U.S. market. While momentarily disastrous, the loss of the U.S. market might have been manageable if the postwar crisis had not also brought the closure of European markets and led to a drastic contraction of world market share for Swiss producers. Cutthroat competition and intense rivalry within the watch complex ensued (Bolli, 1957; Bumbacher, 1992).

The early 1920s were a period of great instability in the watch industry.[4] Disruptions in the watch market presented the Swiss with new and different problems (Knickerbocker, 1976). In the previous two decades, significant capital had been invested to meet the U.S. manufacturing challenge. Firms were larger, and the industry contributed a more significant share of the country's gross national product (Landes, 1984). The severity of the crisis forced family businesses to take drastic steps simply to reduce inventory. Opportunism, price cutting, and increased exportation of movements and parts further destablized the industry (L. Tissot, personal communication, 1990). This unprecedented threat resulted in a call for industry regulation and the formation of a cartel.[5]

During the 1920s, various associations were created to represent the interests of industry members. In 1924, the Fédération de l'Industrie Horlogère Suisse (FH; Federation of the Swiss Watch Industry) was organized to govern firms assembling watches from component parts and the few firms with integrated manufacturing operations. In 1926, 17 manufacturers of ebauches were organized into a trust known as Ebauche SA. Manufacturers of components other than ebauches (e.g., balance wheels, assortments, hairsprings) were organized into the Union des Branches Annexes de l'Horlogerie (UBAH). In the late 1920s, members of the various associations agreed to set levels of output and prices, and explicit rules were designed to restrict exportation of manufactured parts (Knickerbocker, 1976). The merger of two powerful watch companies, Tissot and Omega, into the Société Suisse pour l'Industrie Horlogerie (SSIH) further consolidated the industry's decision-making powers.

When this degree of collaboration proved insufficient to control opportunistic firms, a third convention was held and the planks originally proposed in 1928 and 1931 were adopted and codified by the national government. Despite consolidation over the previous 20 years, the industry was still dominated by small firms (Table 7.5). In conjunction with industry and banking leaders, the Swiss federal government created the massive holding company Société General de l'Horlogerie Suisse SA (ASUAG; which included Ebauche SA, the large parts manufacturer, and other leading component producers). The following activities were subject to government regulations:

1. The export of parts, spare parts, and movements—the rules were to be enforced against any manufacturer, whether or not he was a member of any of the industrial organizations; and
2. The opening of new firms as well as increases in size and structural changes such as mergers—the purpose was to prevent overproduction (Uttinger and Papera, 1965, p. 203).

This final merger halted the exportation of parts and components to competitor countries (Knickerbocker, 1976). By the mid-1930s, the federal government completed the consolidation of the industry by legally recognizing the cartel and establishing a permitting process to regulate its actions (Bumbacher, 1992).

The Statut de l'Horlogerie and the Codification of the Swiss System

The Statut de l'Horlogerie (Protection of the Watch Industry Order) of the early 1930s established a regulatory system that governed Swiss watch manufacturing for the next 30 years. Through a combination of cartelization and government ownership, the Swiss industry was regulated to control vertical integration, foreign sourcing, and offshore production. Swiss manufacturers could buy only from Swiss component producers, and component producers could sell only to Swiss firms (Learned et al., 1961). To further limit competition, the government regulated the sale of machinery. The Statut de l'Horlogerie regulated

TABLE 7.5. Number of Persons Employed in Swiss Watchmaking Industry by Size of Enterprise, 1929

Number of employees per enterprise	Number of enterprises in category (% of total)	Number of persons employed (% of total)
1	421 (15.7%)	421 (0.7%)
2–9	1,268 (47.3%)	6,331 (10.7%)
10–49	764 (28.4%)	16,868 (28.7%)
50–99	134 (5.0%)	9,209 (15.7%)
100–199	56 (2.1%)	7,827 (13.3%)
200–499	30 (1.1%)	8,626 (14.6%)
500 and more	10 (0.4%)	9,517 (16.2%)
Total	2,683 (100.0%)	58,799 (100.0%)

Source: Suisse Recensement Fédéral des Entreprises, Année 1929, II, 2–5.

the volume of Swiss watch production by requiring permits for the construction and expansion of production facilities (Chapuis and Jaquet, 1956).

The resulting industry structure consisted of the parts manufacturers, who sold their output to assemblers; the assemblers; and the brandname manufacturers. Under the law, the Société General de l'Horlogerie Suisse SA (ASUAG) could sell only to firms recognized by the Swiss government. It could not export parts or technology. Manufacturers fabricated complete watches but were restricted from selling movements and other parts to assemblers, thereby eliminating their competition with parts suppliers. They also were restricted from setting up production in other countries. Assemblers were prohibited from establishing production facilities outside of Switzerland, and they could buy parts from non-Swiss manufacturers only if prices were 20 percent below Swiss levels. The law's greatest effects were in regulating who was allowed to produce, and what and how much they could produce. The new law gave the industry tighter control not only over its own firms, but also over the exports of independent Swiss machine manufacturers who were in no way connected with the watch industry. The latter were no longer permitted to sell watch manufacturing equipment to foreign buyers. By requiring export and manufacturing permits, the government essentially held supply below world demand and ensured Swiss firms handsome profit levels.

The five most difficult years of the industry's history, 1928–1934, gave way to a period of industrial stability. Having traded off free markets for government control, the Swiss industry had largely recovered by 1936. The industry experienced a renewed period of fantastic growth, further strengthened by the Swiss declaration of neutrality during World War II. While the Statut de l'Horlogerie covered employers, labor leaders were not central to the discussion. Their actions represented a defining moment in the ability to fully implement the cartel, however. As with the industry structure of the past, skilled and obedient workers were essential to the success and full recovery of the industry. While labor had made several overtures to employers, it required strikes in 1936 and 1937 to effect a labor peace. The state once again stepped in, not wishing to see its investments of the previous two years go up in smoke because of labor unrest (Jequier, 1991). After a long series of negotiations, labor signed an agreement to give up the right to strike in return for increases in wages and better working conditions. This policy, coming on the heels of the Statut de l'Horlogerie, was to have profound implications for the industry's long-term ability and willingness to change in future years.

Summary

The return to peace and prosperity was enticing enough for both economic parties in the industry to give away large measures of control. In the case of labor, without its stimulus and prodding to ensure continuous innovation within the industry, watch manufacturing firms entered a period of innovative complacency and lethargy disguised by 20 years of growth enjoyed largely in the absence of any serious competition. By giving up the right to strike, workers yielded any measure of influence they had over the strategy of the major firms. Even as members of the labor union saw major firms resting on their laurels, they were basically restricted from activity. Union members were well aware that firms were passing up the opportunity to make needed investments in new technology. Labor relations had become so sour that workers regularly withheld ideas and knowledge that could have increased productivity within firms. As long as the firms delivered on continuing wage increases and modest improvements in working conditions, they were essentially free to make whatever investment decisions they saw fit.

For some 20 years this period of relative stability satisfied both parties to the agreement. With the rise of new competition, increasing costs, and growing economic uncertainty, both firms and labor grew concerned. Labor had known for quite some time that employers had access to new innovations but chose not to invest in them. At the same time, workers were not keen to embark on a period of experimentation that could jeopardize the relative good times of the postwar years. Thus, competitive cooperation, the critical element identified by Lazonick (1991) and Best (1990) that acts as a basis for innovation in vertically specialized and horizontally fragmented industries, was eliminated, with dire consequences for the future.

With this new strategy, the Swiss took the lead again in the 1930s and 1940s, rising to dominate all markets—even the United States—and accounting for up to 80 percent of worldwide production in the late 1940s (Table 7.6; Bolli, 1957). In 1943 and 1944, "a third of Swiss watch exports went to America, 4,750,000 pieces in 1943 and almost 4 million in 1944. In those years, watch exports made up three-fourths of Swiss exports to the U.S." (Meier, 1970, p. 294). Unlike World War I, World War II did nothing to diminish demand for Swiss watches. The Allied powers, in particular the United States, were shocked at the Swiss ability to continue exporting all through the war even as their own armies commandeered most of their domestic watch products. In 1943, for example, the U.S. jeweled watch industry produced *no* output. Watch demand was met by importation of Swiss watch movements and parts (U.S. Department of Commerce, 1950). Swiss willingness and ability to service almost the en-

TABLE 7.6. Imports of Swiss Watches and Movements, 1934–1942

Year	Number of watches and movements	Foreign declared value
1934	920,393	$2,834,092
1935	1,201,986	3,668,645
1936	2,228,606	5,877,676
1937	3,127,274	8,183,117
1938	2,386,226	6,562,570
1939	2,919,147	8,057,789
1940	3,536,982	10,220,772
1941[a]	4,400,000	13,700,000
1942[a]	5,700,000	20,700,000

Source: *Boston Traveler*, June 1, 1943. Cf. *New York Times*, May 22 and May 23, 1943.
[a] The figures for these years are obviously not on the same basis as for the other years.

tire U.S. market greatly benefited the Swiss in the short run (Table 7.7; Bolli, 1957). However, by devastating the American market for U.S. firms, this initial gain became a serious loss when only 10 years later the United States enforced severe trade restrictions on watch product im-

TABLE 7.7. U.S.–Swiss Trade during World War II

Year	American products imported to Switzerland (millions of Swiss francs)	Swiss products exported to the United States (millions of Swiss francs)
1931/35	102.8	60.1
1936/40	131.0	108.6
1941/45	120.2	177.8
1939	132.7	129.7

		(a) Raw materials, (b) manufactures, (c) foodstuffs (millions of Swiss francs)				(a) Watches, (b) chemicals, (c) textiles (% of total export)		
		a	b	c		a	b	c
1940	199.2	136.5	41.7	21.0	139.9	50.8%	16.7%	15.4%
1941	151.3	28.9	91.6	30.7	108.0	53.7%	23.7%	10.5%
1942	235.3	18.8	8.9	207.6	102.2	73.6%	11.8%	7.5%
1943	56.4	14.3	3.9	38.2	152.8	74.5%	5.2%	12.5%
1944	21.2	12.2	1.5	7.4	140.8	71.7%	5.6%	16.8%
1945	136.8				385.3			

Source: Bolli (1957, p. 123).

ports. Further cutting into the domestic jeweled watch market was the public's change in preferences and tastes for inexpensive watches. While U.S. jeweled watch producers made fuses and chronographs for the war effort, inexpensive watchmakers such as Timex made serious inroads into the U.S. market and the Swiss share of it. Nevertheless, the Swiss hung on to the higher end of the market, went head-to-head with Timex for European markets, and produced movements for some watches assembled in the United States under the trade names of Bulova and Longines (among others).

8 An Unexpected Competitor

In the middle of the 20th century, the major contenders in the battle for the world watch market were incapable of taking seriously the emergence of a new competitor outside the sphere of influence of the Western Industrial Revolution. Like other industries called into duty in support of World War II, U.S. and European watch manufacturers were seemingly poised to enjoy a period of relative prosperity after more than a decade of economic uncertainty. In the United States watch companies had been commandeered to produce for the nation's defense. This left the domestic market open to the Swiss, who gladly came to dominate the national supply of watches. Switzerland, as a neutral, produced war materiel based on watch technology for all parties to the conflict. While the U.S. industry was left virtually flat on its back and badly in need of revitalization after its complete redirection for the war effort, the Swiss manufacturers' equipment had been seriously run down. In both cases, each country's watch firms were desperate to get back to the business of watch manufacturing. Both labored under different constraints in attempting to return to their core industrial activity.

Thousands of miles away, Japan had been devastated. The country's production system had been bombed into oblivion; to most observers, it seemed unlikely that Japan would regain any measure of economic strength for decades (MacIsaac, 1976; Rosovsky, 1966). Recovery was expected to be further retarded simply because Japan had been so far behind the rest of the industrialized world when the war began. According to knowledgeable observers in the years leading up to World War II, Japan's industrial system was incapable of sustaining a war effort of any scale. In truth, Japan's misjudgment about the length of the war led to its defeat.[1] In light of this, then, why would anyone think that this small island economy, which had so drastically misjudged its capability to wage a war against the world's great economic powers, suddenly prove capable

155

of rising in a span of 20 years to become the leading manufacturer of a product that was once the premier symbol of technological supremacy?

In 1868, Japan was inwardly oriented, having purposely isolated itself from Western influences. Even at the turn of the century, the country was thought to be backward and lacking a modern perspective on industry and economy. Japan was presumed to have few of the prerequisites for industrial takeoff. It is not surprising, then, that the Europeans and Americans remained unaware of a new competitor who would redefine the world watch industry. As one Swiss industrialist said of the Japanese, they were not even irrelevant because to be irrelevant implies that they had been noticed. It was simply unimaginable that the fight for control of the world watch market would turn on an industry that did not exist in Japan much before the late 1890s (L. Tissot, personal communication, 1989).

The Japanese did not set out to create a watch industry that could challenge the craftsmanship and artistic qualities of the Swiss industry. Even today, the capacity to produce beautiful, refined, elegant, and exquisite watches remains almost exclusively the domain of Swiss watch companies. Instead, the Japanese aspired to produce a watch product for the masses.

The Japanese watch industry, bolstered by a strong, highly protected domestic market and a virtual lock on other Asian markets, benefited from the ongoing redefinition of the watch market itself, which began in the middle of the 19th century with the creation of a watch for the common person: the Roskopf. Purposeful Japanese planning and persistent execution yielded the world's largest watch industry by the 1970s. What was it about the foresight and institutional tendencies of the Japanese system of production that led to Japanese dominance of the world watch industry?

A brief account of factors leading up to the emergence of the Japanese watch industry sets the stage for its phenomenal development in the 20th century. Japan's rise to industrial preeminence has been chronicled in a vast literature, but not without controversy. Here, I highlight only a subset of important themes that assist in this story, and leave to others the many unresolved debates (Allen, 1963; Kemp, 1993).

In this chapter I review the history of Japanese industrial development, beginning my story with the Meiji Restoration, which took place in the middle of the 19th century. I recount the development of the watch industry and the emergence of the major watch producers at the beginning of the 20th century. I then trace the development of the industry through the years of World War II and suggest that this period of rapid technological progress and focused industrialization spurred the development of Japan's competency in the production of watches.

The Meiji Era and the Revival of Engagement with the Outside World

In the middle of the 19th century, the 200-year-old Tokugawa regime began to unravel quickly. The previous two centuries had been dominated by an economic and political system that was decidedly inward in orientation. Built around regional shogunates representing rival political factions, and supported by peasant agriculture, the clannish system began to fall apart under the sheer weight of its own size and expense. The stimulus for change was entangled with the growing threat of Western influences, which began to be unavoidable. The fear of external takeover began to weigh heavily on the ruling elites. In 1868, with the overthrow of the aging and increasingly dysfunctional Tokugawa feudal system, the leaders of the Meiji Restoration began the rapid transformation of Japanese society and economy.

At the time, little was known about Japanese economic and institutional capacity or the other ingredients needed for industrialization. Already in place, however, was a domestic craft and artisanal manufacturing capability that supplied domestic trade. Compared with conditions found in developing countries today, by 1868 Japan already had in place many of the necessary prerequisites for industrialization. The country enjoyed stable population growth and a growing agricultural surplus that could be used in support of the state. Japan exhibited an emergent state system with the ability to control and direct its nascent industrial economy. Japan's new leaders demonstrated a strong interest in the technological standards of the day, and in particular a recognition of the benefits of modern machine-based manufacturing of armaments. Of overarching importance to Japan was its strong nationalistic sentiment and its visceral fear of being ruled by others (Allen, 1963; Kemp, 1993).

The path toward industrialization between 1868 and 1900 was conceived of and rested on the recognition that to move ahead rapidly, Japan would have to rely on outside influences while building the requisite institutions to establish a foundation for industrialization. Banking, communications, and transport systems were developed in rapid succession. Educational institutions also were established. The acquisition of technical knowledge was accomplished through massive and expensive imports of foreign expertise and equipment combined with extended trips abroad by Japanese industrialists (Allen, 1963).

In the early stages of its industrial development, Japan lacked the effective business enterprises needed to foster economic growth. Encouraged and often financed by the state, new businesses were formed from old feudal family enterprises. The state built factories and then sold them off to former samurai and other members of the emerging business elite (Hirschmeier, 1964). State-sponsored manufacturing emphasized production for trade, particularly with Asian markets.

State transfer of its factories into the hands of the newly emerging business class brought about the formation of the great business groups of the 1890s. These first gestures toward the centralization of economic power laid the foundation of the Japanese industrial structure. Nonetheless, reorganization of industry into large-scale enterprises would take another 30 years to complete. In the intervening period, home-based manufacturing practices and industrial factories developed side by side. The dualistic structure of production—small businesses alongside behemoths—would become the hallmark of the Japanese system, characterizing her industrial structure throughout the 20th century.

Turn-of-the-Century Industrialization

By the 1890s, factory-based industrialization had barely begun to take hold in Japan (Takeshi, 1996). Few power-driven factories existed. The Sino–Japanese War of 1894–1895 was a deciding influence on the evolution of factory-based manufacturing. After defeating China, Japan enjoyed the luxury of virtually uncontested control over abundant natural resources commandeered from China. Liberated from the restraints of former trade laws such as the "Unequal Treaties Act," Japan was free to pursue trade policies in support of the national economy (Allen, 1965).

After 1895, trade exploded due to the rise of demand for silk exports to the United States. Japan captured and controlled markets in China and Korea and established economic enclaves in those countries that supplied Japan with food and raw materials (Glasmeier et al., 1993). By this time, industrialization had taken hold. Trade, however, was extremely specialized, consisting primarily of the export of textiles. While yarn manufacturing was becoming increasingly capital-intensive, weaving still took place in small factories. A hybrid system of manufacturing characterized the emerging structure of industrial organization in Japan.

The Japanese system of industrialization, based on small home-based producers coexisting alongside large firms, evolved out of the need to satisfy two very distinct forms of consumption (Rosovsky, 1966; Uyeda, 1938). The small-scale cottage industries arose out of the form of industrialization characteristic of the Tokugawa era. These manufacturers produced products for domestic consumption and reflected tastes and interests that were specifically Japanese. The large firms were created in most instances through government investments that were then sold off to members of former samurai families.

Prior to World War I, Japan was substantially ahead of other Asian countries in the pursuit of industrialization. Yet, by Western standards, her economy was still based in agriculture. Foodstuffs and textiles accounted for 42 and 38 percent, respectively, of the country's industrial capacity. By 1913, foodstuffs as a share of output had declined to 22 percent and textiles and machinery had increased to 45 and 8.4 percent, respectively. During the war years, the number of people working in factories increased by 70 percent and factory output doubled (Allen, 1963). The postwar depression of the early 1920s, followed by a devastating earthquake in 1923, set the economy back to pre-1920s levels. By 1929, however, factory output had returned to former high levels and was once again overwhelmingly dominated by textiles, which had trebled from their 1923 level. Textiles increased in share of exports from 53 to 65 percent over the intervening years.

From 1929 to 1932, financial problems plagued the economy. An overvalued currency led to state intervention. Deflation followed by severe wage cuts produced social unrest. The seemingly limitless demand for Japanese textile products from the United States came to an abrupt halt with the imposition of U.S. tariff and quota restrictions on Japanese textiles (Glasmeier et al., 1993). Government economic policies were severe and absolute and contributed to a reassertion of military control over the economy.

The dominance of the textiles industry overshadowed the emergent metals and engineering industries. Forced out of its former comfortable position and abetted by the scaling up of industry for war, Japan experienced an explosion of growth in output and employment in these strategic industries. Factory employment in engineering and metals goods increased by 43 percent between 1929 and 1937 (Takeshi, 1996, 1997).

In his detailed treatment of the rise of the Japanese system of production, Michael Best (1990) asserts that the Japanese industrial economy was undeveloped until after World War II. Numerous other treatments, however, suggest that, to the contrary, the Japanese industrial economy was quite competent in the production of machine-based goods prior to 1935 (Ohkawa and Rosovsky, 1973; Rosovsky, 1966). The history of the watch industry also accords with this latter finding. The difference in viewpoints may be due to the concentrated form of trade practiced by the Japanese after 1920, which emphasized trade with Asia as Japan became increasingly locked out of Western markets.

By the 1930s, Japan's industry was competent in a wide range of manufactured goods (Takeshi, 1997). The push toward armaments production only strengthened this capability. The political unrest resulting from Japan's expansionist tendencies served to further concentrate trade in Asia. The war buildup distorted the economy toward heavy industry as the need for weapons, ships, and other elements of a military arsenal crowded out domestic industries (Allen, 1928, 1940, 1963; Kemp, 1993; Okhawa and Hayami, 1973). National policies served to forestall the emergence of new businesses in some established fields and to curtail production for domestic consumption and international trade, by restricting access to the foreign exchange needed to finance exports. A vast number of new economic institutions were established to regulate elements of trade (Allen, 1940). To produce exportable goods that required the importation of foreign technology and raw materials, Japanese firms had to generate output for export equal to or greater than the needed imports. The concentration of development rewarded the powerful business houses that had grown to dominate heavy industry. Although the smaller industries were not destroyed, and existed on the fringes in support of the war effort, nonetheless the large business groups, suffered great hardship. Many failed (Allen, 1928, 1940; Uyeda, 1938).

The history of industrial redevelopment after World War II pertains here inasmuch as the reemergence of the Japanese watch industry occurred only nine months after the armistice was signed. Despite the incredible devastation inflicted on Japan during World War II, a surprisingly high share of Japanese industrial capability survived intact after the war. Watch manufacturing, which had been removed from Tokyo to the Suwa region for strategic reasons, resumed production and flourished even as the wreckage of Hiroshima and Tokyo was still smouldering.

With this industrial history as a backdrop, I turn now to the history of the Japanese watch industry and the rise of the Seiko Corporation.

What Did They Know and When Did They Know It?

A reading of the various histories of the world watch industry fosters the belief that the Japanese were latecomers to the world of manufacturing sophisticated personal timekeepers. Indeed, the histories of U.S. watch companies refer to the generosity of U.S. firms and their selfless transfer of U.S. technology to Japan to facilitate the growth of the Japanese watch industry. While U.S. firms did sell their technology to Japan, it was the sale of clock technology at the beginning of the century and the subsequent sale of very early watchmaking technology that established the seeds of a future technological competency in the Japanese watch industry. Both the Americans and the Swiss underestimated the technical capacity of the Japanese industry; both paid insufficient attention to the development of quartz technology and how profoundly it would reshape world competition for markets.

Early History of the Industry

The Japanese watch industry emerged from domestic clock manufacturing, which was established to eliminate imports from Europe and the United States. There were three centers of timekeeping device manufacturing: Osaka, Nagoya, and Tokyo. Between 1912 and 1925, the wall clock industry was the dominant industry in Nagoya. By the 1930s, due to mismanagement and the small firm structure that dominated the industry, clock manufacturing in Nagoya died out and was replaced first by textile equipment manufacturing and then by automobile manufacturing (Toyota, personal visit to corporate headquarters and corporate museum, 1994).[2]

Osaka was an original site for watch manufacturing. This early location is not surprising given that Japan's metal-working industries were concentrated in the Osaka region. In 1889, the Osaka Clock Company was established by 11 clock distributors in the city. Although production had built up to a respectable level by the early 1890s, the company was not achieving profitability. By 1894, after one year of profitability, the company once again sank into the red in the face of competition from clock-making firms in the Nagoya region, which began to dominate the domestic market. The establishment of the Osaka Watch Company was the result of a convergence of several factors: the intensifying competition in clocks, which reduced prices to an unprofitable level, the influx of new members to the board from outside the Osaka area, and the availability of watch manufacturing machinery from a bankrupt watch company in California.

A Roundabout Way into Watches

The history of the modern Japanese watch industry actually begins in Chicago, Illinois, with the establishment of the Cornell Watch Company, later relocated and renamed the Otay Watch Company of California. One of America's watch pioneers and the original entrepreneur behind the Elgin Watch Company, J. C. Adams, known in the trade as "the great American starter" (Crossman, 1885), along with Paul Cornell, an early Chicago land speculator, started the Cornell Watch Company of Illinois. Cornell had purchased prime land in the city and sought to maximize his returns from it by establishing a watch factory. Watchmaking was very much the high-tech industry of the day (Harrold, 1993). Communities spent considerable sums to woo watch manufacturing plants.

In 1870, the Cornell Company came into existence with Adam and Cornell's purchase of the Newark Watch Company's watch manufacturing equipment. Like many watch companies of the time, the Newark Watch Company had underestimated the difficulty of producing watches. After producing only 200 or so watches, the company failed and sold its machinery to Adams.

Like numerous predecessors, Cornell completely underestimated the difficulty of running a profitable watch manufacturing operation. By now, Adams had receded into the background and Cornell was stuck with a losing proposition. He cast about for a way to dispose of the capital equipment. Hearing that immigrant Chinese labor was cheap and abundantly available in California, Cornell sought financial backers on the West Coast. The major influence of the Cornell Watch Company on Chicago was neatly summed up soon after the plant's relocation to California: "While the Cornell Watch Company cannot be counted a success from every point of view, it had the desired effect of raising the price of real estate in the vicinity, and goes to prove the adage that, 'it is an ill wind that blows nobody any good'" (Crossman, 1885, p. 76).

Moving to California proved to be a short-lived respite in the company's life. In 1874, after moving his company's capital equipment along with some 60 workers (who traveled to California at their own expense simply because of the promise of a job), Cornell set up the watch factory in an abandoned warehouse of a bankrupt furniture company. Management of the persistent problem facing all watch companies—very high operating costs prior to achieving a marketable product—hinged on access to very cheap Chinese labor for final assembly. A labor strike and employee sabotage mortally crippled the firm. After operating only a year in California, the Cornell Watch Company collapsed.

Following yet another attempt to redeploy the capital equipment in the manufacture of watches, this time in Berkeley, California, Cornell finally gave up and sold off the company's assets. There is no complete record of where all of the equipment went once the Berkeley factory shut down, but some of the company's equipment was shipped to Japan, where it was matched up with equipment from the Otay Watch Company and its later reincarnation, the San Jose Watch Company (Bailey, 1975; Ishihara, 1988).[3]

In 1894, Japanese pocket watch manufacturing was initiated in a factory using the Cornell and Otay machines recently imported from the United States. Forty-five engineers were hired from the United States. The American negotiator was particularly cunning and secured very high salaries for the U.S. watch workers. Delays in receiving materials to begin watch manufacturing seriously strained the new company's finances. The first watches were no doubt assembled from kits purchased in the United States and shipped with the used equipment. The delay in starting up the factory put the company in financial jeopardy. After training the Japanese workers for just a year the Americans were fired. But by the end of 1895, watches comparable to those manufactured in the United States were being produced by Japanese mechanics.

From the firm's beginnings, the Osaka Watch Company had difficulty finding and retaining qualified workers. There was already a shortage of labor in Japan due to the overall expansion of the Japanese economy. Moreover, watchmaking required much more sophisticated and precise skills than clock manufacturing. The need for highly skilled laborers led to a policy of paying very high wages to watchmaking workers and to the establishment of a company training system to generate employees with the needed skills. It also acted as an early rationale for increasing the amount of capital employed in watch manufacturing to compensate for labor shortages.

Unfortunately, the machines purchased from the United States were for the production of a kind of watch that was quickly going out of fashion. Consequently, almost immediately the company had to retool to produce a smaller watch, a difficult and costly transition. As the Osaka factory switched from one watch dimension to the next, it began to fabricate its own machine tools to work at increasingly smaller tolerances.

After four years without profits, the initial investors were highly disillusioned about the prospects for the firm. The need for continuous investment to purchase raw materials and new machines, pushed the company further into the red. In addition, there was continuing resistance by Japanese consumers to domestically made watches: it was more fashionable to buy American. After a major reorganization in which some of the original investors were forced out and the value of Osaka's capital stock was reduced, the company finally began to make a profit in 1898. Profits held for two years, but the Boxer Rebellion in China, coupled with a general economic downturn that reduced the market for Japanese watches, eventually returned the company to a condition of insolvency. After a protracted stockholder suit against the board of directors, the company was forced to reorganize a second time. By 1905, the company had downsized. It eventually moved out of watchmaking all together. By 1909, the company was again close to bankruptcy; this time it was dissolved. So ended the first attempt at factory-based watch manufacturing in Japan.

The most important location for the production of Japanese timekeeping devices was Tokyo. Factories in Tokyo specialized in manufacturing pocket watches, wristwatches, and table clocks. The emergence of a watch industry in Tokyo was linked to established wholesale and retail firms that purchased watches from abroad and provided sales and service to watch customers. Tokyo's watch factories, which came into existence in the 1890s, followed lines of production organization similar to those found in America. The time in which watchmakers began new firms was long past. New firm formation had passed into the hands of retailers who knew much about sales but much less about the fundamentals of manufacturing watches. Thus the establishment of the great Hattori Seiko house of watch manufacturing arose out of the service end of the business.

Hattori Seiko

The Seiko group is one of the few remaining Japanese family-run industrial conglomerates (Boyer, 1984, p. 44). It started in 1881 as K. Hattori & Co., a clock and watch distributor and repair shop in Tokyo. In the early 1900s, the company divided into two manufacturing operations, Seikosha (for clocks) and Daini Seikosha (for watches). K. Hattori retained control of the distribution and marketing of watches and clocks.

The Seiko Corporation began as a watch retailer. Retailers who wished to sell watches also had to be able to repair them. The first Seiko watch was assembled in 1895 out of imported parts. In 1899, Seikosha introduced its second watch, one based on parts supplied by the Waltham Watch Corporation. The first Seiko factory was established in 1902; by the early part of this century, the company was producing and selling pocket watches, as well as alarm and table clocks.

Seiko moved quickly into manufacturing wristwatches. The company's first wristwatch, the "Laurel," was on the market in 1913, almost seven years earlier than wristwatches made by major U.S. firms. The reception of the Laurel watch was so enthusiastic that it contributed to the company's decision to sell branded products using the name "Seikosha" in 1924 (Seiko means precision).

Seikosha (Seiko) excelled in the operation of large vertically integrated factories. When new competitors, such as the Citizen Watch Company (1926), entered the industry, they adopted a production structure similar to that pioneered by Seiko.

Seikosha's vertically integrated production system was in sharp contrast to the structure of Japanese wall clock manufacturing. As illustrated by the clock industry in Nagoya, much of Japan's precision manufacturing was executed by a highly vertically disintegrated system of small firms. The early selection in Tokyo of vertical integration over small firm-specialized production was in response to the recognized need for high quality and precision, and therefore the need for skilled labor to control the production process. It also apparently drew on the experience of Hattori's visit to Hayashi of Nagoya. Hattori had visited the Hayashi factory that pioneered the application of machine manufacture to the production of wall clocks (Uchida, 1985). Hattori also was a board member of the Osaka Watch Company, where he witnessed the application of machine technology to watch production. Although there also is some indication that Hattori was a serious admirer of the Swiss, having imported his early watch technology and watch components from Switzerland, nonetheless, the overriding influence on the selection of a vertically integrated model of production may have been the powerful role played by the Japanese government.

From the turn of the century onward, Japan's national government pursued a policy of restricting competition. It was recognized early on that sustaining hypercompetitive markets would require huge sums of scarce foreign exchange and would ultimately limit the level of investment firms were willing to make in productive operations. At the turn of the century, the national government was busy allocating markets to able producers. Given that watches were an essential item in military endeavors and that Japan was in a frenzy to purchase foreign technology to ensure its economic independence, we can only speculate that the watch industry was deeply influenced by national policies that encouraged the formation of business groups capable of producing at large scale, low cost, and high quality.

Although we do not have quantitative evidence of the growth of the Japanese watch industry and its three primary companies from the 1920s through the 1940s, we do have anecdotal evidence that suggests that insatiable domestic demand and national policies of self-sufficiency contributed to the rapid growth of Seiko and its early competitors (Learned et al., 1961). The company produced a line of products that did not intentionally compete with the Swiss industry, but rather followed the strategy of U.S. watchmakers by promising precision, accuracy, and price accessibility.

Other Competitors

Established in the late 1920s, the second major Japanese watch producer, the Citizen Watch Company, grew more slowly than Seiko and emphasized the low-end market for watches. The company's initial strategy, unlike Seiko's, was to produce products for other brandname distributors. The company became a serious competitor of Seiko's much later by forming foreign joint ventures with other watch producers, including the Bulova Corporation, based in the United States.

Another smaller company, Ricoh, began in the watch business at approximately the same time as Citizen, in the early 1930s. The company tried a number of strategies to penetrate the watch market, including, like Citizen, teaming up with foreign companies based in the United States.

Like most manufacturing companies, particularly those having to do with machinery, Seiko, Citizen, and Ricoh were called into service during the war. Seiko was commandeered during the war to produce bearings and other small machined parts for the aircraft and armaments industry (MacIssac, 1976). Seiko also was engaged to produce chronographs and fuses for the Japanese navy. This critical role allowed the company to maintain production during the war, even as firms in other metals industries were forced to contract due to a scarcity of foreign exchange and restrictions on the importation of foreign technology and materials.[4]

World War II and Relocation Out of Tokyo

During World War II, much of Daini Seikosha's manufacturing capacity was moved to the rural Nagano prefecture to avoid bombing raids (see Figure 8.1 for the location of modern and historical sites of watch production; Japan Clock and Watch Association, 1994). The Nagano region was selected for two reasons. First, a much-trusted former employee and confidant of Hattori Seiko's was living in the city of Suwa, in the Nagano prefecture, at the outbreak of the war. Second, Suwa had a tradition of precision machining, which dated back to the days of silk production. Its climate and geography were ideally suited for silk worms—and not much else. The weaving of silk required complex machines; hence, the population developed a broad base of skills to repair weaving equipment. For example, it developed parts manufacturing in order to repair reeling machines.

During World War II the silk reeling industry was displaced as industries moved from Tokyo to the mountainous Nagano prefecture in support of the aircraft and armaments industries. Wartime production efforts brought such varied industries as cameras, microscopes, and watches to the region. The *Strategic Bombing Survey* (MacIsaac, 1976) from World War II identifies two plants producing bearings for the aircraft industry, both located in the Nagano prefecture and both operated by Seikosha. When the silk trade died out after World War II, the region found itself developing a new economic base (Takeshi, 1996).

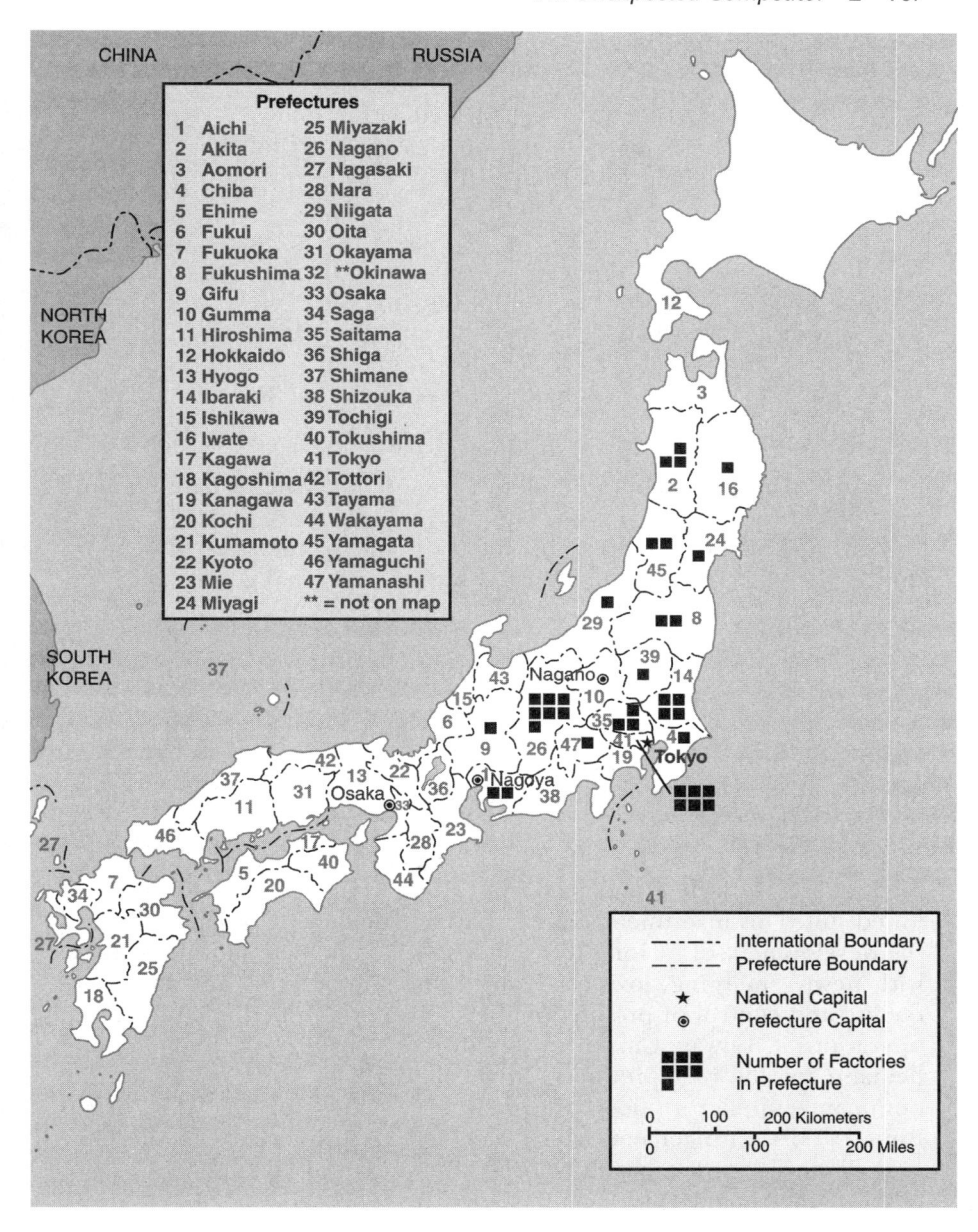

FIGURE 8.1. Distribution map of major watch and clock factories in Japan. *Source:* Japan Clock and Watch Association (1994). Cartography: Erin Heithoff. Sources: Map-Art; JCWA.

Restarting Business after the War

Seiko's watch production was reestablished within four or five months of the war's end. The Nagano operation was spun off as Suwa Seikosha. Soon, it was competing with Daini (still in Tokyo) to design and supply watches to K. Hattori for sale (Neff, 1984, p. 21).

Following the war, Seiko, as well as Citizen, began producing watches almost solely for domestic consumption. In addition, a new firm, Casio, emerged as a major competitor, moving into watches from a base in electronics. Casio had developed its know-how building war-related materials during World War II. Ricoh's competency was also enhanced by the war effort, but the company emerged from the war more dependent than the other companies on defense spending. It subsequently became a lesser player in the watch industry.

Citizen and Casio both developed production systems based around methods of vertical integration. There is some evidence that this strategy was dictated by the high precision needed in the manufacture of weapons during the war. Then again, after the war, the national government continued to rationalize production and to ration foreign exchange, thereby limiting the prospect of other, more fragmented competitors coming into existence. It also may be that the two firms modeled their production systems after the then-leader Seiko.

The next major development in the industry was associated with the conversion of the national economy to an export-led production system. Starting in the 1950s, the buildup of watch manufacturing led to domestic cost competition, which forced Japanese firms to pursue relocation strategies to overcome the effects of rising domestic wages. Seiko began buying cases, bracelets, and glass from Hong Kong. Citizen and Casio followed suit. This investment surge in parts made outside of Japan boosted the development of a Hong Kong industry. To compete more effectively with newly emerging, low-cost Hong Kong competitors, and to ensure quality and control of production, Seiko also established manufacturing operations in Taiwan. Citizen chose a different strategy, deciding to build demand for low-cost products. In the late 1950s, Citizen became the dominant supplier of low-cost pin-lever watches for the U.S. brand Bulova. This arrangement, which included exclusive rights to the Japanese domestic market, basically locked Bulova out of Japan for the sale of other, higher quality products imported from abroad.

The rapid reestablishment of watch production occurred in time to take advantage of the substantial buildup in the domestic demand for watches (Tables 8.1, 8.2, 8.3). The profits from growth made it possible for the Japanese to attempt to close the gap between their product and the Swiss product in terms of styling. Considerably enhanced machining

TABLE 8.1. The Japanese Watch Industry: Production, Shipments, and Stocks of Wristwatches during the Postwar Period (Number Unit: One Piece; Amount Unit: Million Yen)

Year	Volume of product (in units)	Amount of product (in yen)	Volume of goods shipped (in units)	Shipments (in yen)	Stocks (in units)
1945	25,650	1	—	—	—
1946	115,208	24	—	—	—
1947	280,723	152	—	—	—
1948	489,546	458	—	—	—
1949	712,648	875	677,471	887	—
1950	685,713	875	697,333	883	—
1951	916,572	1,285	922,442	1,297	—
1952	1,194,917	—	1,194,203	1,884	1,503
1953	1,584,964	—	1,584,942	2,690	1,835
1954	1,966,010	3,559	1,955,549	3,541	12,296
1955	2,200,546	4,300	2,209,020	4,296	6,103
1956	2,644,127	5,740	2,635,004	5,722	14,990
1957	3,372,049	8,198	3,362,205	8,162	24,832
1958	4,256,185	11,397	4,200,253	11,240	57,228
Jan, 1959	370,560	1,007	369,947	1,004	55,432
Feb. 1959	411,626	1,047	407,832	1,035	56,587
Mar. 1959	83,750	1,055	406,506	977	72,557
Apr. 1959	413,069	1,049	393,925	1,004	84,986
May 1959	448,152	1,105	437,615	1,069	92,671
June 1959	468,580	1,173	469,186	1,181	88,935

Source: Dai-Ichi Bank.

capability had developed during the war years; this capability could be fed back into the design and manufacture of higher quality products. A goal was set to improve styling and bring it up to world standards. An emphasis on unique watch movements for each watch model was abandoned and replaced with a program that reduced the number of unique movements while still maintaining superficial variation of watch cases. The adoption of this farsighted strategy of superficial product variation beat the Swiss and Americans by almost 20 years. This gesture enabled complete reorganization of the industry and led to major investments in new technology to reduce both direct and indirect labor costs.

The Development of Micromechanical Capacity in the Industry

Despite Japanese machine producers' major effort during the war to achieve self-sufficiency in machine tool manufacturing, they still lacked the ability to manufacture machines with the tolerances needed to pro-

TABLE 8.2. The Japanese Watch Industry: Watch and Clock Factory Shipments, 1951–1959 (Unit: Million Yen)

Year	Total clocks and watches	Kinds of clocks					Kinds of watches				Watches as a % of total
		Table clock	Alarm clock	Wall clock	Electric clock	Total	Watch	Stop-watch	Wrist-watch	Total	
1951	3,275	223	684	1,040	—	1,947	13	18	1,297	1,328	40.5
1952	4,504	403	759	1,348	59	2,569	24	27	1,884	1,935	43.0
1953	5,934	509	954	1,609	95	3,167	36	41	2,690	2,767	46.6
1954	7,244	500	1,173	1,764	175	3,612	66	25	3,541	3,632	50.1
1955	8,273	584	1,300	1,784	208	3,876	58	42	4,296	4,396	53.1
1956	10,804	738	1,787	1,923	526	4,974	50	59	5,722	5,831	54.0
1957	14,037	836	2,147	2,043	729	5,755	48	71	8,162	8,281	59.0
1958	16,605	797	1,799	1,670	972	5,238	52	75	11,240	11,367	68.5
Jan. 1959	1,456	67	133	127	115	442	4	6	1,003	1,013	69.6
Feb. 1959	1,528	70	161	126	126	483	4	6	1,035	1,045	68.4
Mar. 1959	1,504	85	167	136	130	518	3	6	977	986	65.6
Apr. 1959	1,524	84	159	131	135	509	4	7	1,004	1,015	66.6
May 1959	1,568	83	155	123	128	489	4	7	1,069	1,080	68.9
June 1959	1,661	70	159	122	119	470	5	5	1,181	1,191	71.7

Source: Ministry of International Trade and Industry.

TABLE 8.3. The Japanese Watch Industry: The Record of Export and Import of Watches (Unit: Thousand Yen)

	Export			Import		
Year	Watch	Clock	Total	Watch	Clock	Total
1950	75,334	295,766	371,100	8,233	4,501	12,734
1951	89,835	431,849	521,684	681,212	6,683	687,895
1952	89,772	331,504	421,276	405,145	26,774	431,919
1953	97,009	315,295	412,304	1,004,317	83,830	1,088,147
1954	69,320	367,292	436,612	605,366	61,102	666,468
1955	125,298	490,307	615,605	612,631	22,400	639,031
1956	486,088	669,449	1,155,537	858,297	22,775	881,072
1957	122,136	558,479	680,615	1,137,163	12,940	1,150,103
1958	139,040	438,758	577,798	906,050	8,984	915,034

Source: Statistics of the Custom Clearance by Ministry of Finance; includes watches' accessories.

duce watches. As the market for machine tools was burgeoning after the war, machine toolmakers demonstrated little interest in producing products for a very narrow market. This forced Seiko to pursue its own machine tool manufacturing capability. Seiko developed a separate engineering department to provide the firm with the necessary equipment to manufacture watches. The two other major producers of watches—Citizen and Casio—were forced to follow suit. This initial bottleneck generated considerable spillovers, however, as all three companies began to diversify even as they moved to satisfy their own internal need for precision machine tools.

It seems reasonable to ask at this point the following question: Why didn't the Japanese simply purchase the equipment needed to manufacture watches? Such a strategy had limited options. As I detailed earlier, the Swiss industry had virtually outlawed the exportation of equipment for watch manufacturing. As for the American industry, although American watch tool companies were capable of producing adequate equipment, the industry suffered significant decline as a result of the war effort, which prevented U.S. watch firms from manufacturing their primary product in order to serve the national government. Thus, while tools were no doubt available in the United States, the Japanese found considerable uncertainty regarding the time frame for their acquisition. In addition, Japanese producers did buy whatever machines were made available from the United States. At one point U.S. firms sold technology to the Japanese as part of that country's rebuilding effort.

Unable to satisfy their need in volume terms, Seiko and Citizen were forced to develop their own machine-building capabilities. What started out as a problem led to an exceptionally fortuitous outcome: the diversification of Seiko into a number of lines of business, including the production of micromechanical machine tools.

Seiko's decision to switch to assembly-line techniques in the production of watches, in contrast with the more labor-intensive model practiced before World War II, was a deliberate result of visits to American auto production facilities in the 1950s (Smitka, 1991). The company also maintained and increased its level of vertical integration, producing most components in-house. A tapered model of integration was followed for those parts purchased externally. All parts were produced either by Seiko or an affiliated small firm located near a main plant.

Internal Family-Based Competition Leads to Product Innovation

To increase the rapid development of products that could compete on world markets, Seiko fostered an intense interdivisional rivalry between its two main plants. This rivalry was expected to produce internal innovation and ensure that Seiko would produce products that met with the approval of export markets (Johnstone, 1999). This same type of rivalry was employed in Seiko's sales programs. While initially maintaining an independent sales force to counter resistance from jewelers (the prime sales outlet), who were predisposed toward U.S. and Swiss brands, postwar demand was so high that Seiko was able to use distributors to reach growing markets.

Pursuit of New Marketing Strategies

At the end of the 1950s, Seiko still sold only 5 percent of its total output abroad. The United States was by far the largest market in the world, but Japan's watch manufacturers had barely penetrated this huge market because most Americans believed that Japanese products lacked the quality and styling of either Swiss or U.S. products. To overcome this resistance, Seiko established a major sales operation in the United States, one that quickly established a reputation for excellent service. As the product had high reliability, the need for on-site or regional service centers was limited. Nonetheless, Seiko established repair centers on both coasts and maintained a program of quick turnaround to allay customer fears of lengthy times for repairs.

From the early 1960s on, the Japanese, and in particular Seiko, were producing products with world-class levels of technology and accuracy. In the early 1960s production volumes increased tenfold and Seiko began to drastically increase its worldwide distribution (Table 8.4). While Seiko watches did not enjoy the same world renown for design and fashion as those produced by the Swiss, nonetheless the company created a product that the Japanese and increasingly the world's middle-class consumers could afford and trust for high quality.

Through massive investments in new technology, Seiko was able to reduce the share of labor costs in watch production to 35 percent by the late 1960s (Hoff, 1985). Vertical integration allowed Seiko to enjoy rapid turnaround in the development of new products. Seiko set out to bring advanced products to market in small volumes, intending to introduce refinements in the production process before large volumes could be achieved. Experimentation was supported and enthusiastically encouraged as the company attempted to position itself closer to the quality and design of Swiss watches. Some evidence suggests that by the late 1960s, Seiko maintained a 15 to 45 percent variable cost advantage over the Swiss for jeweled watches (Hoff, 1985; Knickerbocker, 1976; Table 8.5).

Summary

Watch manufacturing typifies the conventional model of Japanese industry, in which vertically integrated firms maximize the benefits of size to increase market share and eventually dominate world markets. The watch industry, and particularly the Seiko Corporation (the same can be said for Casio and Citizen, which generally followed Seiko's production model), demonstrates how a vertically integrated system of production based around a core technology can yield strategic advantages that enable an enlightened firm to expand beyond the limits of a single technology.

But no firm exists in isolation. Seiko, Casio, and Citizen subsist within a production system that allows firms to maximize the benefits of size while gaining the advantages of small firm flexibility. Flexibility in this instance arises from a system comprised of large firms that deploy resources most advantageously and a vast network of subcontracting relationships that incorporates new technologies and is therefore capable of self-renewal (Smitka, 1991). In the case of the watch industry, however, the spatial location of this interacting, partly integrated, partly disintegrated system of production has occurred in conjunction with an elaborate spatial division of labor between Japan and Hong Kong. Since the creation of the Japanese industry, Hong Kong and Japan have been intimately linked in a supply and assembly chain.

It is important to see how national export policy and the dramatic expansion in watch production were interlinked. Starting in the 1960s,

TABLE 8.4. Evolution of Production and Exports of Watches and Clocks (Value units: million yen; Quantity unit: 1,000 pieces)

Year	Production Watches Value	Watches Qty	Clocks Value	Clocks Qty	Total Value	Total Qty	Exports Watches Value	Watches Qty	Clocks Value	Clocks Qty	Total Value	Total Qty
1950	885	694	1,023	1,634	1,908	2,327	57	39	281	345	338	384
1951	1,298	934	1,929	2,126	3,227	3,050	45	52	418	341	463	394
1952	1,925	1,217	2,656	2,586	4,581	3,803	35	27	312	235	347	262
1953	2,774	1,617	3,160	3,056	5,939	4,673	35	43	289	289	324	332
1954	3,649	2,002	3,619	3,596	7,269	5,599	13	43	281	281	349	495
1955	4,399	2,240	3,857	3,559	8,256	5,798	34	214	443	449	480	483
1956	5,848	2,686	4,873	4,206	10,721	6,892	378	198	562	925	940	1,123
1957	8,318	3,418	5,925	4,898	14,243	8,317	36	17	438	402	474	419
1958	11,524	4,303	5,238	4,232	16,762	8,534	50	60	362	353	412	413
1959	13,821	5,454	6,005	4,801	19,826	10,256	91	72	545	791	636	863
1960	19,163	7,141	7,921	6,681	27,084	13,822	162	145	745	1,033	907	1,178
1961	23,576	9,185	9,962	8,008	33,538	17,193	627	535	952	1,258	1,579	1,793
1962	28,288	10,641	12,115	9,358	40,404	19,999	913	710	1,090	1,593	2,003	2,303
1963	34,622	11,668	12,837	9,807	47,459	21,475	1,821	1,733	1,200	1,808	3,021	3,541
1964	38,223	13,192	14,661	11,403	52,884	24,595	3,581	2,790	1,466	2,266	5,047	5,056
1965	42,538	13,658	16,980	13,528	59,518	27,186	5,817	4,234	2,210	3,080	8,027	7,314
1966	44,504	15,580	17,664	13,705	62,168	29,285	9,082	7,012	3,059	3,991	12,141	11,003
1967	51,386	16,510	21,790	15,237	73,176	31,747	11,377	6,305	4,414	4,584	15,791	10,889
1968	60,616	17,748	24,215	17,917	84,831	35,665	17,282	7,190	4,509	5,647	21,791	12,837

Year									
1969	77,115	21,426	28,074	20,611	105,189	42,036	24,603	9,362	5,459	6,795	30,062	16,157
1970	86,657	23,923	39,786	25,772	126,443	49,695	32,873	11,310	8,977	9,420	41,850	20,730
1971	94,505	24,327	45,020	29,065	139,525	53,392	40,818	12,598	12,471	11,964	53,289	24,562
1972	102,038	25,464	46,287	32,722	148,325	58,187	48,956	14,030	17,762	17,028	66,718	31,058
1973	122,906	28,046	62,499	40,628	185,405	68,674	54,205	15,364	22,000	20,273	76,205	35,637
1974	167,173	32,369	73,964	36,446	241,137	68,815	83,573	17,961	27,482	19,213	111,055	37,174
1975	176,113	30,227	53,368	26,562	229,481	56,789	95,896	16,579	21,846	15,213	117,742	31,792
1976	216,132	34,001	74,170	36,965	290,302	70,966	139,633	21,280	30,347	21,363	169,980	42,643
1977	281,677	44,738	79,894	40,065	361,571	84,803	191,738	28,541	31,455	22,076	223,193	50,617
1978	302,907	49,192	91,411	39,575	394,318	88,767	200,813	30,709	26,643	20,790	227,456	51,499
1979	314,664	59,575	85,969	36,034	400,633	95,609	193,809	33,633	28,422	22,065	222,231	55,698
1980	382,711	86,233	100,196	46,506	482,907	132,739	257,211	47,939	46,063	34,402	303,274	82,341
1981	416,741	103,389	112,323	52,952	529,064	156,341	286,132	58,964	55,488	37,513	341,620	96,477
1982	319,149	101,080	89,860	45,153	409,009	146,233	234,169	62,619	44,776	32,846	278,945	95,465
1983	315,382	116,939	96,583	56,606	411,965	173,545	249,165	74,752	49,212	42,739	298,377	117,491
1984	349,835	148,507	103,332	73,399	453,167	221,906	291,115	94,071	51,592	47,307	342,707	141,378
1985	369,181	182,752	98,763	74,602	467,944	257,354	298,530	125,162	50,012	50,280	348,542	175,442
1986	311,091	200,819	98,905	80,885	409,996	281,704	232,673	133,705	39,232	52,665	271,905	186,370
1987	262,676	200,159	90,353	84,913	353,029	285,072	201,782	149,983	30,008	50,234	231,790	200,217
1988	292,040	266,299	95,250	86,353	387,290	352,652	212,971	182,007	27,366	45,653	240,337	227,660
1989	291,261	288,179	91,380	83,185	382,641	371,364	208,508	200,667	26,261	43,686	234 7683	244,353
1990	322,551	344,398	91,966	87,221	414,517	431,619	242,618	256,904	25,992	48,496	268,610	305,400
1991	336,288	389,789	92,186	87,427	428,216	477,216	254,819	278,814	25,302	51,592	280,121	330,406
1992	281,486	374,146	82,187	71,699	363,673	445,845	237,602	305,567	20,871	40,087	258,473	345,654
1993	244,361	390,596	66,323	57,097	310,684	447,693	197,494	322,708	15,872	32,685	213,366	355,393

Source: Ministry of Trade and Industry, Ministry of Finance (quoted in Japan Clock and Watch Association, 1994).

TABLE 8.5. Per-Unit Value of Watch Movements Imported into the United States (Calculated from 1970 Data)

Movements	From Switzerland	From Japan
0 to 1 jewel	$2.08	$1.15
2 to 7 jewels	$3.65	$3.06
8 to 15 jewels	$12.55	$10.85
16 to 17 jewels	$7.50	$4.23
Over 17 jewels	$28.80	$6.63

Source: Knickerbocker (1976, p. 20).

the Japanese government placed heavy emphasis on the export of watches as a technology that had both direct and indirect benefits for Japan. First, in a global market the undervalued yen proved to be an enormous advantage relative to the Swiss franc. Second, the technological complementarities were significant and proved to be more important as time went on, particularly for the electronic components industry which used sales to watch companies to achieve economies of scale (*Japan Economic Journal*, 1981).

The strategy to build up watch exports privileged three firms—Citizen, Casio, and Seiko—by giving them preferential access to capital markets and foreign exchange, to the detriment of other firms. Even as Japanese companies attempted to team with foreign producers to acquire new technology (e.g., Ricoh with Hamilton in the 1970s), restrictions on foreign exchange inhibited the success of other players and served to further concentrate the market for watches (Knickerbocker, 1976).

Government regulation and assistance (e.g., in the form of R&D grants) accelerated Japanese penetration of the world watch industry. The Japanese watch industry continued to rationalize further by pursuing vertical integration that streamlined operations and reduced inefficiencies.[5] The government encouraged vertical integration to minimize the proliferation of marginal watch producers and to minimize the drain on foreign reserves caused by the importation of watch machinery (Knickerbocker, 1976).

The Japanese also had the advantage of a huge protected home market. Because other watch manufacturers were effectively locked out of their home market by a 70 percent tariff on imports, Japanese producers could set high domestic prices and hence enjoy high profits. As with other industry segments, domestic prices for watches were kept artificially high to cover fixed costs. Thus, in international markets watches could be sold close to or at marginal costs (Hoff, 1985).

As the Japanese watch industry gained in technological competency, companies like Seiko began to circulate their watches more widely, displaying their products and designs at the International Horological Institute in Switzerland. The Swiss no doubt found their innovations novel, but showed no signs of being concerned with Seiko as a serious threat until the advent of the electronic watch. When Seiko was selected as the timekeeper for the 1964 Winter Olympics, held in Sapporo, Japan, its reputation was greatly enhanced. In full view of the entire world, the Japanese watch company demonstrated what had heretofore been the exclusive purview of the Swiss: to flawlessly keep, record, and display time at exceptionally high levels of accuracy.

Although Japanese market share eventually declined in key markets such as Hong Kong, early declines were relatively small and in the low-value segment of the industry. However, examination of Japanese watch and clock statistics clearly shows that Japan made an effort to become a major player in the world watch industry after 1950 (Japan Clock and Watch Association, 1994; Learned et al., 1961). From 1950 through 1980, production and exports tripled every decade, rising from 600,000 to 86,000,000 pieces over four decades. It helped that at the time of the transition the global market was growing; hence, neither Swiss nor U.S. manufacturers felt much effect from Japanese growth in market share.

It seems reasonable to ask how the Japanese were able to overtake the Swiss watch industry in only 20 years, to become the world's largest producer of watches. This unexpected outcome is all the more surprising given that the U.S. industry was in a state of disarray, with watch manufacturers locked in acrimonious battle with watch assemblers over the continued import of Swiss watch parts. Thus, the U.S. market was always in jeopardy. While the Swiss had regained much of its lost share of the U.S. market, it quickly yielded at least some of it to a new, surprisingly agile competitor that appeared to have overcome its most difficult problem: to produce a high volume of watches reliably, at low cost, and with great variety. Although the U.S. watch manufacturing industry had the capacity to produce in a vertically integrated fashion, it never achieved the same level of integration as the Japanese industry. Nor did it enjoy the agility that the Japanese industry appeared to have in introducing new products in rapid succession. The Japanese would succeed in producing watches with a much lower labor content through the internalization of watch technology. The willingness of the Japanese to forego short-term profits during the period in which they developed automated production paid off when shortsighted competitors assumed market-based competition would produce the desired outcome. It failed to emerge, however. This lack of vision will be seen again in later chapters when we examine the development of the electronic watch—the watch that changed the world.

9 Only the Young Survive

The U.S. Watch Industry between the World Wars and after World War II

It is now necessary to return to a discussion of the U.S. industry from after World War I through the 1960s. Making sense of the U.S. story also requires considerable reference to the Swiss industry after World War II and into the early 1960s.

From the beginning of the twentieth century, the U.S. watch industry suffered a number of major setbacks that ultimately led to its decline as a major player in the world watch industry. The experiences of the three primary companies at the turn of the century—Waltham, Elgin, and Hamilton—combined with those of two primarily postwar firms—Timex and Bulova—demonstrate the progression of the U.S. industry from one comprised of mass producers with highly differentiated product lines relying upon a market-based model of competition, to one with market regulation that resembled an oligopolistic structure.

Unlike the Japanese industry, however, in which competition was effectively controlled by government policies designed to limit the destructive elements of unbridled and undirected competition, the U.S. industry was effectively restricted from collaborating in any way by the Sherman Anti-Trust Act. Although some market regulation was introduced in response to the federal government's needs during World War II and the Korean War, these gestures were only temporary diversions from a structure of competition based on interfirm rivalry. Some degree of control over competition, like the "Watch Trust" created at the turn of the century, could have provided the protection the U.S. industry needed to regain a firm footing in the world watch industry. However, the historic independence of U.S. watch firms, the opportunistic cluster of small watch assemblers, and the strong anticollaborative ideology governing U.S. industry proved insurmountable in the constant change that defined the world watch industry after World War II.

In this chapter we revisit the trials and tribulations of U.S. watch firms as they faced growing international competition, first from their old rival, the Swiss, and then from a new upstart, the Japanese. Here we examine the range of options pursued by different firms who clearly saw the contested market terrain very differently. At this point, several U.S. firms were being driven as much by structure as by strategy, an unenviable position to be in. Waltham, Elgin, and Hamilton retained production systems originally formulated at the turn of the century. Although some updating had occurred, the outlook of these firms was strongly influenced by the historic moment in which they came into existence. A very important part of their outlook was shaped by their original product market strategy, the sale of fine watches. The two newer major domestic competitors, Timex and Bulova, pursued different strategies that relied heavily on international operations. This made matters all the worse for the older firms created in an era in which production and markets were almost entirely domestic.

The decision by both Timex and Bulova to sell in international markets, supported by a strategy that recognized the importance of the growing global division of labor in watch parts manufacturing, allowed the two to take the lead from the more traditional firms. This failure to act internationally in the face of growing domestic and international competition proved too great for the older firms and ultimately forced them to abandon the market for domestically made watches. The U.S. market became the single most important market for Swiss watchmakers at the end of World War II. Unhappily, it was rapidly closing. The Swiss firms' inability to cope with changing circumstances was due in part to the Statut de l'Horlogerie (see the discussion in Chapter 7, pages 150–151), which had dictated how the Swiss watch industry worked since the 1930s. Originally designed to preserve the system, the Statut eventually served only to tie the industry to an outdated production system. What initially constituted a strategy designed to preserve Swiss dominance of the world watch industry ultimately resulted in slow strangulation as the outmoded system of production and marketing restricted the industry's ability to make changes after the late 1940s.

During the interwar period (1919–1941), Elgin led Waltham in watch sales, 18 million to 8.3 million. Meanwhile, a new manager at Waltham pursued a profit-making strategy that involved decapitalization and progressive wage cuts.

A glance at the history of the Waltham Watch Company provides an ominous picture of greed, mismanagement, and sheer ineptitude. The factory was run chaotically and workers were paid by the piece. When the market contracted and prices fell, the manager simply cut workers' wages, fueling intense resentment and an unwillingness to contribute to the firm's survival.

By the end of the 1920s the company's market position had been dangerously eroded, in part due to serious competition from the Elgin and Hamilton Watch Companies. But a more serious problem had arisen with Waltham's products. During the turbulent years of the 1920s, the company continued to cut wages in an effort to reduce costs. It also failed to make the necessary investment to maintain the capital-intensive production process. This strategy backfired. Watchmaking required solid relationships between employee and employer (Moore, 1945). The level of precision needed to make a watch required a joint commitment to quality by capital and labor. As management cut back on piece rate–based wages, workers simply skimped on quality. It did not take long for this type of information to hit the street.

Over the next 10 years a series of managers were brought in to straighten out Waltham. This revolving door, however, further destroyed morale and worker competency. Even when one administrator managed to return the company to the black, his effort was the result of penny pinching and working tired equipment to utter exhaustion. The milking of assets, combined with an unwillingness to hire new employees from the outside, left the company stale, rigid, and unable to compete.

Elgin suffered over the long term from its sluggish movement into wristwatches. Commandeered by the federal government to make products for the war effort during World War I, the company had a difficult time returning full-time to watch manufacturing. Cost-cutting was the order of the day, but this proved to be of limited usefulness in the longer term, since product quality eroded and workplace discontent increased.

More than anything else, however, both companies were hurt between 1930 and 1960 by the expansion of the Bulova and Timex Watch Corporations and the increasing worldwide dominance of the Swiss watch. This problem will be examined later in this chapter.

The U.S. Industry after World War I

U.S. firms attempted to return to watch manufacturing after having been seriously diverted by the war effort in World War I. While watch manufacturers profited handsomely from producing armaments in the short run, wartime industrialization drove up manufacturing costs and diverted skilled labor from watch manufacturing. Unlike the Swiss, who maintained a functional balance between watch and armaments manufacturing, U.S. firms turned their entire attention to making fuses and other war-related material. During the war years, the lack of sufficient U.S. domestic capacity in watch production opened the U.S. market to significant Swiss penetration, and the composition of Swiss exports to the United States changed dramatically. By the 1920s, although the *volume* of Swiss exports had stayed the same, their value went up by as much as 50 percent. Before the war, Britain and Germany were Switzerland's largest export markets for watches. But, by the end of the war, the United States had become the largest importer of Swiss watches and watch parts. Although domestic watches were still being produced, their movements were imported almost exclusively from Switzerland (U.S. Senate Committee on Finance, 1951).

The recession that followed World War I prompted cries from U.S. industries in general for trade protection. In 1922, the Fordrey–McCumber Tariff Act was passed, increasing tariffs on many manufactured goods as much as 50 percent. Although U.S. watch manufacturers and watch assemblers were divided on the necessity of and benefit to themselves of burdening their foreign competitors with increasing tariffs on watches and watch parts, the major producers, including Waltham and Elgin, won out and tariffs within the watch industry were raised in the early 1920s.

In 1929, in response to economic chaos of the Great Depression, tariffs were again raised on imported goods. This time Swiss watch manufacturers appealed directly to the U.S. Congress, indicating that increased tariff levels were destroying the Swiss national industry. This appeal prompted a brief return of tariffs to 1922 levels. The momentary respite against higher tariffs was quickly overwhelmed by the passage of the Smoot–Hawley Act of 1930. Tariffs on almost all industries, watch products in particular, were raised by as much as 50 percent. Within two years, Swiss exports fell by first 20 percent and then by 40 percent from 1929 levels. From 1929 through 1935, total Swiss exports to the United States, including watches, fell 77 percent. This drastic contraction in legal exports set off a dramatic increase in watch smuggling.

Rising tariffs on imported watches and watch parts unleashed considerable conflict between U.S. watch assemblers and whole-watch manufacturers. While the whole-watch manufacturers charged Swiss producers with unfair trade based on low wages, U.S. watch assemblers argued that the real problems threatening the whole-watch manufacturers were the superiority of the Swiss technology and the manufacturing acumen of the Swiss (Meier, 1970).

Efforts undertaken by U.S. watch firms to recapture control of the domestic market after the war were impressive. By 1934, the effective tariff on Swiss movements was 81 percent, severely reducing that country's share of the U.S. market. Having enjoyed considerable control of the domestic market prior to World War I, U.S. firms were determined to return to the days of limited Swiss imports.

The U.S. government conceded that its tariff actions had harmful effects on both the Swiss watch industry *and* on the U.S. industry. In 1934, the Department of State notified the American public that it desired to enter into negotiations for a reciprocal trade agreement with the Swiss. In 1936, two years of negotiations eventually produced the "Trade Agreement of American and Switzerland" (Meier, 1970). In the agreement, the Swiss made concessions on numerous products and accepted the "Declaration Regarding Smuggling." This gesture was designed to reduce the illegal importation of Swiss watches (Meier, 1970). American concessions covered 59 items, approximately one-third of which referred to tariffs on watch and clock imports, previously set at high levels by the 1930 Smoot-Hawley Act. The agreement also put in place a quota system setting import levels of certain items, including finished watches and watch parts.

The reduced tariff (still a whopping 33 percent), combined with rapidly rising wages during the late 1930s and devaluation of the Swiss currency, sent employment levels plummeting by one-third as U.S. firms sought various avenues for shoring up market share (Learned, Christensen, and Andrews, 1961).

Despite the remaining effects of the tariffs, within a few years, Swiss market penetration was back to previous levels, albeit in modified form. Locked out of the jeweled market segment by high tariffs, the Swiss turned around and began massively exporting pin-lever movements that were much cheaper and more fashion-oriented. Assemblers in the United States enjoyed the best of both worlds: a supply of cheap but high-quality watch movements and a market that welcomed low-priced, high-fashion watches (U.S. Senate Committee on Finance, 1951). The Swiss successfully recaptured a portion of their lost market share by following a strategy of low-cost, high-quality design, a strategy first used in the last century to take the lower price market from the British (Landes, 1979). However, as I discuss later, this push to maintain a position in the low-priced segment by "cheapening" the Swiss watch resulted in a slow degradation of the public's perception of what constituted a "good" watch (Yankelovich, 1969).

Hamilton from the 1920s to the 1960s

After the Great Depression, the two major U.S. watch firms, Elgin and Waltham, were in shambles. The difficulty of reconverting after World War I, combined with the long-felt effects of the depression, severely weakened these once powerful competitors. Hamilton, which had technology of a newer vintage and a more fleet-footed management, emerged from the depression weakened but still intact and viable as an operating business. At the end of the 1930s, two additional developments completed the reorientation of the firm: reorganization of its administrative operations and the pursuit of a more rational production process. Of the two, the former was far more important in realigning the company's cost structure for the sake of greater competitiveness.

But before Hamilton could reorganize its administration, the company had to reorganize its production system and streamline activities, which it began as a result of its mobilization for World War II. In that mobilization the company had to redirect its entire watchmaking effort, grafting onto its preexisting system countless additional production activities to manufacture war materials. To do this, many parts of the domestic watch business were neglected. Remedying that neglect was costly.

Sauer (1992), in his history of the Hamilton Watch Company, provides a 1947 *Fortune* magazine description of Hamilton's manufacturing process of the 1930s:

> Each foreman arranged for the tools and dies he needed to conduct his department's phase of the work, and progress depended to a great extent on his personal relations with the machine shop foreman. Ultimate success depended on the talent in the foreman's hands and the experience stored in his skill. There were no tight specifications to follow. At best there might be a sketch on a small file card. Occasionally, a tiny part might turn out to be ten thousandths of an inch too large or too small, and it took a master craftsman to reassemble the various parts after gathering them from the respective departments. (p. 92)

This archaic system was slowly but firmly changed to produce standardization and more orderly operations. Efforts at achieving replicability and interchangeability led to major developments in human resources management and skills development.

World War II had an important impact on watchmaking firms around the world. The micromechanical capabilities so deftly developed in the course of watch manufacturing were in great need by war industries. All watch companies, whether Swiss, Japanese, British, or American, participated in the production of war materiel. Unlike other watch companies, however, Hamilton's success in the production of war products during World War II fundamentally altered the company's long-run future. Its reputation for producing the watch that kept American railroads on time positioned the firm as a major contributor to the war effort. Some of the products needed, such as the chronograph, had never been manufactured in the United States. During the early stages of the war, the United States imported chronographs from Switzerland. When that country could not produce enough products to meet United States demand, however, the War Department put out a call to U.S. companies to manufacture the critically needed device. Hamilton not only produced one that excelled in quality, but also was able to make other products needed by the armed services. The superiority of Hamilton products was both a blessing and a curse. The company made millions of dollars on military products, but military production diverted the company from its focus on watches.

Distracted by the war effort, all U.S. watch companies emerged at the end of 1945 with smaller shares of the U.S. market. Of some seven million watches purchased in the United States in 1947, only two million were made in the United States (Bolli, 1957; Sauer, 1992). Due to a better distribution system and a better product, however, Hamilton emerged from the war better able to take up the business of watchmaking than its domestic competitors. A new strategy to lift sales relied on a heavy dose of advertising emphasizing the quality, accuracy, and beauty of Hamilton watches. It worked. The company dropped all references to railroads from its advertising as it moved into the age of science and electronics.

Unlike Waltham and Elgin, Hamilton continued to pursue new product development and innovation. Technological developments arising out of World War II research efforts were used in developing products through technology transfer. In the mid-1950s, the company introduced a self-winding watch, followed quickly in 1957 with the introduction of its electric watch, which was simply a mechanical watch powered by a small battery. This development was truly momentous and for a time propelled Hamilton to a technology leadership role in horology. But the product proved unreliable and hard to repair. Thus its early introduction failed to achieve mass appeal. Also, despite considerable innovation and new product development in watches, the company's fascination with defense work reemerged with the Korean War. Its reentry into defense production was one of Hamilton's many steps toward diversification that would eventually take the company farther and farther afield from its watchmaking roots.

Hamilton began the 1960s a bigger company but one that was less profitable and more vulnerable to external events (Sauer, 1992). The company had grown domestically and internationally through nonwatch or micromechanical acquisitions and was a larger and more diversified corporation. It had a much bigger stake in defense manufacturing and other product areas only loosely associated with the company's core competency: micromechanics. Forays into these other product areas did not bring bountiful profits. Defense production was expensive and the other endeavors simply did not generate the same level of profits (minus costs) that came from watches (Sauer, 1992).

In a business environment with growing markets, companies with a broad product range make large profits. Like the Swiss, Hamilton had a watch for every occasion, a watch of every conceivable size and movement type, a watch to fit every customer mood. This diversity, while advantageous in a stable or growing market, proved problematic for Hamilton when the low-priced market began to be flooded in the late 1950s and early 1960s by cheap imports from areas such as the Virgin Islands, where Timex had set up assembly plants. Hamilton's attempt to respond to the influx of low-cost watches hurt the company. When Hamilton stopped following a pricing policy that respected the overhead costs of the jewelers who were selling the watches, these traditional distribution channels revolted and abandoned the company's products. The company's large product line, long supported by jewelers willing to carry a full line of the company's products, was no longer economical. Now variety meant an inventory excess of parts that had to be written off when Hamilton's market share contracted (Sauer, 1992).

By the end of the 1960s, Hamilton was in trouble. A series of late-to-market product introductions had hurt the company's reputation. Internal problems associated with rapid management turnover and overdiversification in unrelated product lines left it seriously wounded and in need of an outside champion. Major declines in the value of the company's stock led to the sale of divisions and the end of movement manufacturing in the United States—measures deemed too little too late by investors (Sauer, 1992).

Ironically, the company's savior was once its fiercest competitor, the Swiss. The 1974 purchase of Hamilton by the Swiss consortium SSIH (later SMH) saved the company from outright closure.

Hamilton's Pulsar: The Electronic Wonder Product

Although its future looked bleak, Hamilton continued to innovate, bringing to market the first electronic watch. Starting its research in the mid-1960s, Hamilton pursued the development of "Pulsar," "a solid state wrist computer" (Sauer, 1992, p. 247). This revolutionary device came on the heels of the development of microelectronics, a period in which transistors were being replaced by integrated circuits and the dawn of the computer age was just around the corner. The premature announcement of the Pulsar in 1970 placed the company in the indelicate position of advertising a nonexistent product. Rapid development was achieved, however, because parts for the watch were purchased from various divisions of the company outside of watches, and from vendors in allied industries such as glass, audio, and batteries. Of particular significance was the decision to go with a light-emitting diode (LED) technology developed by a small company in Dallas, Texas. To see the time, the user had to push a button. While this technology was ultimately considered inconvenient and replaced by liquid crystal display (LCD) technology, Hamilton went with it in order to be the first to market an electronic watch (Sauer, 1992).

Pulsar came to market in 1972. This space-age product was an immediate success. Hamilton priced the product at the high end, selling it to celebrities, politicians, and collectors. The original Pulsar was followed by several model variations. For a short time, the revolutionary device breathed new life into the company.

Bulova from the 1950s to the 1980s

Although Waltham and Elgin were no longer serious competition for Hamilton after the 1950s, the company was not alone in the market. The Bulova Watch Corporation had been making watches since 1874, but did not become an important player in U.S. watchmaking until the 1960s. The company almost went bankrupt in the 1950s, in part because of Timex's success with low-end watches and in part because of the continuing expansion of Swiss watches in the middle and upper ends of the market. While total watch sales in the United States rose by 25 percent between 1954 and 1958, Bulova's sales fell by 17 percent (Bulova Watch, Annual Report, 1968, p. 3).

In 1960, the company reorganized and expanded from its traditional "middle" niche into both the low and high ends of the market. On the low end, Bulova entered into an agreement with Citizen Watch of Japan. Citizen was to produce jeweled movements for Bulova's "Caravelle" watch. On the high end, Bulova engineers designed a tuning-fork-regulated movement used to drive the "Accutron" watch, and until the development of quartz technology in the early 1970s, the Accutron was the most accurate watch in the world. By 1968, sales of the Bulova brand—the company's traditional, mid-priced model—comprised only 55 percent of total low-, mid-, and high-end watch sales (compared to 80 percent five years earlier) (Bulova Watch, Annual Report, 1968, p. 2).

The accuracy of the tuning-fork watch came at a cost. The product was delicate and hard to maintain—it did not take well to normal wear and tear. Bulova had invested heavily to produce the Accutron watch, and had to continue to spend money to keep the watches already in circulation working. This placed an added burden on the company and helped drag it down in the 1960s.

By the late 1960s, Bulova began using semiautomatic assembly lines to manufacture its watches. Recognizing the importance of producing for all its market niches while maintaining low costs, the company touted its flexible production technology (Bulova Watch, Annual Report, 1968, p. 11):

> Not to be overlooked is the unique elasticity of this sophisticated production process. For example, last year it absorbed the manufacture of the new Accutron calendar model (after two years of production engineering, tooling and processing of a pilot run), plus the new Caravelle transistor electric and a new 23-jewel, thin, self-winding Bulova calendar model, and also *simultaneously* digested the manufacturing of 759 domestic and 4,802 international styles of timepieces, plus 46 varieties of electronic, electric, and conventional clocks and 29 radios and cordless clock-radios. This diversified mix is typical of the flexibility which technology imparts to our operations.

Although Timex came close, Bulova led other U.S. watchmaking firms in marketing expenditures through the 1960s. In distribution, the firm expanded from its traditional placement in jewelry stores to large jewelry chains and general merchandising outlets. By 1968, the company had over 20,000 distributor outlets in the United States.[1]

The two leading American firms, Timex and Bulova, presented significant problems for Japanese and Swiss manufacturers. Both corporations followed the American system of mass production. Employing a combination of sophisticated production technology and labor flexibility (through the internationalization of production), Bulova produced a range of products spanning all price categories. The firm became the industry leader by scattering production around the world and trading off wage levels for labor skill. As part of this strategy, the company also subcontracted with Japanese component producers for low-priced movements (Knickerbocker, 1976; Landes, 1983).

The medium price range was Bulova's strong suit. It offered consumers literally hundreds of different styles manufactured in the company's Swiss factories. According to some estimates, at one point Bulova was the largest manufacturer in Switzerland. At the high end, with its aggressively marketed tuning-fork technology, Bulova was in a class by itself.

The company's international production system maximized site-specific advantages such as skill levels, technology, and markets. For example, Accutron was made in the United States where technology levels were high despite less-qualified workers in terms of manual labor skills. Medium- and low-priced mechanical watches were manufactured by high-skilled Swiss workers, whose wage levels at the time were low by global standards. The company's international orientation provided important opportunities to test-market new products. By having a strong brand policy and by aggressively marketing its products, Bulova moved into markets worldwide.

Bulova's connections with overseas watch companies helped it expand. It purchased Universal Genève, a Swiss watchmaking concern, in the mid-1960s, using it as a platform for manufacturing watches for sale in Europe. Bulova also entered into a licensing agreement in 1969 with Ebauches SA for making tuning-fork movements; Ebauches SA would in turn sell these movements to Omega and Eterna, both members of the Société Suisse pour l'Industrie Horlogerie (SSIH) group, later to become Société Suisse de Microélectronique et de l'Horlogerie (SMH), the dominant Swiss watch holding company. In 1970, Bulova expanded into Japan, setting up a joint venture with Citizen Watch to manufacture and market tuning-fork watches there.

Sales and profits expanded for Bulova until 1975, when a series of events coincided to force the company to the brink of failure once again. The end of the Vietnam War resulted in smaller defense contracts, while a worldwide recession dampened consumer spending. The Swiss franc appreciated by 19 percent, causing the cost of finished Accutron watches—by this point produced almost entirely in Switzerland—to rise dramatically (Knickerbocker, 1976). The highly dispersed production system, which at one time afforded the company great flexibility, became an albatross when the market for watches destabilized in response to the worldwide recession. The company was plagued by having too many production sites with widely varying costs of production, high wastage and reject rates, and a vast number of watch models at varying stages of maturity. Managing this far-flung and differentiated system proved costly and inefficient. In a drastic effort to remain competitive, the company shut down many of its overseas operations, but the gesture came too late.

Far more devastating than an inefficient production system, however, was the effect of the quartz revolution, which had begun to eat away at Bulova's high-end markets. Bulova had devoted some R&D money to quartz technology, but the success of its tuning-fork movement probably encouraged more caution than the situation required. The quartz technology Bulova did develop was devoted mostly to the Accutron, not watches at the low end of the market where Japan would soon vault into the lead.

Why did tuning-fork technology fail as a viable technical alternative to quartz technology? The accuracy it afforded was unsurpassed. It was the first truly nonmechanical watch movement innovation to challenge the status quo, which was dominated by the Swiss. Bulova's worldwide control over its own technology was the problem. Because the company maintained tight control over the technical process for the manufacture of the tuning fork, the innovation did not diffuse widely. Eliminating the vicelike grip on the control of the technology architecture could have meant a rivalry with quartz movements, at least in the early stage of the introduction of the new electronic technology. As a representative of the Seiko Corporation commented, "The tuning fork could have been far more important than it eventually became. But the technology failed to disperse widely and was overtaken by quartz. Our strategy with quartz technology was to see that it filtered rapidly around the globe" (S. Aota, personal communication, Seiko Horological Institute, Tokyo, Japan, July 19, 1994).

In 1976, C. P. Wong, a Hong Kong businessman, took control of Bulova, bringing with him the head of Mostek Corporation, a manufacturer of semiconductor products. This gesture marked the beginning of the ascendancy of Hong Kong as a site for electronic watch production. Despite this, quartz efforts came too late to save the company from losses in 1977 and 1978. In 1978, Bulova was sold to Loew's Corporation, which ended all U.S. manufacturing and changed everything about the company. Loew's also phased out some manufacturing facilities in Switzerland and France. After plunging to the brink of bankruptcy in 1984, the company began to generate modest profits by making specialized watch products. It entered into marketing agreements with Benetton and Club Med for the manufacture of midrange special purpose watches. The Accutron disappeared in the shuffle, and the Bulova brand once again became the company's "flagship," selling watches in the United States through specialty distribution channels.

Timex from the 1940s to the 1970s

One key to the decline of Waltham and Elgin was the creation of the Timex Watch Corporation in the late 1940s. Timex, born from the ashes of the moribund Waterbury Clock Company, did little to revolutionize the making of watches (at first), but did a great deal to revolutionize their marketing. From the start, the Timex watch's claim to fame was durability and low price; it would "take a licking and keep on ticking." When a Timex finally broke, you just threw it away and bought a new one. Timex could trace its bold marketing ploy back to the famous American "dollar watch." Dollar watch advertising touted easy and cheap replacement rather than costly repair should the watch fail.

Before World War II, the Waterbury Clock Company still manufactured the $1.00 Ingersoll pocket watch, but it had nearly gone bankrupt in the late 1930s. In 1942, the company was bought by a group of Scandinavian businessmen led by Jaokim Lemkuhl, an immigrant from Norway. The group bought Waterbury specifically to produce fuse timers for bombs and artillery shells, and immediately began converting the company's watch manufacturing plant to wartime purposes. It soon became the largest fuse-producing firm in the United States.

When the war ended, fuse sales naturally slumped. Lemkuhl decided to return to watch production. By 1950 he had a new kind of watch on the market that was priced at under $10. Wartime technology had produced a new alloy that Waterbury engineers substituted for jewels in the movement. This advance made the watch technologically competitive with jeweled-lever watches, which cost much more to manufacture.

In the early 1950s, Timex brought out a line of fashion watches for women. The company promoted the watch line as an accompaniment to a woman's wardrobe—a watch for every mood, dress, sport, and general use. The transition from marketing a watch as a piece of jewelry to marketing it as a fashionable yet utilitarian device was to have major implications for the industry in years to come. This promotional strategy would also be used to rescue the Swiss watch industry.

Timex relied on mass production, relatively unskilled labor, and constant attention to production efficiency and quality control. The company followed a policy of (1) rigid standardization with fully interchangeable parts, (2) maximum mechanization to reduce human error and to minimize labor costs, and (3) centralized quality control. These goals were achieved in large part thanks to the utter simplicity of the product. It had far fewer parts than most other watches. Whereas an imported watch contained more than 120 parts and required over 100 operations, the Timex had 98 parts and required only six operations. An imported watch had 31 screws while the Timex had only 4. Production was highly regulated and standardized thanks to the work of 500 toolmakers who designed most of the company's equipment (Uytterhoeven and Knickerbocker, 1972, p. 4). The highly automated and standardized system virtually eliminated hand adjustment, a costly procedure that was a constant source of trouble for the more traditional manufacturers.

Timex was an early adherent of strict inventory control. The company tracked parts, finished watches, and sales, enabling it to make highly accurate forecasts of demand. Parts and work in progress were tracked on a daily basis, giving the company great flexibility in keeping overhead down to a minimum.

By the end of the 1950s, after coming out of nowhere, Timex operated production plants around the world. Though the main plant was in Connecticut, production took place in Texas, Arkansas, England, Scotland, West Germany, and France (only miles from the Swiss border). As if intentionally tweaking the Swiss, Timex produced the world's cheapest watch in the shadow of the world's leading watch industry in Switzerland.

Timex prospered in the late 1950s thanks largely to advertising and distribution policies. The company launched a major advertising campaign in 1956, using a series of "torture tests" in which the Timex watch "took a licking and kept on ticking." Unlike most manufacturers, who sold their watches through jewelers and/or department stores and gave a 50 percent margin to retailers, Timex retailed its watches in drugstores where the margins were only 30 percent. This break from convention was initially viewed as a huge risk because traditional retailers threatened to boycott Timex watches. This proved to be an idle threat: within a decade Timex watches were available in over 20,000 outlets. What the company lost in per unit profits it more than made up in overall profits due to high volume. Although the Timex watch did not enjoy the aesthetic appeal or intricate craftsmanship of many Swiss watches, it excelled in self-promotion and availability. These two factors would become the most important prerequisites for success in the low and middle range. Such strategies resurfaced again years later as major competitors tried to regain market share. By 1962, after only 13 years of operation, Timex was selling one-third of all watches sold in the United States.[2]

In the 1960s, Timex gradually worked its way up the product line, introducing jeweled watches in the early 1960s and eventually producing inexpensive electronic watches. Using the same "torture test" ad campaign, the company expanded its operations to Europe and South America.

Having learned how to use local sensitivity to penetrate smaller markets in Latin America and Africa, Timex went after one of Switzerland's exclusive domains, Germany. This time, instead of following its traditional path of explicitly mining the low-cost, simplistic qualities of the Timex watch (as a throwaway), Timex went into Germany to sell customers on the idea that a good watch did not have to be expensive. By offering German retailers high margins, strong advertising, and good service, within two years Timex had grabbed 10 percent of the German market. This made the Swiss sit up and take notice. A similar strategy was applied in the Far East, the other bastion of Swiss dominance.

Early on, Timex used U.S. trade policy to its advantage. A key element in Timex's ability to monopolize the low-end segment of the watch market was its decision to carry out watch assembly in the U.S. Virgin Islands. Time and time again, through much of the 1960s, the company presented testimony to the U.S. Congress regarding the national importance of maintaining a domestic watch industry. By arguing in favor of the developmental benefits of watch assembly in the Virgin Islands, and by making nostalgic references to the nation's history as a watchmaker, the company received remarkable tariff advantages that effectively restricted access to the U.S. market.

Given the product's mass appeal, it was only a matter of time before the higher quality watches worked their way into traditional distribution channels. Jewelers could not ignore the jeweled Timex watches because customers wanted them. But Timex also yielded some ground to jewelers to ensure the successful introduction of its more expensive watches into jewelry stores by offering them higher markups.

Meanwhile, during the 1950s, both Waltham and Elgin continued their slow but inexorable declines. Hamilton and Bulova did somewhat better, but even their profits slid precipitously as both the Swiss and Japanese made continuing inroads into the U.S. market and Timex caused further cost pressures.

U.S. companies had sought protection from foreign imports before. They stepped up this campaign after the Korean War. Although U.S. companies still had the advantage of tariff-free assembly in the Virgin Islands, this labor cost reduction was not sufficient to offset the ability of both the Japanese and Swiss to manufacture cheaper movements (which accounted for the bulk of a watch's cost). Moreover, the Japanese also had access to cheap assembly—in Asia. The U.S. watch industry intensified its

Washington lobbying, but did not receive the reception it desired. The watch companies, frustrated by inattention from Washington, went so far as to trot out the national security banner. This gesture had no effect, however, as critics of protectionism showed that the U.S. war effort did not rely solely on U.S. sources for timekeepers, fuses, and other micromechanical devices. President Lyndon Johnson leveled a devastating blow to the domestic watch industry when he effectively canceled tariffs on watches, a gesture that would eventually destroy the U.S. movement industry.

In the 1970s, Timex began to falter. Although the firm jumped on the quartz bandwagon as early as 1972, its lowest priced quartz watch, when assembled, was still more expensive than those made by the Swiss company Ebauche SA. While the company excelled at marketing, it was a dismal failure at the management of new technology. Furthermore, the falling costs of solid-state technology pushed digital watches into the price range Timex had long dominated with its pin-lever watches. As the major semiconductor producers introduced their own quartz watches (coupling them with LCD technology), competition intensified further. Other major traditional watch manufacturers, such as Hamilton, developed an in-house capability to make integrated circuits long before Timex did.

Simultaneously, Timex suffered managerial problems. Lemkuhl retired in 1973. A former shipping company manager from Norway, Ole Martin Siem, was appointed to replace him. This caused long-simmering tensions to boil over between the heads of marketing and the heads of engineering, and the marketing head resigned.[3] In the years that followed, Timex struggled to compete with producers better situated to make quartz watches. In the early 1980s attempts to diversify into home health care products and personal computers failed. The former operation was spun off within a few years, and the computer operation drowned in the fierce competition of the day. Meanwhile, watch assembly began moving to export platforms in Asia. Foreign plants had long been important to Timex's manufacturing strategy, and by the early 1970s Timex qualified as an "American" watch only by virtue of the location of its corporate headquarters in Connecticut.

Thus, aside from Timex, the United States—in the course of the 20th century—went from being a major watchmaking powerhouse to near insignificance. As manufacturers moved their operations to cheaper labor locations, and less craft-intensive production dominated the low end of the watch market, watchmaking practices disappeared from both Elgin and Waltham. U.S. watchmaking, never a significant producer of high-quality and highly priced watches, was overwhelmed by changes in technology—especially the advent of quartz.

Abandoning Industry Regulation: Instituting Industrial Change

The effect of the decline of the U.S. watch industry was felt around the world in the other dominant watch manufacturing regions, particularly in Switzerland. The U.S. industry's slow death precipitated numerous short-term, industry-led interventions, all made in an effort to save the industry. Although the most dramatic decline in the industry took place after World War II, nonetheless, retaliatory gestures designed to sustain a domestic industry forced other watch industries, including the Swiss, to respond. By the end of the Great Depression, Swiss firms were heavily dependent on the U.S. market for the majority of their sales. As a consequence, any decline in access to the U.S. market was viewed as a significant threat. Already, by the 1930s, U.S. firms were using national trade policy to their advantage as a way of limiting Swiss encroachment on the U.S. market. In addition, U.S. firms chose to constrain market access in other ways that were relatively novel. U.S. firms forced Swiss producers to change their competitive behavior by using nontariff barriers to entry such as regulating imports based on the number of jewels. The Swiss were also forced to pursue small, often erratic markets to fill in the gap created by U.S. market instability. Regulating on the basis of number of jewels set up later problems for the Swiss industry that will now be reviewed. Through forced product variation, Swiss firms found themselves overburdened with massive inventories and significant sunk costs, which laid claim to resources needed to confront the quartz revolution.

Starting in the mid-1920s, the Swiss industry was becoming increasingly dependent on the United States for markets. U.S. firms were clearly concerned about market encroachment, having just barely recovered from the effects of World War I. Partly in response to strategic moves by U.S. watch firms and partly in reaction to growing U.S. protectionism, the Swiss industry was forced to make a series of concessions and product design measures to counter changes in what had become its largest and therefore its most vulnerable market. Starting in 1936, the Swiss had reoriented a major share of production to satisfy the U.S. demand for watches with more jewels, an unnecessary and superficial product augmentation driven by U.S. firms' marketing strategies (54 percent of U.S. imports had 15 to 17 jewels in the period 1941–1945, compared with 3 percent in 1929 and slightly above 20 percent in 1931 and 1932 [Bolli, 1957; U.S. Senate Committee on Finance, 1951]). With this gesture, the Swiss industry became increasingly tied to the fate of the U.S. market, a position of dependence that would wreak havoc on the Swiss industry. Ironically, U.S. firms demanded more jewels as a means of recovering market share; then, when the Swiss followed suit, their actions were viewed as a hostile and aggressive act. That same year, the Swiss franc was devalued, further aggravating U.S. watch manufacturers, who were able to claim that Swiss production was now threatening U.S. production through unusually low wages. A series of legislative debates and customs office claims sent chills up the spine of the Swiss industry, as the United States threatened to embargo watch products sitting on New York's dock and to enact new product definitions in order to tax them upon entry into the United States (Bolli, 1957).

The reorientation of the U.S. watch industry towards armaments during World War II gave the Swiss an unexpected opportunity to increase their dominance over the domestic U.S. industry. The Swiss share of the U.S. market rose from 10 percent in 1941 to 32 percent in 1945 (Bolli, 1957). Between 1940 and 1945, Swiss exports to the U.S. increased from four to eight million watches and modules. By the end of 1945, 49 percent of the Swiss watch industry output was destined for the United States, a figure never seen before or since. A telling statistic is the fact that only 2.6 million completed watches were exported to the United States. The residual comprised watch movements imported by U.S. assembly firms for later manufacture.

The next 10 years were to be some of the most chaotic and tumultuous in the history of the U.S. industry. The U.S. industry, now in deep crisis after having effectively given up the domestic industry in trade for military defense contracts, called for a virtual closure of domestic markets to foreign imports in an attempt to buy time to revive the domestic industry. Tariffs and quotas were the preferred route, with the number of jewels in a watch serving as the basis for protection. In an effort to preserve what was left of a huge portion of the world's total market share for watches, the Swiss responded by manufacturing watches with every conceivable configuration of jewels. This led to the production of uneconomic watch series and only served to exacerbate an already serious problem of bloated inventories.

The level of turmoil unleashed by protectionist threats made by the U.S. industry was catastrophic. Although rumblings for a strong stance in support of the U.S. watch industry could be heard from the late 1930s on, after World War II the din became loud and clear. In 1945, the U.S. government responded to fears raised by watchmakers by convincing the Swiss to abide by a voluntary quota system restricting Swiss watches and parts imports to 1945 levels (Table 9.1; Bolli, 1957). There was no assurance that this level would be maintained. Congress indicated that trade negotiations could be reopened at any time if the U.S. industry failed to recover. The Swiss also were responsible for ensuring that illegal smuggling was stopped. In the past, stopping such smuggling had proved virtually impossible. The Swiss production system consisted of vertically integrated parts producers selling watch parts to literally thousands of watch assemblers. A carefully concealed packet of high-grade Swiss watches could be smuggled into the United States on a person's body. In a quota system that the Swiss rightly feared, small producers would only be driven toward more illegal trade.

Upon finally reaching an agreement to limit exports to 1945 levels, the Swiss were obliged to establish enforceable penalties for the illegal export of watches. The country also was "encouraged" (code for "required") to provide American watch firms with the necessary parts and tools to rebuild the domestic industry. Of course, this gesture was in direct violation of the Statut de l'Hologerie, passed into law in the 1930s. In these very important ways, the Swiss lost control over their markets and over the distribution of manufacturing technology for the production of watches. For the next three years, this reluctantly accepted policy regulated trade between Switzerland and the United States.

As the Swiss share of the U.S. market dwindled, smaller and more unpredictable outlets increased in importance (Table 9.2; Uttinger and Papera, 1965). Other markets in the Americas, particularly Brazil and Mexico, began to demand placement of assembly operations in their countries in order to retain market access in conformance with import substitution promotion policies. According to one observer, these rules were not applied fairly. The Japanese were allowed to distribute goods without comparable assembly restrictions. So onerous became individual country negotiations, all under the auspices of the General Agreement on Tariffs and Trade (GATT), that at one point the Swiss watch trade negotiator said, "It [access to markets] began to be driven by reciprocal trade arrangements where the number of watches allowed into a country was a function of the Swiss market being able to absorb some comparable [in value or symbolically] good such as 20,000 cases of Chilean wine" (R. Retornaz, vice president of the Fédération de l'Industrie Horlogère Suisse from 1945 to the early 1980s, personal communication, 1991). Just when the Swiss badly needed stable environments to counteract the unilateral actions of U.S. trade negotiators, everything seemed out of control.

The U.S. market was not the only one to close unexpectedly. Starting with the Communist takeover in China in 1948, the international markets of the Swiss were rocked by instability. Hong Kong, the gateway to Asia, had been by far the largest outlet for Swiss products sold in Asia. Since many of these products were resold in China, the contraction of the Chinese market proved to be devastating. Then, in 1958, the Indian government "prohibited the importation of watches into India in an effort to conserve already low foreign exchange reserves" (Learned et al., 1961, p. 230).

TABLE 9.1. Swiss Watch Exports to the United States

Year	Totals Value (Mio. Fr.)	% of total watch exports	Watches and movements Quantity (1,000 pcs.)	% of total exports	Value (Mio. Fr.)	Watches Quantity (1,000 pcs.)	Value (Mio. Fr.)	Movements Quantity (1,000 pcs.)	Value (Mio. Fr.)	Cases Quantity (1,000 pcs.)	Value (Mio. Fr.)	Parts Value (Mio. Fr.)	Misc. parts Value (Mio. Fr.)
1926	58,3	22,6	3,775	22,0	54,3	967	13,9	2,809	40,4	176	0,53	3,3	0,14
1927	60,3	22,1	4,118	22,3	56,3	1,387	17,2	2,731	39,1	181	0,49	3,4	0,13
1928	47,7	15,9	3,142	15,6	42,6	995	11,9	2,147	30,7	311	0,46	4,2	0,36
1929	64,9	21,1	4,599	22,2	56,9	1,891	21,4	2,708	35,5	428	0,54	7,0	0,45
1930	30,7	13,1	2,043	12,6	27,2	801	9,1	1,242	18,1	234	0,34	2,9	0,25
1931	13,2	9,2	757	6,6	11,4	154	3,0	603	8,4	11	0,05	1,6	0,13
1932	6,9	8,1	441	5,4	6,1	74	1,3	367	4,7	—	—	0,9	0,03
1933	7,4	7,7	483	4,6	6,1	40	0,7	444	5,4	—	—	1,2	0,02
1934	12,6	11,6	1,003	8,0	10,2	68	1,0	936	9,2	9	0,04	2,4	—
1935	17,2	13,8	1,378	9,1	12,8	75	1,4	1,303	11,4	—	—	4,3	0,02
1936	28,1	18,5	2,455	13,8	22,7	188	2,6	2,267	20,1	7	0,05	5,3	0,03
1937	50,1	20,9	3,463	14,5	40,8	231	4,5	3,232	36,3	7	0,12	9,2	0,04
1938	38,8	16,1	2,525	11,6	32,0	189	4,1	2,336	27,9	9	0,09	6,7	0,04
1939	45,3	23,2	2,895	17,2	37,9	223	5,0	2,671	32,9	11	0,16	7,2	0,05
1940	71,1	33,2	4,188	27,5	57,2	288	7,2	3,900	50,1	11	0,16	13,6	0,16

1941	58,0	25,1	3,471	24,0	51,3	342	8,1	3,130	43,2	8	0,05	5,9	0,75
1942	75,2	26,4	3,950	28,3	73,2	699	19,7	3,252	53,5	44	0,27	1,4	0,38
1943	113,9	33,7	4,747	32,7	111,6	1,145	44,3	3,602	67,3	52	0,55	0,5	1,21
1944	101,0	33,3	3,829	32,2	96,9	794	34,7	3,035	62,2	28	0,44	2,8	0,83
1945	241,9	49,1	8,369	44,5	223,6	2,665	104,9	5,704	118,6	4	0,08	13,7	4,56
1946	263,6	43,6	7,980	38,6	241,6	2,171	109,4	5,809	132,2	16	0,23	17,7	4,01
1947	260,9	33,9	7,555	31,5	227,8	1,267	77,3	6,279	150,6	26	0,36	28,6	4,12
1948	267,1	35,9	8,179	33,6	228,7	1,496	68,3	6,682	160,4	21	0,34	33,5	4,48
1949	229,2	32,6	7,505	31,9	194,4	1,728	60,0	5,777	134,4	29	0,28	29,8	4,70
1950	256,2	35,1	8,939	36,9	229,1	3,218	92,1	5,721	137,0	93	0,61	21,9	4,53
1951	317,8	31,5	11,477	34,2	287,2	4,220	111,8	7,257	175,4	131	0,95	24,0	5,68
1952	357,2	33,0	12,487	37,5	325,3	4,400	114,5	8,086	210,8	160	1,03	25,2	5,62
1953	402,9	36,4	13,517	40,9	372,9	3,994	116,7	9,523	256,2	204	1,03	23,1	5,80
1954	299,9	28,8	10,191	32,8	278,8	3,483	93,1	6,708	185,7	116	0,54	16,6	3,94
1955	298,2	27,7	10,933	32,4	273,6	4,180	96,6	6,753	177,0	238	1,18	18,8	4,63
1926/28	55,4	20,0	3,678	19,8	51,1	1,116	14,4	2,562	36,7	223	0,49	3,6	0,21
1931/35	11,5	10,3	812	7,0	9,3	82	1,5	731	7,8	4	0,02	2,1	0,04
1936/40	46,7	22,4	3,105	16,2	38,1	224	4,7	2,881	33,4	9	0,12	8,4	0,06
1941/45	118,0	35,8	4,873	33,1	111,1	1,129	42,3	3,745	69,0	27	0,28	4,9	1,55
1946/50	255,4	36,0	8,032	34,4	224,3	1,978	81,4	6,054	142,9	37	0,36	26,3	4,37
1951/55	335,2	31,5	11,721	35,6	307,6	4,055	106,5	7,665	201,0	170	0,95	21,5	5,10

Source: Statistiques du commerce extérieur de la Suisse; Annuaire statistique de la Suisse.

TABLE 9.2. Ten Principal Customer Nations of the Swiss Watch Industry, 1950, 1955, and 1960 (Million of Swiss Francs)

Country	1950 francs	Country	1955 francs	Country	1960 francs
U.S.	256	U.S.	298	U.S.	272
Hong Kong	41	Italy	63	Germany	81
Italy	40	Hong Kong	58	Hong Kong	77
Canada	36	Spain	47	Italy	74
Brazil	32	Canada	46	Canada	60
Germany	28	Germany	44	England	51
England	24	England	35	Singapore	39
Tangiers	24	Singapore	35	Brazil	38
France	19	Tangiers	30	Mexico	32
Spain	17	Sweden	30	Spain	31

Source: Uttinger and Papera (1965).

Once the recipient of almost 50 percent of Swiss watch exports, the United States's share of Swiss watch exports declined to 28 percent, or 3.6 million units by the mid-1950s. Of those, two-thirds were watch movements to be assembled into final products by U.S. watch firms. As America's domestic watch manufacturers cut back their production, only Timex and Bulova continued to present problems for the larger Swiss watch producers. Great damage had already been inflicted on the Swiss industry, however. The buildup to supply the United States provided a false sense of security for many small Swiss firms that had few reserves to cope with unforeseen and rapid change in market conditions (Uttinger and Papera, 1965).

In the early 1960s, foreign competition ended the Swiss monopoly on mechanical watch production and the country's quasi-monopoly on the world watch industry. The slow erosion of the Swiss share of world exports met with demands from industry members to change the laws that had regulated the industry for 30 years.

Reasons for industry discontent were numerous. The more profitable and better run firms (those that produced medium- and high-priced products: Tissot, Rado, Omega, Eterna, Longines, Certina, Rolex, Patek Philippe) lobbied against the cartel, arguing that it protected firms producing low-quality watches (L. Tissot, personal communication, 1990; Landes, 1984; Uttinger and Papera, 1965). They criticized current laws for fixing the level of Swiss production at a time when other countries were making substantial inroads into the Swiss share of world export markets. By 1957, major assemblers banded together and challenged the fundamental fabric of the cartel (Learned et al., 1961). Together with the dissenting assemblers, the Fédération de l'Industrie Horlogère Suisse (FH) studied the problems and recommended that the Statut de l'Horlogerie be significantly changed. In 1961, the Federal Assembly of the Swiss Confederation ratified a new decree eliminating the regulation of output and encouraging rationalization of the industry. While calls to revise the original legislation were almost unanimous, there was still a widely shared belief in the need for some type of industry protection. Thus, the new law did not entirely eliminate industry-wide regulation. Parts prices were still determined on an industry-wide basis, but incorporated flexibility for individual circumstances. Once the dike had been breached, however, remaining attempts to regulate the industry through industry legislation would prove ineffective. While the new law took effect in 1962, it was not until the early 1970s that restrictions on watch manufacturing were entirely eliminated.

Although the law regulating the industry structure was changed, labor law was not revised, removing an important and badly needed check on industry behavior. Back in the 1930s, as part of the Statut de l'Horlogerie, in return for job security the labor unions agreed not to strike. As the industry fell into chaos, the unions were bound by the old contract to resist attempts at labor actions. The union leadership, having been brought up on the notion of labor peace, thwarted worker discontent and resistance, both of which intensified as industry fortunes evaporated. During the period of greatest instability, home work once again became an important component of the industry's operations. As one informant noted, "Part of maintaining the industry output was to continually reduce prices in the face of growing competition. My husband was a foreman. As things got worse more work was sent home with women. Because no one knew how to precisely estimate the cost of the product, prices continually fell. Women's wages continued to drop as output was increased. It was a bad time for everyone" (C. Veya, personal communication, 1991; Table 9.3).

The problem of pricing was crucial. The Swiss watch industry, even after having undergone some degree of consolidation, still churned out thousands of watch models each year with little true understanding of real production costs. For medium- and high-priced watches, parts had to be produced for repairs. As parts prices were effectively subsidized by the more efficient producers, inventories were extraordinary in scale. Because of the fragmented production system, a major segment of the industry never developed cost-conscious record-keeping standards. Production volumes were based on past practice rather than on long-range projections. Very few members of the industry really knew whether a watch cost a dollar or a 10 dollars to produce. Thus, when the crisis came and firms were forced to write down inventories to prices below actual

TABLE 9.3. Number of Firms and Workers in the Swiss Watch Industry, 1960–1980

Year	Number of firms	Managerial employees	Industrial workers	Home workers[a]	Total employees
1950	1,863	10,052	43,119	7,068	60,239
1955	2,316	11,847	50,312	7,867	70,026
1960	2,167	13,510	51,617	9,089	74,216
1965	1,927	14,588	58,012	11,322	83,922
1970	1,618	17,307	58,738	13,403	89,448
1975	1,169	16,744	39,210	6,613	62,567
1976	1,083	15,314	34,677	5,191	55,182
1977	1,021	14,950	34,872	5,003	54,825
1978	979	14,692	33,613	4,364	52,669
1979	867	13,685	29,911	3,120	46,716
1980	861	14,155	30,018	2,825	46,998

Source: Fédération de l'Industrie Horligere Suisse (Federation of Swiss Watchmakers).
[a] Home workers were self-employed. They were paid on a piece-rate basis for the crafting of a watch component or the assembling of a watch, under a contract with one of the larger Swiss watch firms.

cost, the number of firms forced from the industry was unprecedented (Hoff, 1985; Table 9.3).

As a consequence of all the turmoil, the watch industry underwent a series of unprecedented mergers. The healthier and larger establishments joined forces to match the size of their Far Eastern and American rivals. Within two years three firms were producing 32 percent of Swiss exports. The leading vertically integrated manufacturer, Société Suisse pour l'Industrie Horlogerie (SSIH; formed in the 1930s through the merger of Tissot and Omega), became the third largest watch manufacturer in the world (behind Timex and Seiko). In 1971, the Société General de l'Horlogerie Suisse SA (ASUAG) expanded beyond component manufacturing by creating the General Watch Company, a holding company for several brand name watch and component manufacturers (Knickerbocker, 1976). A third holding company, Société des Garde-Temps (SGT), was created primarily to manufacture low-priced and electronic watches.[4]

In addition to the three holding companies, there were a number of other important groups. Rolex, although privately held, had 1972 sales estimated at 200 million Swiss francs (almost one-fourth of Swiss exports by value) (Knickerbocker, 1974). There also were four middle-sized groups, including two subsidiaries of U.S. companies, Zenith and Bulova, and the prestigious brand names, including Piaget, Patek Philippe, and others. The remainder of the industry was made up of hundreds of small companies assembling and selling cheap watches primarily for retail channels.

Swiss dominance of the low end of the worldwide market came to a screeching halt with the appearance in the late 1960s of quartz technology. Quartz gave craftsmen little to do; in Landes's (1984) terms, science had replaced art. The Swiss might very well have jumped onto the quartz bandwagon; the technology, capital, and connections were all there. But they balked. Early quartz models required users to push a button to see the time; they cost much more than a good mechanical watch; and they sometimes proved less accurate than mechanical watches. Quartz watches also were initially hard to manufacture and had serious problems with their batteries. At the time the Swiss industry lacked risk-takers, acting much like British watchmakers in their stubborn refusal to change. In the critical years of the early 1970s, the Swiss failed to recognize either the growth potential or the true threat of quartz. Neither the Japanese nor new entrants to the American industry were so short-sighted, and so the Swiss were badly hurt.

Signs that the Swiss industry had become too rigid began to appear as early as the late 1950s. A Swiss watchmaker, Max Hetzel, developed a tuning-fork regulator that would control the U.S. Bulova Accutron watch. Hetzel offered the device to Swiss watchmakers, but he was turned away. When the Swiss finally recognized the device's merits, they had to *buy* the right to use it (Landes, 1984, p. 50). Stung by the experience, Swiss watchmakers set out to make more accurate timekeepers. In 1967, they developed several quartz chronometers that doubled the accuracy of even the most accurate mechanical timepieces.

Hence the Swiss dominated the watch market for a long time by making an inexpensive product that told time accurately, and by finding and building markets. However, when the product changed suddenly, Swiss industry either could not or did not change rapidly enough. Even if it had, there would have been little need in the new firms for the highly skilled craftsmen who formed the heart and soul of the Swiss industry.

Summary

Up until the mid-20th century, watch manufacturing and competition within the watch industry took place between two competitors with two very different but overlapping production systems. Neither proved invulnerable to the remarkable events that characterized the first half of the 20th century. The constant vying for market share and market supremacy between U.S. and Swiss manufacturers must be understood in the context of wars, periods of protectionism brought on by deflation, the addition of new competitors, and the effects of incomplete institutional transformation. Was one model of industrial organization superior?

Retaining a foothold in a highly competitive international industry proved to be difficult for most watch-producing countries. When established positions are continually contested, false moves often lead to lost leadership and drops in market share. Why did different countries' industries stumble? While there is little need to recount the complex sets of factors leading to the failure of the British industry, the discussion presented in this chapter suggests that an institutional context that is marked by flexibility at one point in time may promote inflexibility at another point in time. As new competitors enter the market, often unencumbered by sunk investments and rigid cultures and institutions, turmoil is unleashed in an industry, usually to the detriment of former leaders. Bad decisions are only recognized in hindsight. Mistakes made by leaders can often seem trivial if taken out of context (e.g., changes in distribution channels, the failure to adopt a new technique). This is because any change can be magnified by unexpected events that occur in the external environment and that cannot be anticipated or accounted for in advance. These unexpected changes often reduce the degrees of freedom firms face as they attempt to navigate periods of rapid change.

A great deal of recent academic research has focused on the benefits that accrue to firms situated in context, such as a city region, in which endogenous technical change arises in tandem with industry localization. While there are voluminous treatments of the effects of exogenous technical change, there are surprisingly few recorded examples of how exogenous technical change challenges the basis of a localized industrial complex. In the case of the world watch industry, the evidence suggests that exogenous technical change can have profound effects on an industrial complex that enjoys very high levels of external economies through localization. Exogenous technical change exerts an unusually powerful effect on highly integrated complexes, particularly those in which past rounds of technical change have been based on internally derived information. As the next chapter suggests, norms of behavior that manifest in organizations and are conventionally thought to help localized complexes remain supple and therefore capable of change, are in fact creations of the local context itself and often have only limited capability, even if aware of the threat of a pending exogenous technical change, to transform the system.

10 Going Electronic, Moving to Hong Kong

Two major developments in the world watch industry began in the 1960s and reached maturity by the end of the 1970s: the commercialization of quartz watch technology and the emergence of Hong Kong as the world's most formidable manufacturer of low-priced watches. These two developments are linked in time and space. All major producers of watches, including U.S., Japanese, and Swiss firms, were vying to bring the first electronic watch to market. All knew this would be the next frontier. Each faced many of the same technological problems, but with very different competencies to resolve them. U.S. and Japanese firms had acquired considerable know-how in the field of early microelectronics during World War II. As a neutral, Switzerland lacked a domestic armaments industry to drive the development of derivative technologies such as microelectronics. Switzerland also lacked an extensive system to draw upon for needed complementary technologies, such as liquid crystal and light (LED) emitting diode displays.

The link between the two seemingly unrelated events—quartz technology and the rise of Hong Kong—was at the heart of the electronics revolution itself: the redefinition of the meaning of skill and the variable value of labor and its cost in modern manufacturing. For the Swiss, skill meant everything. Their existence and the identity of their industrial culture were completely bound up in the creation of precision timepieces. This competency required mechanical skill. Other relevant watch-producing regions were not burdened by this reality. For firms in Japan, and to a lesser extent the United States, skill was an enabling influence but not an end in itself. Thus, the electronics age simply had a more profound impact on the Swiss system of production.

The dawn of the electronics age occurred coincidentally with wage inflation in both the United States and Japan. Hence, rising labor costs would have fueled the search for cheap labor locations, even had there been no link to electronics. But there were direct links to electronics: during the initial development of electronics technologies, product life cycles were shortened as new technical capabilities evolved rapidly. High development and production costs and the instability of early products encouraged hand assembly rather than automation, though automation arrived eventually (Dicken, 1998). The age was characterized by firms' pursuit of strategies that matched specific pools of labor with the specific needs of manufacturing selected products. Watch producers followed this trend. However, not all watch firms enjoyed the same spatial flexibility. For Swiss firms, the creation of a spatial division of labor had long been off-limits, due to regulations promulgated in the 1930–1950 period that severely limited their ability to pursue such a strategy.

The emergence of Hong Kong as a major site of watch production did not occur suddenly or unexpectedly. Indeed, the country's link with the industry was part of larger worldwide economic trends. By the early 1960s, Hong Kong had become a cheap labor haven for all assembly-intensive industries. Foreign firms in the West seeking a low-cost production location often turned to Hong Kong. The British colony was not totally dependent upon the foreign import of technology, however. Since the closing of China in 1949, Hong Kong had become the center of the Asian textile and apparel industry. Entrepreneurs from China, and especially Shanghai, began transferring capital and textile production technology out of China when it became clear that the communists were going to outlaw private capital ventures. China's indigenous watch industry had been concentrated in Shanghai and Guangzhou. Watch firms there presumably felt the same need to relocate to Hong Kong as textile-based firms.

In the first part of this chapter I focus on the most important development within the watch industry during the last quarter of the 20th century: the development of quartz technology. This revolutionary product technology—solid-state watches with no moving parts—differed in almost every respect from conventional mechanical-movement watches. Quartz watches could be fabricated without the expensive labor required to assemble and test mechanical movements. Moreover, quartz watches were more accurate than mechanical watches. Given the development of quartz technology in the 1960s, the world watch industry would never be the same again. Ascertaining how and why first Japan and ultimately Hong Kong came to dominate this new product, even as the primary innovators largely remained U.S. and Swiss firms, is a core element of this chapter.

After recounting the development of the quartz watch, I explore the emergence of the Hong Kong watch industry. Unlike any of the previous watch production systems I have detailed, the Hong Kong industry chose a model of industrial organization that relied heavily on imported parts and labor-intensive assembly. Very little capital investment was required and only minimum human capital needs emerged. In Hong Kong, shortages of land and an ample and cheap labor supply yielded a production system organized around small workshops. The organization of watch manufacturing simply followed the model of the larger system of relatively high-precision hand assembly that characterized Hong Kong manufacturing. As it did in so many other industries, Hong Kong became the world watch industry "assembly shop." This dependence on a low-wage model of production was inextricably linked to the fates of both the Japanese and the Swiss industries. For both countries, Hong Kong represented the assembly node in the global spatial division of labor for watches.

The emergence of a low-wage competitor at the low end of the price range is reminiscent of previous eras of competition, for example, the competition between the British and the Swiss in the 1800s. But there are fundamental differences. To be an effective competitor in the 1800s required an indigenous competency to produce skill-intensive parts for labor-intensive manufacture. In the case of Hong Kong, at the end of the 20th century, being a global leader requires little more than maintaining a ready supply of low-wage labor, as critical parts can be sourced from around the world. The links between global competitors in this instance now take the form of a dependent supply chain in which movement producers such as the Japanese and the Swiss must sell parts to the Hong Kong industry in order to achieve low costs that can be passed on to their middle- and upper-sector products. Autonomy has diminished across the spectrum of firms and long-term stability has probably faded away forever.

The Challenge of Quartz: Now Anyone Could Compete

The history of the watch industry had been governed by the search for precision and the ability to manufacture a high-quality and beautiful product. With the advent of quartz technology, the romance of the watch evaporated. Now virtually anyone could make watches and find a market for them in the world economy. All the established players in the industry rose to the occasion and tried to compete in this new market. In the final analysis, however, only one country's industry would emerge as the winner, and its success would turn on strategic decision making and gut instincts.

Over the course of 20 years, several precursor technologies to quartz were developed. In 1952, Elgin produced the first electric watch that replaced a watch's usual mainspring power source with a battery (Johnstone, 1999). Five years later, Hamilton introduced a second electric watch, based on a mechanical watch design but again powered by a small battery. These early products all suffered from the same problem: a lack of sources for power supply parts. The third intermediate technology, the tuning-fork watch, did away completely with the balance-wheel regulatory mechanism. Hetzel's desire to extend the length of the superprofit stage of his new technology prevented his tuning-fork invention from becoming the basis for a whole new generation of watches. Other factors also impeded its ascendance. First, the product was finicky and difficult to keep running accurately. More importantly, because Hetzel chose to retain control of the technology through licensing agreements, it remained off limits to other companies. Other watch producers took the clue about the need to move into the electronics era, and set about searching for another regulating system—one in the public domain. They found it in a quartz crystal—the basic building block of sand.

The Japanese were the first major players to move commercially away from the confines of the mechanical watch toward the potential of quartz technology. Their foray into electronic watches began with Seiko's construction of a quartz clock devised to keep time at the 1964 Summer Olympic games. It took another five years of focused effort to bring a quartz watch to market. The first watch, Seiko's "Astron," was introduced in 1969. This first electronic watch was a serious blow to both the U.S. and the Swiss industries. The pressure was building as Japan made a play to become a leader in the world watch industry.

Digital Delay

The dilemma facing the Swiss industry was not only that a new and revolutionary technology had been developed halfway around the world. The Swiss lacked the technical competence to commercialize the new quartz technology. Also, more than one option was open regarding the new product's configuration. On the surface, quartz was just another time regulator, but how such timekeeping would be displayed was another matter. The development of quartz was quickly followed by the innovation of digital display technology that presented time like a calculator, first using light-emitting diodes (LEDs) and then liquid crystal display (LCD) technology. Thus the problem was not just with the device that regulated time, but also with how the time ultimately was displayed.

This new innovation presented a peculiar problem to the U.S., Japanese, and Swiss watch industries. Both the time-regulating device and the external display technology were different. There were also a number of questions that needed answering. How big would the quartz market be? What price point would the product settle at? How would the introduction of the new technology affect the structure of competition? The answers to these questions were quickly forthcoming.

The introduction of quartz watch technology fundamentally changed the nature of timekeeping. But the problem was more than just a new technology; it meant new competitors, new products, and new distribution channels, none of which the Swiss industry was prepared to handle.

Semiconductor Firms Search for Markets

On the other side of the ocean, U.S. semiconductor firms were prowling product markets in search of niches in which to apply their newfound technologies. This search was unexpected by both the Swiss and the Japanese watch industries. Having just witnessed the tremendous business first gained and then lost in developing the calculator industry, firms in the semiconductor industry were quick to jump to watches. Parts could be mass-produced and were easy to assemble. Semiconductor firms also could appreciate the ease of using the same labor pool to assemble watches and to assemble semiconductor chips into cases. As one giddy editorial in *Business Week* exclaimed in 1975, the U.S. semiconductor industry was writing another chapter in its book on how to win in world markets. This time the product was the solid-state digital watch, which was taking over the market from the conventional mechanical timepiece. Without the hundred or more small parts that made the mechanical watch a labor-intensive product, the digital watch could be assembled in minutes at low costs. As a result, the newfound U.S. electronic watch industry took back ground the mechanical watch industry had gradually surrendered to Swiss watch factories (*Business Week*, 1975, p. 128).

Were watches like calculators? By the mid-1970s, more than 60 U.S. firms were competing in the digital watch market. Owing to their novelty, the market for digitals grew rapidly during the early 1970s. Why were chipmakers moving into watches? The answer was simple: factory capacity for chips in the early 1970s exceeded the demand for components. While the United States was a pioneer in the calculator industry, it lost the industry to Japan just four short years after introducing the electronic calculator. Neither the Swiss nor the Japanese were prepared for the entry of U.S. semiconductor firms into watch manufacturing. Before long, however, complete bedlam ensued and the market for digitals collapsed.

Why was the U.S. presence in the industry so devastating to conventional watch manufacturers? Though initially they both shared the problem of being a new entrant in an already unstable market, now Japan and Switzerland faced different problems and have separate stories.

The U.S. Strategy for Digitals

The first players in the digital watch market were small firms exploiting the growing availability of relatively cheap chips. Some of the companies that sold digital watches also made calculators, while others were former mechanical watch manufacturers trying to retain market share. Component producers such as Texas Instruments and National Semiconductor quickly overtook these first movers.

In 1975, Texas Instruments, the world's largest producer of semiconductors, began shipping watches in order to become "the world's largest consumer electronics firm" (*Business Week*, 1975, p. 80). Having briefly dominated the calculator market only to stumble and ultimately lose out to the Japanese, Texas Instruments was determined to dominate watches. It had set up an automated production line to assemble watch modules.

Texas Instruments was a relative latecomer to the watch industry. National Semiconductor had entered the market the year before with a two-pronged strategy: take the high-end jewelry segment to department stores and the low-end novelty-watch segment to mass merchandisers. It hoped to dominate digital timekeeping by "buying" watch industry professionals.

Other semiconductor entrants included Intel, Hughes Aircraft, and Fairchild—all chip producers searching for market outlets for their primary product. All but Hughes ventured into actual watch manufacturing. Hughes resisted this step, wisely predicting that acquiring the necessary experience to sell into consumer markets could prove too costly and risky.

Other chip producers lacked such humility; this would ultimately be their downfall. Intel, for example, wishing to make a major, rapid foray into the market, acquired existing watchmakers, but then struggled to bring a reliable product to market. Such a gesture was costly. Fairchild went even further in acquiring product capacity by buying component producers, including display makers, case makers, and watchmakers. All these efforts were high risk relative to potential rewards, especially given the rapid decline in product prices and rising competition, but their lack of experience in commercial markets blinded electronic component producers to the problems of consumer marketing.

When both the Swiss and the Japanese introduced their first digital watches in the early 1970s, the going price for such a watch ranged from $1,500 to $2,000. Leading companies in both countries initially priced the watch high because of their respective philosophies, which equated "new" with high technology and high price. Within five years, a digital watch with an LED display cost just $50. This dramatic decline in price drove the structure of the market and ultimately wreaked havoc on everyone, including the electronics firms.

Technical dilemmas were another source of trouble plaguing all players in the digital watch industry. The first watches were heavy, unsightly, and inaccurate. These quality problems pushed the Swiss to the sidelines. Initially, traditional watchmakers could not see much future in digitals. But the Swiss, more so than the Japanese, were vulnerable to waiting, a problem they had recognized with the introduction of the quartz.

Going digital required technology that the Swiss neither owned nor had a reasonable opportunity of creating on their own. Fearing the worst, the Swiss plunged into an expensive arrangement with Hughes to produce digital movements. It quickly became apparent that competing in digitals required more than control over the production of movements. Here, the Japanese were better prepared than the Swiss. Already a pioneer in electronic gadgets, the Japanese could pursue a cautious, calculated move into digitals. In other words, unlike other competitors, Japanese watch companies were already vertically integrated and poised to produce digitals with limited additional capital outlay. The same conditions did not apply to the Swiss.

Component Producers Drive the Price of Digitals through the Floor

The ease with which electronics companies could acquire watchmaking capacity intensified competition within the industry. From 1974 to 1975, the price of a digital watch dropped from $125 to $50. The next big step downward was forecast for 1977: some industry watchers believed that prices would fall to $20. Then Texas Instruments stunned the market by introducing a plastic-encased digital watch for $9.95! The actions of Texas Instruments were quickly followed by other semiconductor makers who hoped to destabilize and thwart further actions by watch assemblers who could not buy components cheaply enough to compete at the $20 level.

No one doubted Texas Instruments' intent. Dropping prices so low would discourage new entrants and run weaker competitors out of business. But Texas Instruments also had its eye on the market niche dominated by Timex. The largest watch company in the United States was caught flat-footed when Texas Instruments slammed through the $20 barrier, once Timex's exclusive domain.

As prices fell, forecasts concerning the eventual size of the digital market grew wildly. In 1975, 3.5 million units were sold. By the middle of

1976, forecasts had already been adjusted upward twice to 13 million. U.S. movement and watch producers had a combined capacity of 36 million, perhaps the single largest reason for such steep price declines. U.S. producers, flush with capacity, were forecasting a 50-million-unit market, an increase that would come at the expense of the Swiss and the Japanese.

Which Way to Watch a Watch?

Amid massive increases in capacity, the digital industry confronted a critical turning point precipitated by options in display technologies. While the majority of watches through 1976 were produced using a light-emitting diode (LED) display that required the watch owner to push a button to tell time, an alternative technology, liquid crystal display (LCD), was waiting in the background, though still hampered by a history of mechanical failures.

Many of the biggest semiconductor watch-producing companies bet on LED, primarily because this display method was easily fabricated using existing semiconductor technology. Ultimately, LED technology proved to be inferior to LCD technology. In 1972, however, it was impossible to tell which technology would come out on top. LCDs drained battery power, which reduced the life of the watch. Further, the production technology for LCDs was more expensive than that for the LED. Because LED was the technology chosen by the big U.S. companies, such as Texas Instruments, its initial position was ensured as the display technology for American digitals—that is, until the technical problems with LCD were overcome and LCD price fell to levels comparable with the LED.

Thus, it was only a matter of time before LCD, a more visually appealing, less energy-consumptive, and more convenient technology pushed forward into the digital market. Few big companies saw this coming; sunk costs made it difficult for them to respond to LCD. For example, the full commitment by Texas Instruments to LED caused it to build a production line to manufacture displays for both its watch and calculator markets. Whereas in 1975 the ratio of LEDs to LCDs sold was 2 to 1, by 1977 the LCD market share was expected to equal or exceed that for the LED market.

At the deciding moment in choosing between the two technologies, U.S. firms chose LEDs, while Japanese firms chose LCDs. In the United States, LEDs were used in more applications and were more widely available. Thus capacity existed since many competitors were producing marketable products. Since U.S. watch producers were expecting the market to deliver the right product, there was no consideration of the need to look down the road and consider the long-term viability and superiority of LCDs. Japanese firms, on the other hand, steeped in the tradition of developing in-house capacity by seeking the best technology and cultivating it through cross-subsidization, chose to make the initial risky investment in LCDs, perceiving rightly that in the long run LCDs would dominate the industry. They bet on the long-term victory of LCDs, and their bet paid off. Two years after the deciding moment the industry tables were turned and the Japanese leapt ahead, leaving U.S. firms far behind.

The ascent of LCD technology dealt a cruel blow to U.S. manufacturers, but was ultimately just one of many problems beginning to plague semiconductor producers. Few of the U.S. chipmakers had ever really understood marketing. As each company sought to increase volume to bring down price, it forgot to factor in customer preference. Companies like National Semiconductor spent too little on product quality, and their reputations suffered. As an unnamed former National Semiconductor employee commented, "The company never developed an understanding of the consumer market. It saw watches as it saw calculators, simply as packages for selling components" (*Business Week*, 1977a, p. 78) National was not the only company plagued with this problem, one that signified the arrogance and wrongheadedness of the industry as a whole.

As prices fell, chipmakers were forced to go after a mass market. In a head-to-head competition with Hong Kong, they simply could not win. That market turned out to be promotional merchandise, which cheapened the image of digitals further (*Business Week*, 1977a, p. 78).

Digital displays proved to be ephemeral; projections for huge market share never materialized. But the chaos unleashed by the imagined threat sent the Swiss industry in particular into a tailspin. The introduction of the quartz movement was far more destabilizing for the Swiss than for the Japanese or the Americans. While the three countries' industries had heretofore been built around precision machine technology, the Japanese proved adept at taking advantage of the new capability and responding to the new competitive challenge. U.S. firms had the technology, but were in disarray and had already yielded too much market share to put up a fight. The Swiss had to convince too many industry participants of the seriousness of the threat. They lacked the unity of purpose to confront this large challenge. The Swiss also lacked the basic manufacturing capability to produce quartz watches in large numbers, not to mention the other components of the product, such as digital displays. The Swiss were seriously shaken when digital display technology began to make inroads in the medium and then the low market segments.

Meanwhile, the Japanese and in particular Seiko took the new competition in stride, for they forecast (correctly!) that digital display technology would never account for more than a small segment of the industry. Seiko took a risk and bet on quartz analog technology. This bet hit the jackpot, for analogs came to dominate the market. Asked why Seiko bet on analogs, one company official said, "The company figured the world had been reading time by analog for hundreds of years. It was unlikely people would change that much" (S. Aota, personal communication, Seiko Horological Institute, Tokyo, Japan, July 19, 1994).

Success Goes First to the Large and Then to the Nimble

As the market began its inevitable shakeout, vertical integration, once pursued because the increasing power and complexity of integrated circuits had drastically reduced the value that could be added by assemblers, began to reverse. Now, even complex chips were worth so little and were produced at such high volume that massive new markets were needed to keep chip plants at capacity.

Halfway around the world, a new production location was emerging. Hong Kong, long a low-cost case and bracelet producer and supplier for the Swiss and Japanese industries, was quickly becoming the assembly location of choice for digital watch companies. Hong Kong began to assemble digitals in matchbox–sized factories. This situation was fine as long as Hong Kong firms emphasized contract assembly. But when Hong Kong producers began to produce other OEM products at the very low end, they flooded the world market with millions of cheap watches. Given that digital technology was completely self-contained, there were no effective barriers to prevent Hong Kong firms from entry. And, since Hong Kong already had case and bracelet capacity, not to mention plastics extrusion and assembly capacity associated with its toy industry, it was an easy next step for the enclave's watchmakers to pursue their own markets.

The first casualties in "the digital wars" were the very companies that had introduced the new innovation. For example, the original American digital watch, "Pulsar," disappeared. It had been the Hamilton Watch Company's last-ditch effort to remain a viable producer of high-technology products. The first U.S. firm to introduce a widely distributed digital watch, in the early 1970s, Hamilton marketed the product as a luxury, never anticipating that the price of a digital watch would plummet from $2,000 to $10 in a matter of five years. But U.S. watchmakers were not the only firms to find rampant price competition deadly.

The U.S. chip producers became their own worst enemies. Price cutting over five years reduced the cost of a digital watch to $10. At that rate, no one, no matter how automated, could make a profit by selling below Hong Kong watch assemblers. Three companies remained in the ring: Timex, Fairchild, and Texas Instruments.

The dramatic decline of firms in the industry began as Texas Instruments used predatory pricing to stabilize and dominate the market. Timex, long the sole player at the less-than-$25 end of the market, was trying to get the technology right. Were the remaining contenders strong competitors, or simply the last to go down with the sinking ship?

The real demolition of the cheap digital industry occurred with the advent of the plastic watch. Ironically, when watch producers dropped the price of a digital watch below $20 by encasing it in plastic, consumers rebelled. Consumers preferred a metal case, and refused to buy plastic. As soon as the price of a metal-cased watch fell to $20, it drove the plastic watch out of the market entirely (*Business Week*, 1977b, p. 78). It quickly became apparent to all participants that less and less money was available in the watch industry. Even as the number of digital watches sold increased, sales revenues dropped off significantly. Growth began to level off; in some segments, it was already in steady decline.

The drop in prices wasn't just a problem for U.S. firms. The Swiss, still dallying on the sidelines of the electronics revolution, choosing to position the new precision device at the medium and high segments of the market, continued to send mixed signals about a possible presence in the low-end digital watch market. Money was being spent in preparation for such a position, but no real effort was exerted. In the meantime, as a digital movement declined in cost, it approached and then fell below the lowest cost mechanical movement. This was the point of the Swiss industry's greatest vulnerability.

Two Years Later, Not a Chip Maker to Be Found

By 1982, in one of the most rapid and dramatic retreats from a once wildly expanding product market (Williams, 1982), virtually no American semiconductor firms were left in the digital watch industry. Texas Instruments retreated when LCD technology firmly overtook LED technology, leaving it with warehouses full of finished products customers did not want. This was particularly ironic given that Texas Instruments held the basic patent for LCD technology, but had committed earlier to LED technology by sinking millions into a production line for LED displays. Fairchild began losing money in watches in 1977 and retreated from the cutthroat industry when its more important microprocessor markets began to falter. Perhaps the most telling sign that U.S. chip producers, long the technology leaders, had lost their way was their failure to move beyond digital movements with one or two basic functions. The public's interest in digital watches changed as the price came down. Now they wanted more bells and whistles—watches that could take a pulse and read out the temperature. U.S. digital watch firms were too slow to heed the public's desire for gadgetry. At the end of the 1970s no American firms were making digital watches.

In 1980, a *Business Week* story summed up the U.S. problem: "Companies must have technology but also marketing capabilities. U.S. watchmakers either had one or the other but not both. The chip producers had technology but didn't understand the customer; Timex had the marketing acumen but lacked the technology. Neither technology nor selling can do the job by itself" (*Business Week*, 1980, p. 168). The loss of the digital market to Hong Kong pointed to one irrefutable fact: areas such as Hong Kong and Taiwan benefitted from the combination of a wage differential and extraordinary flexibility. The United States could not overcome this advantage with technology. Additionally, the Japanese were able to put technology and marketing on an equal footing and do both successfully. They bet on the long run and accurately predicted the ascendancy of LCD technology. When the market turned, they propelled it forward. Having taken this calculated risk, they eventually emerged the winner.

Timex was down, but not out of the race. Throughout the 1980s, the company struggled to regain its once secure position as producer of one of every two watches worn in the United States. It finally succeeded through careful niche marketing, total offshore production, and a major emphasis on technology. The firm had long ago become accustomed to coming from behind and inventing, sometimes reinventing, itself when necessary. Unencumbered by an elitist tradition and perversely enjoying coming from behind, Timex succeeded in retaining a position in the U.S. market and eventually secured a share of the world market as well.

The emergence of the new technology presented every player in the industry with an array of complex choices. Strategic response options on the part of firms competing in the industry depended on whether the competitor was a technological leader or a follower. The ability to act was also influenced by whether a firm relied upon markets versus vertical relations to allocate resources for the development of new products. Success ultimately turned in part on whether a firm maintained a short-term versus a long-term investment payback time horizon and thus was willing to take risks. The ability to act deliberately was also tied to whether a firm could take advantage of technological complementarities associated with other industries. Significant sunk costs played a role in the strategic choices firms faced. The extent of firms' flexibility depended upon the presence of rigidities within social institutions, including labor relations. Finally, the extent to which a firm controlled its distribution channels could reduce its freedom to act.

In a convincing article, Tsuyoshi Numagami (1996) argues that the clarity of vision demonstrated by the Japanese when they took the lead in the production of LCDs for quartz watch displays, which in turn gave them an advantage in terms of the production scale for overall quartz electronic watch capacity and dominance, was the result of a calculated strategic choice to go with what was considered the superior technology, even though in the short run LEDs were in greater supply.[1] Hong Kong simply followed the leader and benefited from the vision of the Japanese watch industry. It may have been propelled forward thanks to its role as a major supply outlet for Japanese quartz movements and LCD displays. Numagami (1996) argues that meanwhile U.S. firms got caught in a "flexibility trap" (p. 133), a condition he defines as market-induced myopia in the selection of a technology in a period of major technological change. It delays the technological turning point if the important part of the technological evolution is generated by the aggregation of the choice patterns of firms (or the volume demand of one major player).

Whereas in the United States no individual firm was in a position to chart a bold path and follow it to its logical conclusion, in Japan Seiko not only chose a direction, but began the switchover to a new technology standard quite early and stayed with it as the market gyrated back and forth in search of technology convergence. Meanwhile, in Switzerland, most firms sat on the sidelines, unable to decide which way to go when faced with what appeared to be a myriad of uncertain choices. The Swiss culture's "bottom-heaviness" diminished Swiss firms' ability to act quickly and decisively. The lack of a single trend-setting response was partly a result of Swiss unfamiliarity with the technology. Although the Swiss had begun to develop an electronic capacity, at the time they were very much followers rather than leaders. The introduction of the quartz watch was the first weak signal portending the flood tide associated with the impending turn toward electronics. Unlike the Japanese watch companies which, having been rebuffed at the end of World War II by machine tool producers who ignored their need for tools to make watches, and then simply went ahead and made their own tools, the Swiss watch companies were still too fragmented to will into existence the capacity to produce the electronics needed for quartz watches. But perhaps the bigger inhibitor was the cultural view embedded in the industry's most prominent firms: they simply could not believe that the buying public would ultimately select an electronic watch as the product of choice.

Finally, the quasi-fragmented Swiss production system, a lumbering giant, lacked the centralized control and ideological commitment to make a massive new investment and chart a new path. The beneficiaries of such a change were largely mute at the time decisions needed to be made; assemblers were followers rather than leaders; major brands had succumbed to infighting. Part of the resistance came from the fragmented aspect of the industrial system in which no individual firm could appropriate sufficient returns from the needed collective investment to move into electronics to guarantee participation, much less support, for such a gesture.

As I stated in the introduction to this chapter, the ability of U.S. and Japanese firms to quickly and thoroughly control the production of digital quartz watches was intimately linked to their ability to distribute labor-intensive assembly operations to a low-cost location. While assembly could have been done anywhere—and it was carried out in Japan, Switzerland, and the United States—the labor-intensive nature of the activity, in combination with complementarities associated with the electronics industry, made Hong Kong an excellent assembly center for the quartz watch industry. But it wasn't just the island nation's experience with assembly operations that made it such a target of opportunity. Firms in Hong Kong already had almost two decades of experience with the production of watch cases and bands, and had long-standing business relationships with watchmaking firms, both as assemblers and as distributors.

What no dominant country's firms counted on was the ability of Hong Kong companies to rapidly assert themselves as major assemblers of low-cost watches, thus becoming competitors that would eventually take over the low-price end of the industry. No dominant country, except perhaps Japan, fully understood the economics of the newly emerging electronics industry. In this new era of competition, there would never be a period of time when price declines moderated. Producers of watch components faced constantly falling prices for parts. Rampant price competition became a fixed feature of the industry. This inevitably meant that only a few companies could ever make serious money; to do so would require controlling entire market segments. Even with market control, volume parts production and parts standardization would always be required to bring costs down. Thus every major player would have to maintain a position in the low end or forfeit economies of scale in the production of standardized parts.

Hong Kong emerged as a major player in the world watch industry in part because of her early participation as a low-cost assembler and parts manufacturer. Hong Kong's history combined fortuitously with development of the early electronics industry, which needed both a location for hand assembly and an outlet for final products, such as semiconductors.

Entrepôt Economy, Aggressive Economic Competitor

Hong Kong is located at the mouth of the Pearl River on the southern fringe of the Chinese mainland. This tiny city-state consists of a few islands, a peninsula, and a small piece of the mainland (Figure 10.1). Hong Kong became a British protectorate after the opium wars of the

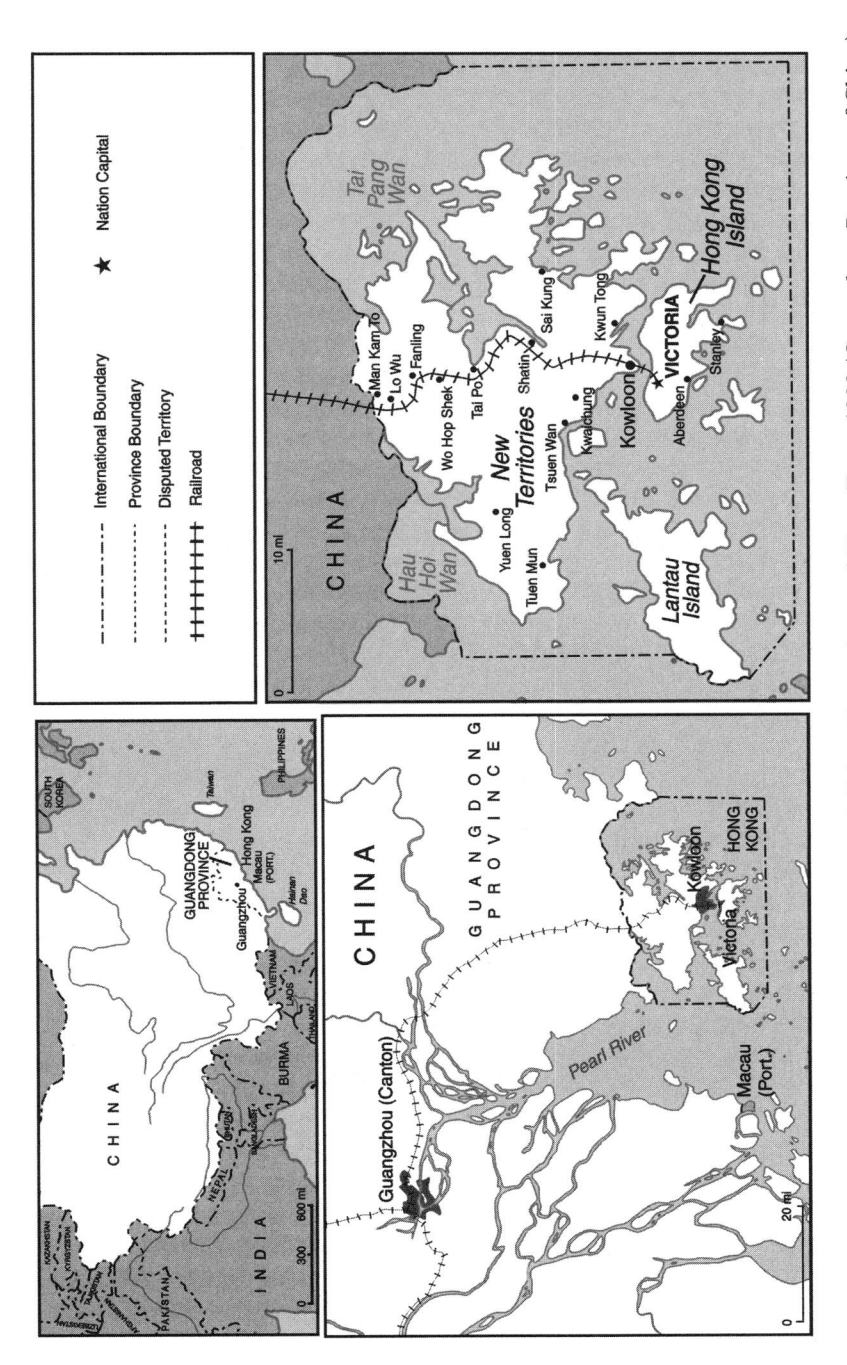

FIGURE 10.1. Sketch map of cities and provinces of China. Guangzhou and Hong Kong, 1988 (Guangdong Province of China). Cartography: Erin Heithoff. *Sources:* www.multimaps.com; Le Defi Horloger, Blanc (1988).

19th century. It was returned to the control of China in 1997. It has always served as an entrepôt, the gateway to and from Asia. Goods from all over the world are transported to Hong Kong, repackaged, and then shipped out to other parts of Asia; goods from all over Asia are transported to Hong Kong, repackaged, and shipped out to the rest of the world.

Prior to 1949, Hong Kong's economy centered on its role as an entrepôt, particularly its identity as a trade outpost for China and other Far East nations (Henderson, 1989b; Hsia, 1978; Riedel, 1974; Wong, 1988; Youngson, 1972). The island's regional prominence derived from its monopolistic position as the gateway to the West.

However, starting in the 1950s, Hong Kong's economy evolved from its former dependence on trade to export-based manufacturing. This transformation was triggered by three interrelated events. First, in 1949, the Communist Revolution drastically curtailed access to the huge consumer market in China, formerly Hong Kong's largest trading partner. Second, during the Korean War, a United Nations–initiated trade embargo on all Chinese goods had a strong negative impact on Hong Kong's economy. Third, an influx of immigrants fleeing China's political upheavals resulted in severe unemployment and a harsh economic environment in Hong Kong. Within five years of the communist takeover in China, Hong Kong's population exploded as more than one and one-half million Chinese migrated to the city-state in search of economic security. Hong Kong was ill-equipped to deal with this development; with its traditional livelihood in trade eliminated, it was confronted with unprecedented economic problems. As unemployment skyrocketed, the peninsula had little choice but to pursue a labor-intensive, export-led development strategy.

Structure of Hong Kong Manufacturing

Hong Kong's manufacturing industry began as labor-intensive assembly operations (Hong Kong Government Industry Department, 1985; Lin et al., 1980; Lui and Chiu, 1994). If manufacturing is defined as the employment of either chemical or mechanical processes to convert raw materials to consumer and producer goods, then few of Hong Kong's sectors qualify.[2] Despite record output growth rates, capital investment levels in Hong Kong industry have always been surprisingly low. Manufacturing establishment size is small. A large number of establishments on the island are sweatshops.

Lack of capital investment in manufacturing stems from a number of interrelated factors. First, in the 1950s, labor was relatively abundant compared with either land or capital. Production was organized around the country's most available resource: its labor force. Second, the economy's dependence on exports made it vulnerable to unexpected declines in markets. To limit risk, manufacturers minimized investments in capital equipment while maintaining the flexibility associated with labor-intensive manufacturing. Third, the largest share of Hong Kong's manufacturing employment was concentrated in apparel production, which continues to resist the application of new technology to reduce labor inputs (Lau and Chan, 1991). Historically, this sectoral structure has exposed production employment on the island to intense international competition from other, lower cost countries (Lui and Chiu, 1994).

Amid the major population influx of the 1950s, Hong Kong's government initiated favorable tax policies that encouraged producers to move their investments—first from China, then from other parts of the world—to the city-state. Japan was particularly well positioned to take advantage of this arrangement, given rising wages and labor shortages at home, and growing trade friction with the United States abroad. But the combination of very low wages, a seemingly infinite supply of docile labor, and a favorable business climate eventually enticed investors from around the world.

International trade policy also influenced the structure of production in the city-state. For example, the garment industry is regulated through an international quota system. Hence, via an "origins" rule, the Hong Kong government has restricted the amount of garment production originating on the mainland but reexported from Hong Kong. As Lui and Chiu (1994) note, "For those products exported to restrained markets, a special outward processing arrangement is administered by the Hong Kong Government Trade Department to ensure that goods manufactured in Hong Kong but partly processed in China only qualify for Hong Kong origin status if they fully meet Hong Kong's origin rules" (p. 62; Hong Kong Government Industry Department, 1985, quoted in Lui and Chiu, 1994). This has helped retain industry to some extent (Table 10.1).

Hong Kong manufacturing persists with its labor-intensive production system for three reasons. First, labor-intensive production is flexible: producers can quickly organize a labor force to meet their needs. Second, an elaborate subcontracting system facilitates the rapid manufacture of a wide variety of products with minimal personal capital investment. Third, because of the subcontracting system, responsibility for capital investment falls on successive levels in the subcontracting hierarchy.

This manufacturing arrangement also determines which products can be manufactured profitably. Because each producer attempts to minimize capital investment, the system remains undercapitalized—and thus labor-intensive. As Lui and Chiu (1994) note, "Automation does not nec-

TABLE 10.1. Watch Exports, Entrepôt Trade, and Imports from 1968 to 1991 (Unit: million Hong Kong Dollars)

Year	Imports	Entrepôt trade	Exports
1968	338	88	69
1969	478	110	108
1970	574	136	135
1971	673	183	175
1972	739	225	202
1973	1,011	392	293
1974	1,376	560	510
1975	1,313	618	644
1976	1,610	642	1,208
1977	2,211	762	1,694
1978	3,179	874	2,734
1979	4,309	1,240	4,354
1980	6,288	2,098	6,288
1981	7,443	2,584	7,104
1982	6,318	2,265	7,168
1983	8,134	2,903	8,259
1984	8,605	3,225	8,875
1985	9,390	2,973	9,237
1986	12,244	3,504	11,323
1987	14,777	4,666	13393
1988	20,424	7,000	16,588
1989	21,716	9,325	16,344
1990	26,050	10,819	18,319
1991	27,282	15,455	15,855
1992	33,618	21,467	16,017

Source: Hong Kong Trade and Development Council (1994).

essarily keep production costs down in the short run. In an industry which is dominated by small capitals, alternative strategies are often formulated within a framework of short-term calculation—these small manufacturers have no control over product markets and have to respond to rapid changes in the market" (p. 61).

Finally, subcontracting helps ensure flexibility because small, dependent firms act as buffers when markets shift unexpectedly. The lack of capital investment ensures that the effects of economic instability are borne almost solely by labor. Although an entrepreneur with a good idea can set up shop and begin production quickly, low barriers to entry in most industries mean that the profitability of even a new product is quickly eroded as competitors enter the market. When markets sour, producers simply stop manufacturing, often shifting to entirely different industries to take advantage of the next wave of economic opportunity.

The watch industry follows this pattern. It has always been comprised of small assembly shops. The period of highest employment occurred in the early 1980s with the advent of quartz technology (Table 10.2). As the industry has evolved, plant size has fluctuated and grown through time.

Watchmaking in Hong Kong

The origins of Hong Kong's watch and clock industry can be traced to trade. The country began as a central location for the shipment and sale of watch and clock products from Europe and the United States (Table 10.3). In the early part of the 20th century, Hong Kong was the largest point of sale for watches and clocks in Asia. Although the production of watches and clocks was more established in Shanghai and Guangzhou,

TABLE 10.2. Watch Manufacturer Numbers and Employees

Year	Manufacturers	Employees
1970	102	4,043
1971	106	4,187
1972	126	4,574
1973	144	5,887
1974	197	7,630
1975	237	9,393
1976	345	12,880
1977	368	15,326
1978	530	21,182
1979	770	31,931
1980	1,187	40,478
1981	1,238	38,531
1982	1,314	36,179
1983	1,394	36,731
1984	1,373	35,657
1985	1,436	31,551
1986	1,191	28,152
1987	1,302	26,850
1988	1,368	26,678
1989	1,335	24,983
1990	1,434	19,270
1991	1,321	19,484
1992	1,323	16,647

Source: Hong Kong Trade and Development Council (1994).

TABLE 10.3. Main International Markets and Exports, 1960–1991 (Unit: million Hong Kong dollars)

Market	1960	1965	1970	1975	1980	1985	1986	1987	1988	1989	1990	1991
U.S.	8	13	55	199	1,671	3,791	3,927	4,570	5,627	5,090	4,413	4,483
Mainland China	—	—	—	—	152	833	1,012	1,369	1,997	2,229	2,831	3,028
German	1	1	4	30	635	590	857	1,049	1,215	1,134	1,398	1,198
Japan	—	2	16	43	266	491	654	874	1,212	1,294	1,192	1,188
Switzerland	—	1	21	99	265	278	324	516	653	679	808	589
Britain	2	4	9	43	477	375	530	616	852	860	836	579
Spain	—	—	2	25	162	186	339	439	531	505	470	445
France	—	—	1	11	299	255	466	507	596	608	612	418
Others	6	9	100	347	2,649	2,774	3,558	3,916	4,663	4,676	5,573	4,799
Total	17	30	208	797	6,576	9,573	11,667	13,856	17,346	17,057	19,133	16,727

Source: Hong Kong Trade and Development Council (1994).

the advantageous location of Hong Kong made it the gateway for watch distribution. As a point of distribution, Hong Kong also served as a watch service center. A capability for the repair of watches was established early and persists to this day. The basic need for repair and adjustment of watches was the cornerstone of the Hong Kong watch industry in its formative years.

During the 1950s, the isolation of much of Asia, exacerbated by the Korean War, resulted in severe shortages of material goods. Watches and clocks were in particularly short supply. While Japan's industry quickly recovered after World War II and took up production within a matter of months after its defeat, insatiable domestic demand swamped production capacity. Japanese watch firms began to shift labor-intensive assembly of relatively low-grade mechanical watches to Hong Kong when domestic capacity had been exhausted (Johnstone, 1999).

Hong Kong became the center of watch retailing in Asia, particularly for Swiss watches. In 1953, 81 percent of the watches sold out of Hong Kong came from Switzerland; another 15 percent came from Germany. Hong Kong was Switzerland's regional distribution center and the largest distribution center for watches in the world.

Being a regional service center required the ability to fabricate watch parts for repair. In the 1930s, domestic workshops formed to produce watch cases. Given pent-up demand for repairs and rising incomes, and therefore growing watch ownership, the repair business was highly lucrative. Several watch repair companies developed offshoots to supply parts for watch cases. Although business was brisk, the production system remained both domestic and small in scale. The growth in watch-case manufacture continued through the 1950s. Worldwide, watch demand was booming. With growth came demand for more variety and the need to satisfy more markets. Products began to change. The use of steel as opposed to copper for watch cases required more sophisticated technology to manufacture high-quality products. Demand for case manufacturing also was boosted when the Japanese began to establish case-manufacturing operations in Hong Kong to supplement domestic capacity. The introduction of foreign technology, in particular metal-forming machine tools, increased the capabilities of Hong Kong manufacturers and encouraged the development of more complex business enterprises. Hong Kong's export industry was becoming so successful that countries in the region began to restrict Hong Kong producers' access to their markets.

Despite its early capacity to make watch parts, especially cases and bands, the watch industry in Hong Kong has primarily emphasized assembly rather than fabrication. From the early days of the industry's development, up until the advent of the quartz crystal watch movement, the most complex and expensive aspects of watches were imported primarily from Japan and secondarily from Switzerland.

The 1950s were boom years for the Hong Kong watch industry. The industry imported almost $100 million in watches and then reexported more than one-third of them to other Asian markets. Switzerland was the major supplier of the industry, with more than 82 percent of the Hong Kong watch movement market (Hong Kong Watch Trade Association, 1994; Table 10.4). Although most watch sales occurred legally, some portion occurred illegally. Thus, from an early stage the Hong Kong industry was plagued by contraband activities.

By the mid-1950s, Hong Kong's role as a reexport location began to wane. A regionwide depression, coupled with the protectionist efforts of neighboring countries, reduced the reexport of watches in the region. While Hong Kong continued to distribute watches made in the region (such as those from Malaysia), the country's former status as the Swiss reexport center was permanently diminished.[3]

Two developments in the 1960s laid the foundations for the explosive growth of the Hong Kong watch industry. By the early 1960s, U.S. electronics and select watch firms were using low-cost locations to assemble inexpensive watches. Timex had already shifted most of its watch assembly offshore to the Virgin Islands. This gesture placed significant cost pressure on other watch producers, including other U.S. firms such as Bulova. The second development, and perhaps the more important, was the declining share of the Hong Kong watch market controlled by the Swiss. Within a matter of two years Japan had doubled its share of the Hong Kong watch industry market for export by moving into low-end watch products. Seiko's entire low-end product line was manufactured in Hong Kong. This loss of market share was almost totally absorbed by the Swiss. By 1962, Japan had begun to move into the global watch industry, using Hong Kong as a place to buy inexpensive, world-class watchbands. It was an important market to export from, free from U.S. tariff restrictions.

TABLE 10.4. Exports and Imports in 1950s and 1960s

Year	Imports (Hong Kong dollars)	Quantities	Exports	Quantities
1953	96,384,121	–	32,548,411	–
1956	144,482,752	3,011,286	16,362,120	478,583
1957	174,946,571	4,021,629	10,245,528	308,827
1958	105,006,315	1,230,972	14,981,131	442,134

Source: Hong Kong Watch Trade Association (1994).

Starting in the 1960s, Hong Kong firms began developing their own watch brands for OEM (original equipment manufacturers) distribution,[4] selling watches and clocks to more than 15 countries. The majority of these products were private-label watches, produced for retail outlets. It was not long before Hong Kong began replacing its regional, Asia-based trading patterns with international trade patterns. In addition to the sale of completed products, Hong Kong began to be the major international producer of watchbands for export.

Throughout the 1960s, in an era of intense price competition among Swiss, Japanese, and U.S. watch manufacturers, Hong Kong continued to operate as an export center. Gradually, however, starting in the late 1960s, the composition of the city-state's exports changed from strictly assembled goods to manufactured goods, including cheap mechanical watches. When electronic watches were introduced at the beginning of the 1970s, within four years Hong Kong became the central assembly location first for LED, then for LCD, electronic watches. U.S., Japanese, and, to a far lesser extent, Swiss firms ran assembly operations for electronic watches out of Hong Kong.

Up until the early 1970s, Hong Kong primarily assembled mechanical watches based on watch modules imported from Switzerland. In the late 1960s, mechanical watches still comprised more than 60 percent of all watches assembled. Assembled products were sold in more than 100 countries, but were predominantly sold in the United States. Japan was Hong Kong's second largest final market.

The Quartz Revolution and Its Link to the Electronics Industry

The advent of quartz watch technology completely transformed the Hong Kong watch industry. Whereas mechanical watch assembly still required some degree of skill to coordinate many parts at very small scale, electronic watch movement assembly was very simple. The simplicity of the electronic watch opened the way for new competitors. By the early 1970s, Hong Kong had already established itself as a sophisticated offshore electronics assembly location, primarily for U.S. electronics companies (Dicken, 1998). Hong Kong assembled and then shipped all manner of electronic gadgets, including toys, radios, calculators, and watches, worldwide. This early acquaintance with electronics manufacturing led the way to local production of electronic displays. Hong Kong firms initially emphasized LEDs, the standard for many early electronic products. But when the Japanese led the shift to LCDs, Hong Kong firms made a relatively smooth transition to the new technology. Unburdened by the need to invest in the fabricating technology required to manufacture the needed components, Hong Kong producers simply relied upon supply sources developed by the Japanese in order to move into LCDs.

Hong Kong's ability to move quickly into digital watch production was linked to the simultaneous emergence of the semiconductor industry, which used Hong Kong as an early outpost for chip wiring and case assembly. Hong Kong firms began to merchandise semiconductors, and like everyone else in the business scrambled for downstream markets, such as watches. Like U.S. semiconductor manufacturers, Hong Kong semiconductor assembly houses desperately needed new product outlets to smooth out the erratic flow of international semiconductor demand. In the early years of semiconductor development, a number of Hong Kong chip assembly firms moved into electronic watch manufacturing to control downstream markets. Their first products were low priced and distributed as private-label products almost exclusively in Asia. This proved to be a replay of the American semiconductor experience.

The chaos experienced in the electronic watch industry in the United States and Switzerland was also experienced in Asia. The shift to LCD technology occurred in a short span of five years as LCD watches became the dominant Hong Kong watch product. The number of LCD assemblers increased from 237 in 1970 to 1,187 in 1980. The shift from mechanical watches to electronic watches, while by no means painless, occurred with a swiftness in Hong Kong that eluded most other manufacturing and assembly competitors. Statistics tell the story (Table 10.5). In the 1970s, mechanical watches comprised slightly less than half of all watch and clock products produced in Hong Kong (valued in Hong Kong dollars). By 1980, the number of watch product units had increased by 10 times, and electronic watches were half of all output. While mechanical watches had fallen to slightly less than one-third of total product, by 1985, watch product demand had doubled again and electronic watches comprised five times the value of mechanical watch products. In little more than two decades, Hong Kong went from virtual insignificance to world leader in total worldwide watch production.

In the late 1970s, the price of an LCD watch fell drastically from $38.00 to $11.00 in Hong Kong dollars (hereafter, HK$), which is from $5.00 to $1.50 in U.S. dollars (Hong Kong Government Industry Department, 1985). Initially, lower input costs, fueled by price competition among U.S. semiconductor producers, drove prices down. Prices fell further still as price competition set in among Hong Kong watch assemblers. LCD manufacturing requires simple and inexpensive equipment. Thus, barriers to entry into the market were minimal. Many firms began assembling watches, largely for U.S. semiconductor producers. Supply from the Japanese watch producers further eroded prices. By 1980, many

TABLE 10.5. Watch Product, 1960–1991 (Unit: million Hong Kong Dollars)

Category	1960	1965	1970	1975	1980	1985	1986	1987	1988	1989	1990	1991
Electronic watch	—	—	—	—	3,557	5,835	7,247	8,821	11,524	11,332	12,841	108,806
Mechanical watch	—	—	90	457	1,648	781	888	936	850	752	698	457
Electronic clock	—	—	—	—	288	400	533	616	512	302	232	201
Mechanical clock	1	2	9	49	59	381	454	427	300	171	172	114
Watchband	12	23	3	152	288	336	343	463	758	731	814	872
Watch shell	4	5	36	139	736	1,840	2,202	2,593	3,403	3,787	4,375	4,277
Total	17	30	208	797	6,576	9,573	11,667	13,856	17,346	19,075	19,133	16,727
Percentage in Hong Kong's total exports	0.6	0.6	1.7	3.5	9.6	7.4	7.6	7.1	8.0	7.6	8.5	7.2

Source: Hong Kong Trade and Development Council (1994).

Hong Kong watch assembly firms were losing money as both competition and the international recession depressed market prices and their profit levels. Thus, even firms in a low-cost assembly location like Hong Kong, unburdened by high fixed costs, were not immune to the trials and tribulations of a dynamic industry segment subject to wide swings in prices and profits.

In the early 1980s, the emergence of quartz analog watches breathed new life into the world watch industry. Quartz analog watch production in Hong Kong relied upon foreign parts. Because analog watch production also required a higher level of capital investment, Hong Kong's production system took some time to adjust to the new product.

But adjust it did. Since the mid-1980s, quartz analog watches have come to dominate the industry. Evidence of their growing importance is striking. In 1980, digital watches accounted for approximately 60 percent of the value of total watch output. Analogues made up only 8 percent, and mechanical watches accounted for the remainder. In 1984, quartz analogues and digitals each made up approximately 43 percent of total output value (Hong Kong Government Industry Department, 1985). By 1989, quartz analog watches dominated the market, accounting for 84 percent of total output by value. Digital watches made up only 6.6 percent. Because of Hong Kong's extremely fluid industrial structure, its watch industry made a rapid and complete transformation in its product mix (Hong Kong Trade and Development Council, 1988).

Hong Kong watchmakers never completely abandoned the mechanical watch. A few firms still assemble mechanical watches from imported movements (Hong Kong Government Department of Industry, 1985; Hong Kong Trade and Development Council, 1989, 1998). But because of skill and quality requirements, industry output has remained low. Capacity limitations also stem from the complicated nature of mechanical watch assembly, which requires higher levels of capital and floor space.

Hong Kong's mechanical watch share has been declining since its 1980 high of 27 percent. By 1989, mechanical watches accounted for only 2.3 percent of total exports. Comparable figures persisted in 1998. Unlike other industry segments, a larger share of mechanical watches are marketed in third world countries. The decline in the low-value mechanical watches' share of total output means that at least superficially, the Hong Kong watchmaking industry has kept up with technological trends. However, this misrepresents the relationship between advances in assembled product and those in manufactured inputs. While Hong Kong's industry has been propelled by technological advances in watch movement manufacturing, Hong Kong watch producers continue to respond to innovations rather than develop new process technologies and products.

Industry trade reports still attribute the exclusive use of imported movements to the complexity of manufacturing the inputs combined with significant barriers to entry to achieve cost-competitive output.

Ironically, it is within the external watch industry segment that firms have developed some level of technological leadership and unchallenged market acceptance. Hong Kong's world-renowned case and band assembly industry has little competition. Dailywin, a large Hong Kong manufacturer, indicated that it uses CAD/CAM (Computer Aided Design, Computer Aided Manufacturing) for the design and manufacture of cases and bands. This new computer-based technology allows rapid model changes and facilitates the design and manufacture of complex watch cases and bands. Recent industry reports indicate firms are moving toward original design manufacturing (ODM), whereby customers consult with firms, which then develop prototypes and models in preparation for full production. Hong Kong has amassed both know-how and capital in this market segment.

Organization of the Hong Kong Watch Industry

The number of watch assemblers grew precipitously over much of the 1980s (see Table 10.2). In 1988, there were approximately 1,386 watchmaking firms registered with the Hong Kong government. From 1983 to 1988, employment slowly increased from 25,200 to 26,444. But many more new firms were established during the same period. In the late 1980s, over 90 percent of firms employed fewer than 50 employees. Labor constituted approximately 68 percent of operating costs (excluding materials and other business services), while accounting for only 8 percent of gross output (compared with inputs of 84 percent).[5]

In 1993, although there were 1,477 establishments, employment had declined considerably since five years earlier, falling by almost 10,000 jobs to 17,110 employees (Hong Kong Trade and Development Council, 1993). The decline in employment was not reflected in share of national exports. While the number of jobs in the industry has declined steadily since the late 1980s, watches still contribute approximately 7 percent of domestic exports, a figure that has not changed since the mid-1980s.

Recent Industry Trends

Hong Kong industry output has experienced impressive gains since the late 1950s (excluding recession years). Since 1980, exports have increased at an annual rate of 15 percent—with recent growth trends topping 20 percent. Post-1989 growth rates were projected to be 25 percent annually. In 1998, Hong Kong produced in excess of 700 million watches, the majority of which were reexports from China.

Until the late 1980s, strong growth trends persuaded watchmakers to maintain production within Hong Kong rather than relocating it to mainland China. International trade policy further cemented the assembly of watches in Hong Kong. Traditionally, international trade law restricted the country of origin of a watch product to the country of assembly. This helped to maintain watch assembly activities in Hong Kong. In 1991, however, this rule changed. Hong Kong unilaterally redefined its trade law to indicate that a watch's origin would be determined by where the movement came from (Hong Kong Trade and Development Council, 1993). This allowed Hong Kong watch producers to use the label "made in Japan" or "made in Switzerland" on Hong Kong–assembled watches. Consequently, the reexport trend has only recently been significant (as the absolute level of available labor has significantly declined).

Although watches remain among Hong Kong's four largest export industries, the presence of watch manufacturing in the city-state has changed dramatically since 1997. Recent statistics from the Hong Kong Trade and Development Council indicate that since the mid-1990s the number of firms has continued to decline. In 1998, there were 528 firms and just 5,287 employees in the industry. Such drastic decline over such a short time period reflects the almost complete movement of watch assembly activities to mainland China. Access to low-cost labor and abundant space for the construction of new plants has drawn many of Hong Kong's larger assemblers to China. As capacity builds on the Chinese mainland, reexport figures clearly show that Hong Kong firms are sourcing more and more parts from within China. These facilities now support watch assembly levels equal to or greater than those supported by Hong Kong factories. At the low-price end of the market, China has emerged as the largest site of assembly operations in the world.

Ten years ago, Hong Kong watchmakers operated mostly as OEMs for discount houses and retail outlets. Manufacturers produced models to specification, rarely offering their own ideas. Today, many watchmakers have moved from OEM business toward the development of unique styles for every occasion. Their goal is to convince consumers that a watch is a fashion statement and clothing accessory. This level of product variation is Hong Kong's trademark. The production system operates within this philosophy. As one manufacturer noted, "The Japanese always talk about automation, but it [automation] also restricts flexibility and creativity that a labor force provides. As watches become more fashion accessories, people will appreciate interesting hand-crafted items" (Federation of Hong Kong Watch Trades and Industries, 1989, p. 21).

The Hong Kong industry is highly sensitive to variations in customer preferences across countries. High-end watches sell well in Europe and Japan, while low-priced models dominate the U.S. market. Over the last 10 years, the Hong Kong watch and jewelry industries have begun to merge. Moderately priced watches with semiprecious stones and more upscale cases and bands are particularly successful at penetrating the lower end of the U.S. jeweled-watch market. They are making significant inroads into the medium- to upper-price Japanese markets as well. The combination of the two industries allows for extensive variation. The bottom line, however, is still perceived to be price. As one manufacturer noted, "The benchmark for success or failure in the fickle world of watchmaking does not lie in design. Ultimately buyers make their choice based on price" (Federation of Hong Kong Watch Trades and Industries, 1989, p. 41). Such an outlook pervades the Hong Kong industry, where even the highest quality watches are priced far below comparable brands manufactured in other countries.

Hong Kong watch prices (usually free-on-board; FOB)[6] are extraordinarily low by world standards. The average price of a watch in 1988 U.S. dollars was $3.00. Advertised prices ranged from $4–$10 per watch (with a lead time of between 25 and 60 days). Even jeweled watches cost a fraction of those manufactured in Switzerland. Orders can be as small as 100 watches; in some cases, firms have no minimum lot size. More recent data from the Swiss Federation of Watch Manufacturers reconfirms the low-cost nature of Hong Kong products. In 1998, the average world price of a watch was 59 Swiss francs. The average price for a Hong Kong watch was 8 Swiss francs, whereas the average price for a Swiss watch was 235 Swiss francs. The persistently low price for Hong Kong watches reveals the city-state's difficulty in moving up-market into the more lucrative midprice range for watches.

Markets

In 1993, the Hong Kong watch industry exported 592 million watches (*South China Morning Post*, 1993, p. 3). In 1998, the industry exported more than 766 million watches (Hong Kong Trade and Development Council, 1998). Market shares among Hong Kong's major trading partners have remained quite constant over time. As of 1992, the country's major markets by value and volume of exports were (in descending order) the United States, Germany, Japan, and other European countries (Hong Kong Trade Development Council, 1989; Hong Kong Watch Trades Association, 1993). Approximately 70 percent of this output consisted of electronic watches; of this, 61 percent were quartz analogs

A portion of Hong Kong's watch industry consists of parts exports.

Seventy percent of watch parts are sold to China, and the residual is split between Switzerland and the United States (Hong Kong Trade and Development Council, 1989; Hong Kong Watch Trade Association, 1993). Similar findings hold today. Watch cases and parts constitute 14 percent of total industry exports. Major markets for these products include China (32 percent), India (8 percent), Switzerland (10 percent), and Japan (11 percent).

Competition

Hong Kong dominates the low-end watch market. Until recently, the city-state had no rivals for market share in the price ranges in which the majority of its products compete. Prior to 1987 (when the yen was less than one-half its current value), Japanese producers made a foray into the low-priced end of the electronic watch industry. Anticipating increased market share, two of the three major Japanese watchmakers set up assembly plants in Hong Kong to manufacture their low-cost products. Even technology-touting Seiko uses Hong Kong as a low-wage flexible labor location for the assembly of new products.[7] This has become all the more important as the yen has continued to rise in value.

In the 1980s, devaluation of other Asian and Latin American currencies proved to be a boon for Hong Kong watch producers. Whereas previously a consumer might have bought a Seiko low- to medium-priced watch (because of its quality), price differences between Japanese and Hong Kong products have singularly eroded Japan's market share in the newly industrialized countries (NICs).

The more recent Asian devaluation (1997) has had little effect on the Hong Kong watch industry's market share. Rising standards of living have relegated watches to a much lower level of importance with consumers. Whereas once watches were considered a luxury product, they now stand alongside many other inexpensive consumer goods. Watch buyers today are increasingly buying inexpensive watches, saving their disposable income to buy cellular phones, beepers, and other personal electronics devices.

I lack reliable information comparing the success of medium-priced Hong Kong watch companies with Seiko (the traditional market leader in this segment). A Japanese industry representative indicated that this market is unstable, particularly among customers in NICs. Again, the rising value of the yen has pushed Japanese watches out of reach for many former customers, and they are increasingly turning to Hong Kong watches as a price-competitive alternative (S. Aota, personal communication, July 19, 1994).

The Swiss maintain control of the medium-to-high and very-high luxury markets. Interestingly, as the value of the yen has increased, medium-priced Swiss watch demand has benefited. In Asia, the Swiss name still connotes quality, luxury, and status. Even in Japan, consumers prefer Swiss watches at the upper end of the medium-priced range. Thus the Japanese are caught in the middle: Swiss producers are eroding their middle- to high-segment market share, while Hong Kong producers are walking away with the lower end of the market and making inroads in the medium segment (including low-priced jeweled watches). The possibility exists that new innovations—such as a wrist computer—will substantially alter existing watch manufacturing countries' market shares. In particular, anecdotal evidence suggests that Japan's solution to eroding market share and stagnating demand is the development of new core capabilities associated with information technologies in watch- and non-watch-related products. However, Hong Kong's demonstrated ability to rapidly copy new industry developments may circumscribe Japanese producers' ability to control all segments of potential new product markets.

Up until now, Hong Kong's ability to dominate the low end of the market has been unquestioned. In the late 1970s, perceived profits in watchmaking were low; thus, other NICs chose to ignore watches as a target sector. There is some watch-production capacity in Thailand, the Philippines, Malaysia, Taiwan, and South Korea. But none of these countries has a serious watch industry. In recent years, however, China has begun to prove a serious competitor for the very-low-end mass watch. After reunification, Hong Kong watch producers challenged China's indigenous industry because China's factories produced poor-quality watches. Today, however, this imbalance is beginning to adjust as Chinese factories employ newer technology and place greater emphasis on quality in the production of watches for export. Hong Kong firms have been central to this new capability.

The Hong Kong industry is the epitome of watchmaking today. With increases in the speed of new design and product development, industry emphasis is on fashion and impermanence. Hong Kong watch producers tout their ability to produce batches of watches in 25 to 60 days. This is considerably faster than other competitors such as the Japanese (one year, from design to development) and the Swiss (one to two years, from design to development).

For now, Hong Kong producers occupy a very large market segment: low-end watch sales. Until 1985, Japanese firms competed successfully in this market. Since the yen's appreciation, however, the Japanese have been losing low-end market share throughout Southeast Asia at an alarming rate. Customers are both unwilling and unable to buy one expensive

Japanese watch when, for one-fifth the price, they can own two or three different Hong Kong watches. Needless to say, Japanese firms are concerned about their lost market share.

Since the early 1980s, the Swiss have retaken a position in the low-price segment of the industry. The creation of the Swatch saved the Swiss watch industry. Only a decade ago, it was unclear whether Hong Kong watch assemblers could compete directly with Swatch. Today there is little doubt that Hong Kong producers manufacture trendy plastic watches that are sold all over the world. While Hong Kong plastic watches are not direct substitutes for the Swatch product, Hong Kong's flexible and fashion-conscious watch producers show significant capability to produce low-cost plastic watch products that sell into a market segment significantly below the cost of a Swatch.

Problems Facing the Industry

Despite its impressive share of world watch products, Hong Kong producers have a special vulnerability to competition from lower cost locations. In recent years, the industry has been plagued with skill shortages. Thus any attempt to move up-market has been partly kept in check by a lack of workers. Today, the population of Hong Kong has achieved higher levels of education, and the city-state of Hong Kong has emerged as an important international center of trade and finance. Employment alternatives abound even for less-skilled workers. For example, service industries that generally pay higher wages and provide better working conditions than manufacturing have been successful in attracting former manufacturing laborers (Lui and Chiu, 1994).

A lack of skilled manufacturing workers has always been a problem in Hong Kong. This problem is especially acute in watchmaking, which requires designers as well as manufacturing staff (Hong Kong Government Industry Department, 1985; Hong Kong Manufacturing Association, 1988; Hong Kong Trade and Development Council, 1989). This problem continues to hamper the industry's progress toward the mid-price watch market. Two polytechnics in Hong Kong train students in horological design. These schools have produced a number of skilled graduates, but Hong Kong still suffers from a serious lack of local high-fashion watch designers. The situation has certainly improved in the last 10 years, but the president of the Dailywin Corporation, a major Hong Kong watch manufacturer, indicates that there are really only two or three reputable watch designers in Hong Kong. To overcome the lack of local talent, many companies contract with Europeans for design services. For example, a large proportion of fashion watches are designed in

Europe and assembled in Hong Kong. The vast majority of Hong Kong product designs are variations of those developed elsewhere.

A longer term problem relates to two issues: existing labor shortages (both skilled and unskilled) and the inability to make critical watch components. Until recently, the Hong Kong watch industry defied larger industry trends of moving labor-intensive production offshore to China. As land and labor shortages have developed, however, more Hong Kong assembly is occurring offshore. Certainly, with this development, there is even less incentive for Hong Kong manufacturers to improve production technology through capital investment. As long as China proves a viable alternative for solving the labor force problem, Hong Kong producers will remain reluctant to invest in labor-saving equipment.

The second problem stems from industry dependence on foreign components. Since its origin, the Hong Kong watch industry has relied on foreign suppliers for movements and critical watch components. This dependence is tolerable as long as trade relations among movement-producing countries and Hong Kong remain stable. However, over the last 20 years, there have been several periods when Japanese and Swiss component producers were manufacturing at capacity (Hong Kong Government Industry Department, 1989, 1993; Fédération de l'Industrie Horlogère Suisse, 1998). In 1989, component shortages increased delivery times from 45 to 120 days (Hong Kong Watch Distributors Association, 1989). Although more recent data on shortages are not available, reliance on uncertain supply sources makes Hong Kong watch producers vulnerable to unpredictable changes in global demand for watch components.

The structure of Hong Kong manufacturing has come back to haunt the watch industry. While flexible and able to switch models rapidly, the industry does not encourage the design, development, or manufacturing knowledge needed to introduce reputable brand names and move upmarket. Industry watchers contend that for Hong Kong to make enduring inroads into watchmaking and stave off future competition from other low-wage countries, manufacturers must reduce variety, increase product quality, and create a quality image with staying power.

However, Hong Kong's production system thwarts such development. While the trade association calls for industry-wide improvements, exhortations are largely ignored by all but those large firms capable of investing in new technology (and most likely to benefit from compliance). Because the industry is mainly composed of small firms, larger firms cannot ensure that trade association advice will be heeded. Unless the old system is changed, it seems doubtful that industry aspirations will ever be realized.

There is some evidence that the larger firms are beginning to circumvent the problems of low quality and low prices. In recent years, major Hong Kong brandname firms have pressed to move up-market through the acquisition of Swiss watch brand names. This trend has been of major concern to the Swiss watch industry, particularly given that Hong Kong no longer adheres to international trade law regarding point of origin linked to point of assembly. The significance and effectiveness of this trend are unknown.

Patent Disputes

Skill shortages and limited design capacity lead to another problem: patent infringement and the lack of identifiable brand names. Since its origin, the Hong Kong watch industry has been criticized for copying watch designs from other countries' firms. In the early 1980s, RCA of America accused Hong Kong Manufacturing of infringing on its patent for LCD design and circuitry. During the same period, the Swiss also claimed that Hong Kong producers were essentially copying Swiss watch designs (Hong Kong Government Industry Department, 1985). Even more recently, Hong Kong producers were publicly embarrassed at the Basel Watch Fair when Swiss and French watchmakers openly accused them of copying European designs.

Hong Kong manufacturers have attempted to overcome this problem. For example, the Productivity Council regularly publishes a booklet pointing out the latest developments in watch design, development, and marketing. Included in each volume is a description of important designs and patents to which individuals or countries have original claims (Hong Kong Productivity Council, 1988). This effort is aimed at helping Hong Kong watchmakers reduce problems of patent infringement.

Industry trade representatives acknowledge watchmakers' lack of enthusiasm for government programs. While industry leaders note the importance of improving product design, the myriad small watch producers resist such entreaties. Improving designs means paying more attention to detail and using higher quality inputs. Small firms may not be in a position to do this. Moreover, the elaborate system of subcontracted assembly makes policing such efforts almost impossible. Thus despite concerted efforts, Hong Kong's larger firms are unable to coerce the smaller and more price-sensitive producers to restrict copying. National government efforts are impressive, but the cutthroat nature of the industry reduces the effectiveness of such programs.

The industry trade organization has also initiated a number of programs to overcome the copying problem. In 1984, the Hong Kong Watch Manufacturers Association set up a Patent Depository for logging designs. The depository's records provide the basis to dispute claims made by countries against the Hong Kong watch industry. A committee of Hong Kong and Swiss watchmakers was created in 1984 to improve relations among watch producers from both countries. Ironically, Hong Kong firms now find themselves concerned with the copying of proprietary products by Chinese mainland firms. Hong Kong has worked hard to eliminate patent infringements by local firms. This issue is once again simmering under the surface; many of Hong Kong's bigger firms realize the possibility that this problem might reach uncontrollable proportions. As the largest reexporter of products from China, it is impossible for Hong Kong firms or their trade associations to police the entire mainland's watch production.

What the government hopes to achieve with one set of programs, however, is often diminished by the effects of other policies. Hong Kong trade policy implicitly encourages copying. With no import restrictions, individuals can travel to other countries, buy copies of European watches, copy them, and then sell the copies either in Hong Kong or within other free-trade nations. The reverse is also true. Merchants export copies manufactured in Hong Kong to other countries. The chain linking manufacturing, wholesaling, and retailing is broken in several places, allowing distribution of counterfeit watches. While industry representatives say Hong Kong watchmakers are trying to improve their image, wholesalers are not above buying contraband watches assembled in other Asian nations.

New efforts to improve quality, productivity, and material competency were launched in 1998. The Productivity Council has initiated a program designed to help small manufacturers achieve ISO 9000[8] certification. Of major importance for firms competing at the upper end of the low-price product range are programs for material finishing and high-tech plating. These activities play to the industry's strength in the production of exterior parts for watches. This activity still remains the purview of Hong Kong firms. Regardless of what migrates to the mainland, metal finishing and case production is likely to stay within the confines of Hong Kong.

Industry success is also tied to an incredibly flexible production system. While industry representatives hint that improvements have been made in quality and control, they still acknowledge that the average plant size is falling, and the system remains highly fragmented by rampant price cutting (Hong Kong Trade and Development Council, 1998). Noting that the industry's strength is the result of competition, representatives also acknowledge that the level of competition in the Hong Kong watch industry is extreme. Because of the ready availability of imported parts and ease of assembly of the final product (particularly with digital watches), firms can easily enter the industry. Severe competition and price cutting have created great instability. Thus, although the laissez-faire system has greatly benefited watch manufacturers, competitive pressure has gotten out of hand in the recent past.

Kinks in the Distribution Chain

Viewed from another perspective—that of watch distributors—the Hong Kong watch industry still has some distance to travel to overcome problems of design, quality, and marketing (Federation of Hong Kong Watch Trades and Industries, 1989). Now that the industry has moved upmarket with better designs and higher quality products, industry observers suggest the time has come for firms to develop brand names (Hong Kong Trade and Development Council, 1998). However, customers are still dissatisfied with the quality of Hong Kong watch finishes. Another development on the horizon is the growing number of small orders from fashion boutiques and small retail outlets. Watch companies are receiving conflicting signals from these marketing channels. Boutiques are calling for great variety, while small retailers are asking for higher quality on fewer basic designs.

The fragmented nature of today's production has serious implications throughout the production and distribution chain. Since the industry is generally undercapitalized, few surplus profits are available for sales incentives and training. Little or no coordination exists between producers and the distribution chain. Manufacturers offer price reductions on an ad hoc basis. This pits one retail outlet against another. Furthermore, producers rarely listen to retail market information. Retailers on the front line report that customers want higher quality and greater variation within a product line, not unlimited options on one basic style. However, the production system does not respond affirmatively, and the casual system encourages illegal activities that further erode retail market share.

Future Prospects

The return of Hong Kong to China in 1997 had significant effects on the watch industry. Almost the entire manufacturing portion of the industry has moved away. What little manufacturing is left in Hong Kong itself constitutes short runs, testing, assembly, and administration. While only five years ago, as much as 50 percent of the industry was still based in Hong Kong itself, today almost all of the industry has moved to the Chinese mainland. What little manufacturing remains is now open to question because of major changes in the traditional market for watches.

More than anything else, Hong Kong has always been a trade center. Tourists from all over the world come to buy watches and jewelry from Hong Kong's merchants. The retail market for watches saw constant growth from the 1950s to the 1980s, but that is changing now. Hong Kong has been impacted by the recession of the early 1990s, a change in consumer preferences for new status symbols, and a more conservative middle-class faced with growing expenditures for housing, transportation, and education.

The recession of the early 1990s saw a drop in the number of wealthy individuals traveling to Hong Kong and buying high-priced watches. While it was thought that such a thing would never happen, the market for very-high-priced, diamond-encrusted watches is in decline. Indeed, thanks to the Asian financial crisis of 1997, overall watch demand has been down. Firms are being forced into a position of either moving up-market, which is costly and uncertain, or restrengthening their position in the low-priced end of the market, by moving all production to mainland China.

Quite unexpectedly, consumers have begun to develop a taste for things other than watches as status symbols. The ownership of portable phones has increased dramatically. Other types of electronic devices have increased the range of buying opportunities for the middle class. Moreover, as the cost of living has increased, consumers are spending more on essentials, which has reduced their disposable income. All of these changes have added up to a decline in the retail end of the Hong Kong industry.

As price competition has become more fierce, the old rules of retail behavior have begun to change. Watch distributors can no longer guarantee that their competitors will abide by unwritten rules of discounting. As business has stagnated, retailers have abandoned old margins and cut prices to ensure sales. This has reduced the profit level to retailers and has resulted in retaliatory pricing.

Retailers have also begun to move offshore, increasing competition for Chinese tourists who would otherwise come to Hong Kong to buy consumer goods. As retail capability develops in China, Hong Kong producers, with much higher land and labor costs, are finding it increasingly difficult to compete.

The Hong Kong Watch Trades Association, the oldest watch industry association in the city-state has attempted to institute industrywide agreements to contend with recent developments. Given the movement of both manufacturing and retailing to China, some fear that it is only a matter of time before tourists will have no reason to come to Hong Kong for watches. To counteract this possibility, in 1992 the association proposed to create branded products to cement the industry on the island.

After much discussion, the leadership of the association did not prevail. Both merchants and manufacturers saw little to be gained from creating a branded product. The self-interested nature of various Hong Kong firms thwarted all attempts to develop a united front. Watch manufacturers were concerned about who would produce the branded watch and whose name would be used on the watch. Echoes from the past can be heard in these arguments. Remember how the Swiss faltered on the eve of introducing the 1960s version of the electronic watch? The Swiss were unable to agree on the name of the watch or which watch company would bring it to market. Eventually the idea met delays. Like the Swiss before them, Hong Kong's retailers were concerned that a new competitor would simply reduce the market share of existing brands. The Manufacturers Association saw the effort to build a brand as a means by which the Watch Trades Association could reassert control over the industry.

After months of discussion and disagreement, the idea was dropped. Thus the historic flexibility of the industry may be the basis for its further shrinkage. Without a leading firm or some unifying cause, firms have always gone their own way. As one observer of the industry said, "The Hong Kong watch industry is so flexible and capable of changing that it can't plan for anything. People here are so used to adapting that they cannot conceive of planning for the future" (Peter Lee, President, Hong Kong Watch Trades Association, 1994).

Summary

Given the powerful market position held by Hong Kong watch industry firms, why did these companies remain wedded to a competitive strategy based on low wages rather than moving up the product development cycle toward more advanced aspects of the industry to compete in the medium-price watch segment? Why didn't they choose to displace the Japanese, who by the early 1980s dominated the midrange, much higher profit segment of the industry?

Numerous authors attempting to explain the organizational differences between Japanese firms and Hong Kong Chinese firms focus on the unique cultural histories of the two countries. Many Hong Kong manufacturers are former merchants who take cues not from production processes but from the market. A traditional manufacturer's priority might be to employ state-of-the-art processes to create a higher quality but standard product. In contrast, merchant manufacturers are perhaps more highly motivated to respond to consumer preferences and get product into customers' hands, regardless of process.

Finally, political uncertainty has contributed to a "get rich quick" mentality; thus, Hong Kong entrepreneurs have historically had expectations of very fast payback on investment (sometimes in less than three years). The yearning to own one's business adds to the short-term time horizon of many Hong Kong Chinese. In contrast with the steadfast and highly loyal Japanese worker, who is willing to toil selflessly for the collective good, Hong Kong Chinese workers have a strong desire to strike out on their own. Thus they yield up little of their general human capital to grow the businesses of the capitalist class (Wong, 1988). As Wong (1988) suggests, Hong Kong Chinese businessmen are more able entrepreneurs than managers, willing to invest with the expectation of a reasonable rate of return in a finite amount of time (p. 170). To move up-market and compete in the more expensive realm of the industry would require greater capital investment and include much higher risk. This is something Hong Kong producers have simply chosen not to do.

There is little we can ascribe to the internal workings of the Hong Kong industry that would have presaged its development into a world-class competitor. Starting as a parts distributor and service provider, the country grew into the established leader in the assembly end of the industry. Breaking with all previous models, Hong Kong assumed leadership entirely dependent on foreign import of critical watch technology and cheap, docile labor capable of working at small scale. To this day, no watch module manufacturer has emerged to supply the watch industry in Hong Kong despite the hundreds of millions of watches assembled there each year. The lack of a long-term investment outlook combined with insufficient management capability has led to an underinvestment in the scale of technology to dominate the technical end of the industry. The movement of major elements of the industry to the mainland may assure the low-cost market share of some of Hong Kong's larger firms, but the competitive pressures from mainland firms will remain a constant threat.

11 Can One Man Save an Industry?

Turning now from Hong Kong, and returning to Switzerland, we find that the struggle to retain market share and to recover from the complete loss of the low end of the market created a crisis for the Swiss watch industry unprecedented since the 1930s, when global recession brought the Swiss industry to its knees. As the quartz revolution began, the Swiss industry was self-satisfied, having just experienced 20 years of unparalleled growth. Even small producers and ancillary players in the industry had grown rich. Thus, when the first signals of pending change were noticed, the Swiss acted just like they did when Hetzel showed them the tuning-fork watch—they ignored them.

Not everyone was oblivious to the real potential of this new innovation, however. Engineers in the largest firms, and in particular those in the Société Suisse pour l'Industrie Horlogerie (SSIH) group, which included key brands such as Tissot and Omega, knew nonmetallic watches were coming. Engineering staff had already experimented with such devices. They also knew it was only a matter of time before the electronics revolution penetrated the watch industry. The question was not *whether* but simply *when*.

In previous chapters I recounted the turmoil unleashed by new competitors who viewed the watch as little more than a vessel, a necessary final market for an intermediate input. Why did the Swiss act with such caution, and even with disdain? Why did the once incredibly flexible system find itself flat-footed, unable to respond? Here, I recount the experience of the Swiss watch industry once the effect of the quartz revolution had set in. I emphasize how vertically disintegrated systems, flexible at one point in time, are peculiarly vulnerable to unexpected and system-transforming events, such as a technological discontinuity. The story details the fact that in this case, the "cultural crisis of the firm," a condition that disempowers firms and limits their ability to act on their own behalf, brought the industry to its knees (Schoenberger, 1997). In the case of watches, to succeed in the brave new world of electronics, the Swiss needed to rethink their notion of "precision." In this case, Schoenberger's (1997) meaning is extrapolated to an industrial complex as a whole. We can use her description of culture change as a basis for understanding why the Swiss industry found it so difficult to act. Schoenberger's insight bears repeating:

> Cultural change inevitably involves a struggle over power. By this I do not mean the rather circumscribed one envisioned in, say, organizational sociology over who runs this or that bit of the world, but a struggle over who has the power to construct and defend a specific social order and practices, behaviors, knowledge, and understandings appropriate to it. At the same time, cultural change involves a struggle over identity—over who and what the firm is. When these struggles are sufficiently acute, they amount to a kind of cultural crisis in which competing models of the social order, and the material and human resources and identities tied to them, are threatened with devaluation, and oblivion. (p. 122)

The winners in these circumstances could be anyone. In the case of Swiss watches, it took an outsider, Nicholas Hayek, to revive the industry. But in identifying the role of outsider, it is important to keep in mind that such a gesture is symbolic, acting largely to remind decision makers of their responsibility to make decisions. The underlying ability to respond must be present or else, as happened in the case of the British watch industry, things just fall apart.

Given Up for Dead

The early 1980s caused great angst in the industry, as well as outright regional depression. Within a decade the Jura region and the watch industry as a whole had lost more than 60,000 jobs. Hundreds of companies collapsed and the Swiss banks had to step in and restructure the industry's most influential groups.

The restructuring of the Swiss watch industry has been held up as an example of industrial renewal. Most stories of the region's revitalization emphasize the introduction of a new management team lead by Nicholas Hayek, a self-promoting businessman who supposedly "saved" the industry. Yet there is another more colorful and circumspect story about how the largest firms in the industry achieved a degree of renewal.

The Antecedents of the 1980s Reorganization

As SSIH continued to decline, near to the point of receivership, a regional outsider, Ralph Gautier of the Siber Hegner World Watch Marketing group, recapitalized SSIH, including the Tissot and the Omega brands, and in 1977 began the long process of restructuring and reorganizing the company. After assessing the situation, Gautier forced the company's conversion to quartz movements. The major brands went from only 8 percent electronic to 60 percent electronic in just three years. In addition to moving Swiss watchmaking into the electronics age, Gautier accepted the Swiss industry's loss of the low end of the market as final and concentrated on SSIH's strength: selling high-quality, medium-priced watches with cachet. He rightly saw that the market would split more definitively into three categories, and that in the short run cheap mechanical watches had a limited future. He positioned SSIH to take back market share from Japan's Seiko. This approach was strongly supported by advertising and other marketing efforts needed to reposition the company's products and regain market share.

Gautier, who spent his early career selling watches in Asia, knew what the company was up against. Rather than sticking with tradition and searching for solutions within the region, he made numerous forays into Asia in search of locations for the production of low-cost watches. He also secured semiconductors from German and Japanese companies rather than relying on inefficient Swiss production.

By 1978, the SSIH showed a small profit. While the turnaround attracted praise, problems still plagued the company. Given that the higher end market segment requires little replacement, growth was a function of interbrand competition. Success in the midrange segment required inventiveness. SSIH focused on reducing component sizes to give its designers more freedom. The company also introduced new products that took advantage of mechanics and electronics. Gautier felt that people would prefer to tell time via analog devices and therefore aimed research efforts toward quartz analogs rather than strictly competing in the digital realm.

However, innovation often causes severe headaches regarding inventory. To keep up with the competition, SSIH had introduced 30 new models in the previous 30 years. New models meant more than single variants of a product. For each model, superficial external variation increased the number of options multifold. One model could have 12–20 variations depending on the material used, the gender it was designed for, and other external elements. Each model required parts and service capability—a difficult and costly proposition.

Gautier could see that the future for wrist timekeeping would be very different than the past. His vision of watches as multifunctional and quite complicated was right on the money. He also foretold the era of the throwaway watch, "probably made of plastic and priced inexpensively. When the battery runs out, the watch will just be discarded and replaced with a new one, like cheap cigarette lighters or ballpoint pens" (Gautier, as quoted in *Fortune*, 1980). Gautier did not see this task as something for his company to handle. The reintroduction of a cheap plastic watch that could compete in the low-end segment would wait for the next management team. Prototypes for such a watch collected dust in the SSIH vault, to be brought out by another group.

The Antecedents of the Swatch Watch

The Swiss had been experimenting with plastics for more than 30 years before the introduction of the Swatch watch. Plastic was first considered as a material for bearings. SSIH was awarded a patent for the use of plastics for island bearings in 1952. When it was found that plastic did not have the qualities needed to work as bearings, the company pursued other resin-based materials such as delrayon for parts production.

The first plastic watch was designed in the late 1960s and put on sale in 1970. Several hundred thousand plastic watches were distributed through Tissot and Astronolon. The marketing strategy for this watch included making it cheap and mass-distributing it. Given Tissot's distribution channel, jewelers, it was not surprising that distributors balked at the new watch.

The age-old problem of deciding which SSIH branch would carry the new invention further delayed the introduction of the plastic watch. Both Omega and Tissot wanted to introduce the plastic watch. Omega, the larger firm, exerted its influence, resulting in a stalemate. As a counterstrategy, Tissot bought Roskopf in 1972 from Economic Swiss Time. Roskopf was the Swiss equivalent of Timex; it prospered by selling a cheap pin-lever watch. Tissot thought Roskopf could be the foundation for its plastic watch manufacturing, but the company was made up of traders, not brand sellers: it did not know how to use advertising to sell a brand.

Once again, the inability of a horizontally integrated system to agree on a strategy left a critical innovation gathering dust in the SSIH vault. Almost all the ingredients (including technology and a low-price model conception) that would enable Tissot's return to the market in a low-priced segment were in place. All the company needed was the right advertising company to market its new product.

The Emergence of Swatch

With the Swiss watch industry on the brink of bankruptcy, the federal government asserted fiduciary control and transferred ownership of a number of companies to individuals who had no prior history in the industry. This radical step was deemed essential to the precarious industry's survival. In the early 1980s the Société Suisse de Microélectronique et de l'Horlogerie (SMH), which was given the freedom to reconstitute large segments of the industry, created an entirely new production system that promoted an industry image based on fashion and operated on the basis of mass-manufacturing practices, subcontracting, and new distribution strategies.

With the introduction of the Swatch watch, SMH moved the industry away from its age-old watch marketing concept based on precision and toward a new fashion-based strategy. This reconceptualization led the Swiss industry back into the low-end price segment it had abandoned to Hong Kong and Japanese competitors.

This shift precipitated dramatic changes in the structure of the social relations and production system within the Swiss watch complex. The major innovation was the creation of a vertically integrated, highly automated, continuous process manufacturing system designed to produce Swatch watch parts and execute final assembly. To accomplish this industry reconfiguration, the watch caliber (dimension) was standardized, and the number of parts in the watch was reduced drastically. Most importantly, the stylish watch was made of brightly colored plastic, something never previously used in watch construction. Product differentiation was accomplished by varying the exterior color and (within limits) material of the watch, and altering the watch face to reflect different product moods or messages.

After SMH introduced Swatch it undertook a major reorganization and rationalization of its midpriced brand products (formerly owned by the SSIH Group) and gained further control of the industry's reorganization. This phase of adjustment also included the introduction of new models such as the "Rock Watch" and Tissot's "Two-Timer," a combined analogue–digital watch. The use of new materials was a significant departure from the nearly exclusive use of metal in the past. As part of this reorganization, new exterior parts inputs (cases, bands, etc.) were sourced from all over the world (a reminder of the basis for Hong Kong's entry into the industry), and interior movement parts manufacturing was standardized. As a result, midrange watch movements began to converge technologically, much as has happened in the automobile industry (Dicken, 1998). Yet achieving economies of scale in parts manufacturing resulted in the erosion of technological distinctions between brands. Brand differentiation was now based almost exclusively on characteristics associated with the watch's exterior (e.g., metal finish).

The Internationalization of Watch Manufacturing

The crisis in the 1980s saw the last vestiges of the 1930s Statut de l'Horlogerie (Protection of the Watch Industry Order) melt away as Swiss watch firms took to the international scene with a vengeance. Multinational corporate solutions to the Jura's competitive problems precipitated dramatic regional change. In February 1988, SA Fabriques d'Ebauches (ETA), the largest Swiss watch parts manufacturing conglomerate (and part of the Société Suisse de Microélectronique et de l'Horlogerie SA [SMH] company), closed several factories in the Jura in an effort to compete more efficiently in the world market. Under pressure from the Swiss metalworkers union Fédération Suisse des Travaillers de la Metallurgie et de l'Horlogerie (FTMH), the local government in Saint Imier demanded that ETA substitute a production unit equal to the one it was closing. While public dissent had little effect on the rate of plant closures, this latest wave of closures was met with stiff resistance from local residents, who complained that such actions further eroded the industry's prominence in the region.[1]

Pointing to the Swatch as the savior of the Swiss watch industry distorts the recent history. Even as the Swatch was introduced, the industry was in the midst of a major reorganization. This antecedent reorganization was absolutely essential to its recovery. Without it, the costs of later reorganization would have eaten into the profits of Swatch and other brandname products, and the SMH in all likelihood would have continued to struggle.

Reorganization à la Japanese

Swatch was a tremendously important component in the recovery of the Swiss watch industry. However, even more critical was the industry's reorganization along the lines of the Japanese industry and its massive increase in the sale of movements to watch assemblers in Hong Kong. The first change signaled the triumph of the Japanese, and particularly Seiko's, model of watch production. The second change signified the growing importance of interconnections between the Swiss watch industry and the global watch industry.

With regard to the production of Swiss watches, Swatch, despite its reputation as a huge hit, represents only a small portion of the value of the watches manufactured by the Swiss industry. Even as it passes its 140 million sales mark, the product has weaknesses. The standard models are losing some cachet, though demand is being buoyed by a cult affection for the trendier, collector models. The larger question is threefold: how well are the midpriced Swiss watches doing, for example, Tissot, Certina, and Mido? How important is the continued sale of movements to the overall cost structure of the industry? What efforts are being made to establish new markets? The midpriced watches are for the most part more of the same. Significant technological innovations have not occurred in this segment, as witnessed by the sales promotions that emphasize continuity and tradition rather than revolution and change. Old mechanical models are being reintroduced, playing on nostalgia rather than true differentiation and new product development. Does this matter? The answer depends on how effective competitors' campaigns are in promoting new and different watch attributes.

Second, how important is the sale of movements to the cost structure of the industry? This is difficult to ascertain. Clearly, watch producers with movement capacity have been flooding the world market with their movements for more than a decade (Japan Watch and Clock Association, 1994). This has fueled competition and driven down prices. Is the Swiss share of the movement market secure? In the near term, probably yes—if the phenomenal growth in the worldwide watch industry continues. Watch assemblers buy Swiss because of its presumed higher quality. But growth in the world market for watches is at the low end of the market. Thus, it is unclear whether there will be much further growth in the area of Swiss movements. Factors such as currency valuation will have a major effect on the Swiss share of the movement market.

Third, what efforts are being made to establish new markets? Innovation, or its absence, is a key. The Swatch was not a technological innovation—no matter what SMH's corporate propaganda says. The idea of plastic had already been pioneered; the reductions in parts and assembly steps were pioneered with the "Delirium" and reached the mass-manufacturing stage with the "Two-Timer"; and assembly-line capability was also within the industry's grasp. Product innovations and attempts at diversification were only modestly successful. Even at the high end of the market, watch producers were concerned about becoming vulnerable in the absence of innovation. The problem of a tired product line was clearly evident in recent trade statistics. Starting in 1994, plastic watch sales declined for the first time and continued to decline throughout the 1990s.

Ernest Thomke, the originator of the Swatch, saw this type of watch as a marketing innovation desperately needed to break the stranglehold of the old distribution system (*Los Angeles Times*, 1989). Prior to Swatch, ETA was unable to market watches on its own, but lacked confidence in the existing brands. Survival mandated a change in the historic monopsonistic relationship. Swatch was the ideal vehicle for this strategy. Producing and assembling the parts became one process. ETA decided to do it all. Swatch had two purposes: to create profits and to create a path so that ETA could go directly to the market with a solid rationale and break with past modes of operation.

New Markets

Swiss marketing efforts lack coherence. Ventures into Asia are performed more for the opportunity to employ cheap labor than to secure market access. Swatch continues to push into new markets, but the numbers are small. In the late 1980s, Hayek announced the opening of a production operation in Japan, where the company sold about 400,000 watches annually. Always prone to hyperbole, his vision seemed a bit grand when he announced that the Swiss would produce 40 million watches and movements, a sum equal to the Japanese market (Matsuzaka, 1992).[2] Such a strategy contradicts the de facto marketing plan for Swatch, which is to make it exclusive and keep its numbers down. Other foreign operations in places such as Malaysia and Thailand are strictly low-wage production locations, hardly an effort to meet the Japanese head on in newly emerging markets.

Is the Swiss industry less dependent on global conditions? Not at all. Until the early 1990s, Hong Kong remained the country's largest market. The island country has historically been the tourist market of choice for the world's most discriminating watch buyers. But even Hong Kong merchants are beginning to see this dominance decline as consumers' tastes and preferences for high-value trendy goods change, as incomes in the developed world stagnate or shift downward, and as China itself becomes a locale for consumer spending. The Asian crisis, starting in 1997, saw the major market for Swiss products pass from Hong Kong to the United States. If experience is any guide to the meaning of this shift, the Swiss have little assurance that Americans will continue to buy Swiss watches from one year to the next. Given the peripatetic pace of fashion, what is au courant today is rarely in style tomorrow. The Swatch was designed with the baby-boom generation in mind. These folks no longer rule fashion trends, as tried and true brands like Levi's have painfully learned.

Already new brand names, some established by distributors, have made inroads in the young adult market. Fossil, Inc., a company headquartered in Dallas, Texas, has come out of nowhere to take a significant share of the high-volume, less-than-$100, fashion-intensive end of the market, where 35 percent of Swiss watch revenues are made.

Another irritant arises from the designer side of the fashion business. Untold numbers of clothing designers, such as Liz Claiborne, Tommy Hilfiger, and Donna Karan, now sell accessories including watches. These "designer" watches are almost exclusively assembled in Hong Kong. Thus, the Swiss face significant competition at the high end of the low-priced watch market.

Are the Swiss high-end watchmakers at risk? Some are, given the high costs and contracting or unstable markets associated with increasingly global events. While most companies are not likely to fail, many will require outside capital and the patience to simply see the annual watch production cycle through. The recent consolidation of numerous high-end watch brands suggests that even this segment of the market is no longer the exclusive purview of the Swiss.

The more expensive brands have developed second lines in the hope of luring the middle class. This new plan is contrary to earlier plans for major brands, such as Omega, to scale back access to their products and reassert their uniqueness. The problem is that there is simply not enough volume at the top end of the market to achieve reasonable production costs. Many top brands still manufacture too many different movement modules to achieve significant economies of scale. And even the upper-end watches are having problems. Piaget, the venerable 150-year-old, family-run company, recently sold 68 percent of itself to the watch company that makes Cartier. Instability in the global economy and changing consumer tastes make selling very high-priced watches unpredictable (Ocampo, 1992).[3]

The 1990s and Beyond

During the 1980s Swatch was a Swiss savior both in terms of the Swiss industry's return to the low-priced end of the industry and in terms of industry image and morale. By the early 1990s, Swatch was 10 years old. It had become a brand that retained a certain mystique but that was nonetheless aging. Its importance was as much in saving face as in saving the industry itself. Competing in the low-priced segment, it still accounts for a small share of total watch sales (sales, 5 percent; volume, 35 percent, 1999). This is ultimately the problem.

The world watch market is highly segmented. Rough calculations suggest an inverted relationship between value and volume. High-priced products ($750 U.S. dollars) account for only 5 percent of sales volume, but they account for almost 50 percent of sales value. At the low end, inexpensive watches ($50.00 U.S. dollars) account for 70 percent of volume, but account for only 5–8 percent of sales value. The vital midrange is extremely wide, and accounts for something like 28–29 percent of volume, but as much as 45 percent of value. Missteps in this segment therefore can have major implications for the industry overall (Fédération de l'Industrie Horlogère Suisse, 1998).

The 1990s saw a continuing decline in the Swiss volume share of world markets. Starting in the early 1990s, global crises, rising exchange rates, and growing competition created circumstances that continued to eat away at the Swiss market share based on volume.

The export of Swiss parts and movements was a necessary step in reviving the industry. Long restricted from selling parts abroad, after reorganization in the early 1980s, the industry could finally export movements. Movement sales increased dramatically to 50 million units. This newfound freedom had its own liabilities, however. By the early 1990s, a large share of the overall industry including Swatch itself, faced low-priced competitors who were marketing watches with Swiss movements. Thanks to their inclusion of Swiss movements, many cheap watches were billed as "Swiss" by Hong Kong producers. To combat this problem, the industry passed regulations defining what constituted "Swiss made" (50 percent of the sales value of the watch). The new law was designed to restrict illegal foreign practices. But as the 1990s progressed, this limitation served to hamper domestic producers who faced rising costs and a growing need to import components from low-cost locations. As the decade wore on, to counter sharply rising exchange rates, Swiss producers were being forced to confront the undeniable need to move production abroad. The double-edged sword of "Swiss made" then came back to haunt them. Unable to source abroad, the medium-priced segment of the industry watched cost conditions and market share erode. More and more of the external parts had to be sourced abroad to offset high production costs for the manufacture of movements. Small Swiss firms, once secure in providing the external parts for watches, saw their markets decline as more production was moved abroad. Low-priced watches were exempt from the restriction, which only served to add confusion to the market and to further erode the Swiss reputation.

The reentry into the low-cost segment of the industry proved to be "too little too late" in the fight against foreign competition. As one Hong Kong producer who imports both Japanese and Swiss movements for final assembly said, "They [the Swiss] need us more than we need them. What's the difference between a Japanese movement and a Swiss movement? Not much. So we play the big guys off against one another. We are the volume and we control the access to new markets. Today they need us more than we need them" (Eddie Liu, President, Dailywin Watch Company, personal communication, 1994). Recent statistics on watch movement sales tell a revealing story. The Swiss share of worldwide movement sales has continued its erratic trajectory throughout the 1990s. In 1998, Swiss movement sales were down by 13 percent. This segment is critical to the achievement of economies of scale and hence low production costs. A continuing slide in this product market presents serious problems, particularly as the low-cost, high-volume plastic watch segment of the market continues to stagnate (Fédération de l'Industrie Horlogère Suisse, 1999).

The 1990s saw a return to instability. Up until 1991, the Swiss industry appeared to be recovering from the dark days of the early 1980s. Swatch had become a global brand increasingly well entrenched in the traditional Swiss markets of Europe, the United States, and parts of Asia. The cute, cerebral, fashion-oriented watches took the youth market, a segment Switzerland had never had much luck with, by storm. At first, on the eve of the new decade, even the medium-priced watches experienced some degree of stability, if not growth, in market share. In addition, the Japanese still had not matched the Swiss in terms of the aesthetics of the industry. Reigning supreme and seemingly untouchable, Swiss watch manufacturers owned the high-priced luxury end of the market. And it was profitable. While Swiss industry employed slightly less than 30,000 workers, Swiss artisans ruled the high-end market, a market previously immune to price fluctuations and exchange rates.

The first signs of future instability came as the Swiss franc began to gain strength against other foreign currencies in the early 1990s. The integration of the European Economic Community (EEC) was causing problems in key markets such as Germany and Italy. These markets were sacred—they had previously kept the Swiss afloat in dark times, but were now at risk. The immediate impact was felt in the medium-priced segment of the market, where there were viable alternatives and where consumer preferences are notoriously fickle. Additionally, to an important extent, the bottleneck of design had been overcome by the Japanese as the Swiss midrange product lines began to look more and more like one another.

From 1992 on, the midpriced market segment saw a slowing of demand and faced highly contested market shares. Although SMH, the parent company of Swatch, had succeeded in drastically cutting costs through rationalization and standardization within its medium-priced brands (Certina, Tissot, Mido), it did so by applying Swatch production logic to a more refined product segment. This entailed standardizing parts across major brand names with little regard for their reputation. The strategy economized on production costs while significantly diminishing product differentiation based on presumed grades of quality. Cheapening the product proved to be problematic as customers grew increasingly confused about the difference between a Certina selling for $150 U.S. dollars and a Tissot selling for $350.

One by-product was customer alienation. Other strategies such as the introduction of new materials and functions were momentary hits, belatedly satisfying the burgeoning fashion and technology niche markets. While products like the "Rock Watch" and the "Two-Timer" initially met with market acceptance, they proved to be both expensive and hard to regulate in terms of quality and timeliness of delivery. The "Rock Watch" and its variants proved easy to copy; the "Two-Timer" digital–analog was hard to operate and read. The continuous decline in the price per function boggled the Swiss mind. Still locked into the mind-set of mechanically complicated, the Swiss simply could not fathom customer desire for gadgetry and style and willingness to pay for them (within reason). For example, outdoor enthusiasts were willing to pay $150 (U.S.) for a watch that told them the altitude, regardless of whether it worked correctly as a watch or was waterproof. Timex and Casio took this market segment by storm and built it into a two-billion-dollar market (*Forbes*, 1988).

The low end retained its marketability until the mid-1990s. Then, like a great wave, foreign competition, particularly from Hong Kong, and increasingly from mainland China, began to engulf the fashion segment previously controlled by Swatch. Knockoffs and look-alikes confused and eroded the market.

Once considered invulnerable, the dominance of Swiss watch firms at the high end has steadily eroded since the late 1970s. Fifteen major Swiss firms, some of the most prestigious names in watches, have been sold to foreign interests (Union Bank of Switzerland, 1997). To manage the huge overhead associated with carrying very high material costs, particularly gold and other precious metals, some companies pursued joint production agreements with fashion houses (e.g., Ebel produced for Cartier). But these relationships are highly unstable, particularly for the smaller partner. Cartier took advantage of its powerful position to source gold for its watches in combination with its other jewelry products and set up its own production unit by buying a very prestigious Swiss brand. Even this strategy has proved to be ephemeral for top-end producers.

By 1997, restructuring of the industry was again being called for. The strong growth of the late 1980s was nowhere to be seen. While high-priced watches retained their markets overall, they were increasingly being challenged as new competitors popped up in the heart of the industry. There had always been an implied gentlemen's agreement at the high end regarding who would produce what. Given that volumes were very low, markets were extremely segmented, changes were made, and major companies were gradually accepted in the industry. One important development rocked this seemingly stable, decades-old arrangement. In 1993, Swatch added 13,000 new units of its "Trésor Magique," a watch made of platinum. While the Swatch platinums were snapped up largely by collectors, the Swatch marketing ploy was viewed as a shot across their bow by the sacred high-end producers. As it had earlier done in the medium-priced end of the market, Swatch moved into the high end and turned everything upside down. The platinum Swatch broke ranks with tradition and served to further challenge the Swiss image. The impact of this development was viewed by one observer at the Union Bank of Switzerland as follows:

> The introduction of the Trésor Magique immediately made the SMH [Swatch's parent] the production leader of Swiss platinum watches. Annual production previously amounted to less than 10,000 in number but now one must add an additional 12,999 time pieces of this sort. The entire stock of platinum Swatches was sold domestically and abroad within a few days in a targeted action unprecedented in the watch industry. It remains questionable, however, if such an action can be repeated with similar success. (Union Bank of Switzerland, 1993, p. 22)

As for the medium-priced end of the market, the lack of attention to true new product development and product differentiation seriously threatened this segment of the industry. With the exception of one year in the 1990s, market share continued to decline. In 1997, exports were down 5 percent and prices had to be cut to stem further erosion in market share in the face of intensified competition. The Swiss had no response. The midrange had been fragmenting since the early 1980s. The area of strongest world growth, complicated electronic watches, was completely dominated by the Japanese and Timex. The battle lines were drawn in the segment, and the Swiss retreated. Once again, they sat on the sidelines, locked in by their past.

In the face of blurred meaning in the medium-priced segment, and a recycling of tired old mechanical models strategically positioned to play on nostalgia, new entrants challenged vulnerable firms in the medium-priced segment. Given the breadth of the segment, which includes watches costing from $50 to $750, both new Swiss and new American firms entered the fray. The U.S. upstarts picked off the trendy, low- to medium-priced fashion end by producing knockoffs of the more stylish Swiss models. Given the size of the U.S. market and the growing affluence of the younger crowd, the Swiss "Rock" and "Wood" watches were perceived as stodgy and unappealing—and why not? The Swiss products were more expensive and were sold in department stores and jewelry shops, places infrequently visited by teenagers. The market was increasingly being stratified based on age and lifestyle.

For the over-40 crowd (e.g., husbands buying the first luxury watches for their wives) new market entrants like Raymond Weil satisfied upwardly mobile customers as yet unable to buy a Rolex, but still coveting a watch as a piece of fine jewelry. As a new market emerged for this segment, the staid old-time SMH brands were caught flat-footed with little new to offer.

This turmoil rocked the complex infrastructure. Suppliers to the medium-priced segment felt the pressure. The medium-sized watch firms had the flexibility to shop around for components. Disorder increased as the ability to source abroad expanded. While some of the medium-priced producers had long sourced watch cases and bands internationally (justifiably, because of the lack of a local supply), now they were increasingly sourcing internal components abroad. Medium-priced watches still claimed market prestige from the use of the "Swiss made" label. For some medium-priced firms, the regulation put into place in the early 1990s was seen as a hindrance to achieving low costs and profitability. Regulation once again provided only temporary relief from competition but ultimately became a double-edged sword, keeping prices high and thereby opening the industry to further competition from the Japanese and other lower priced new entrants.

The 1997 financial crisis in Asia has had a major impact on the mechanical watch trade. According the the Fédération de l'Industrie Horlogère Suisse (FH), mechanical watches now account for just 7.9 percent of volume and 44 percent of value, down from 9.3 percent of volume and 47 percent of value in 1997. The strategy of reviving the mechanical watch trade clearly has quite a distance to go if it is to be an important future contributor to watch sales growth (Fédération de l'Industrie Horlogère Suisse, 1998).

The reentry into the low-end segment only 20 years ago was heralded at the time as the rebirth of the Swiss watch industry. For 10 years, the SMH Swatch strategy appeared to defy global economics based on high skills, low wages, and the sourcing of low-cost parts abroad. This revival appears to have come to an end, however. Now even Swatch has unstable sales levels, and Hayek's rhetoric sounds increasingly stale and out of date. Recent figures from the Fédération de l'Industrie Horlogère Suisse indicate that consumers are continuing to move away from low-priced plastic watches toward low-cost metal-cased products. While Swatch produces metal-cased products, its volume system is built on economies of scale associated with plastic watch manufacturing. There are many small producers aside from Swatch in the low-priced metal watch market. Hence competition can be expected to intensify as firms vie for this segment of the low-end market (Fédération de l'Industrie Horlogère Suisse, 1999).

Time-to-market has become the key to economic success. The turnaround time for entirely new collections of Swiss watches is now down to less than one and one-half years—although this is still quite slow by low-end market segment standards. By comparison, Hong Kong watch producers can turn around watch collections in as little as six weeks. Because of the long lead time for bringing out new products, turnover has to be high, very predictable, or at least steady—otherwise inventories build up and unsold goods threaten to swamp reserves with little to offer in their place. This continuing difficulty remains an ever-present threat to small businesses, one intensified by the creation of new fashion-based upstarts with few qualms about going to Hong Kong to contract for market-ready models that can be provided in volumes from 600 to 60,000 in just six weeks. In fact, since most of the world's fashion clothing is produced using facilitators located in Hong Kong, it takes little extra effort to find a watch assembler who will produce a fashion watch to match a seasonal clothing collection. Thus, designer companies and others have done just that. Fast turnaround requires the ability to source widely. Swiss turnaround time has never been fast. Geographic isolation makes sourcing with a short-term time frame difficult, if not impossible.

Under these circumstances, when all else fails, the wise course of action is to rely on your distributors. They represent intangible assets—they need you (the watch producer) as much as you need them. This historic symbiosis embodies high levels of loyalty during good and bad times, and therefore represents one of the few tried-and-true aspects of business where a handshake still seals a deal.

Uncertainty strengthens conservatism, however; in periods of turmoil, all bets are off. The opening up of China and the return of Hong Kong to China have created new challenges for the Swiss industry. Until the late 1990s, Hong Kong served as the largest intermediate retail market for Swiss watches. While it is really little more than an entrepôt for Swiss watches sold into Asia and other parts of the world, many dealers have been forced to abandon Hong Kong for China, disrupting decades-old distribution channels. While the impact of this change is felt most noticeably at the high end of the market, Hong Kong is still a very important market for medium-priced watches and for watch movements. Recent statistics from the Fédération de l'Industrie Horlogère Suisse indicate that Hong Kong sales remain erratic and very much subject to fluctuations in economic conditions in Asia. Demand declined 17.5 percent between 1998 and 1999 (Swiss Watch Industry Federation, 1999).

At the end of 1996, industry observers were portending a major shakeout in the industry:

> In 1997, many firms in the Swiss watchmaking industry will be faced with fundamental decisions: unless demand improves significantly, major structural adjustments will be required. This scenario would be shaped by plant

closures, takeovers, and price erosion. According to the UBS survey, smaller firms in particular are concerned about a significant deterioration in sales and earnings. Reasons given are the growing competition from eastern Asia (with the resulting pressure on prices) and weak export demand. (Union Bank of Switzerland, 1996, p. 23)

Summary

No one man alone can save an industry. Hayek's repenetration of the low-end market was a bold, clever, and risky move, but not sufficient to immunize the Jura-based industry from global economic forces. The Swatch strategy may have reached its limit; foreign competition is increasingly eating away at the low-end plastic watch market segment because here, more than in any other segment, low wages count. While Hayek has tried numerous other ventures in other product areas all built around the mystique of Swatch (phones, radios, timers), none have enjoyed market acceptance at the scale of the plastic watch in all its permutations. As if to prove his invincibility, Hayek's most daring plan to date was a joint venture between Mercedes Benz and Swatch for the production of the "Swatchmobile." The car is designed to appeal to the in-town driver in search of convenient parking in Europe's notoriously overcrowded cities, and to the loyal Swatch buyer with an environmental conscience. When the product came to market, it was seen as high-priced relative to its competition, and had little beyond clever design to add to Japanese or other European cars with a long and successful service record, established distribution systems, lower prices, and an equally small size. While much fanfare has preceded the identification of the Swatchmobile partner, it is curious that Volkswagen, a company with a history of producing a car for the low-priced market segment at mass scale, was approached early on, only to back away from the venture. Although it may make sense for Mercedes to go after this market segment, given its lack of presence there, and while teaming up with Mercedes may have given Swatch added cachet, only recently did the project move from the discussion stage to actually manufacturing the product itself. Its introduction was fraught with engineering problems. Recognizing the difficulty of moving into an industry so far removed from watches, the Swatch group recently sold its interest in the Micro Compact Car to Daimler Chrysler, for an undisclosed amount. Little more was said in the press except that the car was an integral member of the Daimler group.

As for Hayek, few new watch-related revelations appear to be in the offing from SMH. The Japanese and Seiko, on the other hand, continue to pursue an aggressive strategy of complementary product development to reduce dependence on watches. In the case of Seiko, by the middle of the 1990s, watches had fallen from 60 percent to 40 percent of company sales. While watches have by no means been abandoned, they are no longer seen as an end in and of themselves. Seiko conceded long ago that by the early years of the next century, China and other low-cost producers would control the watch industry. Seiko is content to allow this inevitable change to occur. It sees itself as the beneficiary of a hundred years of micromechanical know-how and experience. Its diversification plan, first outlined 20 years ago, is deeply rooted today. Seiko makes printers, other computer peripherals, micromechanical robots, chips, optical scanners, and assembly-line technology. As for Citzen and Casio, they have broad market reach as well. In both cases, watches have come to absorb electronic parts rather than acting as an end in and of themselves. The same cannot be said of the Swiss watch industry and its flagship firm, SMH. Only time will tell whether and when the Swiss also will be forced to accept this emerging reality.

In the meantime, the return of Hong Kong to China has precipitated a complete transformation of the Hong Kong industry. The Hong Kong factories of today look more like semiconductor plants than the dark dingy lofts of old. Gone are the small workshops. They have been replaced with large factories sometimes 10 times the size of a typical Hong Kong plant. Beginning in the 1980s, Hong Kong watch producers began to establish production sites on the China mainland where a combination of cheap land and abundant, very low-cost labor could be found. The job decline in the industry has been dramatic, with more than 10,000 jobs simply disappearing. The decline in jobs has been accompanied by an explosion in output. In the year 2000, China and Hong Kong were expected to be making almost a billion watches a year. Eddie Liu, of Dailywin Watches, a large Hong Kong–based watch manufacturer, indicated that the factories on the mainland have finally overcome the enduring problem of the Hong Kong industry: the inability to produce midpriced watches that could compete with the likes of Seiko and certain Swiss brands (personal communication, 1994). He is confident that in the next 10 years, China will rule the world watch industry. This transformation has not been without costs. Aside from job losses in the industry, old retail and wholesale relationships have been seriously challenged. With production growing at remarkable rates on the mainland, distribution is no longer exclusively controlled from Hong Kong. The decline in Hong Kong market share is causing great hardship and resulting in infighting among old distributors. What does the future hold? Many forecast the end of watch manufacturing in Hong Kong. While defensive gestures such as the development of a

Hong Kong–based brand to differentiate that area's products from those being produced on the mainland are being discussed, no single firm has emerged with the financial power and authority, much less the control of markets, to claim the name of this new watch. And so, like the Swiss on the eve of the quartz revolution, when no one asserted enough authority to claim the brand name that would carry the new technology to market, no tactical response has been made. History does appear to repeat itself.

12 *Success Goes to the Nimble, Regardless of Size*

Over the last 150 years, several countries have vied for leadership of the world watch industry. At different moments in history, factors converged to catapult one nation's industry ahead of another's. In a surprising number of instances it was relative success (and consequent institutional inertia) that ultimately resulted in the demise of a nation's watch industry. Industrial dominance appears to have precipitated a lax attitude toward competitors and obscured the need for new product development. In certain instances, a fragmented organizational structure precluded consolidation of the capital needed to purchase new production technology to meet the competition. In other cases, competitors simply had too much invested in existing operations to shift willingly toward new products and markets. Both sets of circumstances resulted in a significant redistribution of market share among countries' industries.

In this book I have traced the development of one technologically rich industry across three continents and five countries. The purpose of this elaborate exploration was to investigate how such factors as technological change, wars, economic recessions, industrial ideologies, industrial structure, and culture shaped the fate of one industry—watchmaking. At the heart of the story is the contingent nature of development. Over the past 15 years, academic debate has sought a "Holy Grail" in the form of a select set of ingredients, including key industries capable of fostering and promoting continuous economic development. As the example of the world watch industry is intended to show, any number of intervening circumstances seem to limit or promote the successful development of an industry. At the same time, certain reoccurring features of industrial development allow for some degree of generalization about how the process unfolds and changes through time.

260

This story has been manifested institutionally in quite different ways across locations. In the case of Britain, Switzerland, and Hong Kong, the watch industry can be characterized as horizontally specialized and vertically fragmented. In the United States and Japan, the watch industries were vertically integrated, though in varying degrees. The exploration of these core experiences suggests that different forms of technological and environmental change are managed better depending on the structure of the industry. Incremental change may be confronted more successfully by a disintegrated system in as much as multiple attempts to resolve the same problem may produce a superior result. In contrast, there is little evidence that either system is particularly adept at coping with paradigmatic shifts. Given the costly nature of radical change, however, key differences in the capacity to cope come down to the need for a limited set of decision makers and a storehouse of resources required to chart a new path forward and then follow it. This is particularly the case when the nature of the change called-for does not present a clear direction.

Although this book has traced the history of the industry across time and space, I have repeatedly emphasized how the Swiss watch industry coped with change. This emphasis was intentional. No other country's industry has had as many or as extensive brushes with fate and still been able to renew itself. Thus, on this account alone, the Swiss industry is worthy of careful investigation. In this concluding chapter I summarize a set of reoccurring themes that have emerged, ones that in important ways have linked the five locations to one another over time.

The historic success of the Swiss and U.S. watch industries (relative to the British industry) is partly explained by the institutional rigidities that pervaded the British industry. British watch manufacturing was governed by skilled artisan guilds. These guilds mounted great resistance to the development of mass production and the use of interchangeable parts. Instead of responding to market signals that called for low-cost, mass-distributed watches, British manufacturers maintained that the public did not want cheap watches. Other explanations for British failure include the relatively high cost of production in Britain, which forced specialization in high-value watches, where profits were greatest. Moreover, rigid social institutions governing the British industry sowed the seeds of the industry's downfall and led eventually to the complete abandonment of the watch industry. Innovators and entrepreneurs diversified into wholly new industries such as bicycles and automobiles.[1]

Inertia clearly contributed to individual countries' failures to adopt significant watch industry innovations. Throughout the history of this industry, various countries developed new techniques to support the production of ever cheaper and ever more accurate watches. Nonetheless, the invention of a technology was no guarantee of its eventual adaptation and application within a domestic watch industry. Several inventions and innovations developed by British watchmakers were never implemented within Great Britain. The same can be said of the French. The Swiss also suffered a similar fate with machine-based parts manufacturing. At least one U.S. watch producer (Hamilton), aware of the benefits of LCD technology and its potential to displace LED technology, still chose to capitalize on LED technology despite obvious implications for competition with makers of LCDs.

On the other hand, the importance of technological developments as experienced through changes in market share evoked more extensive responses. The Swiss were conscious of market signals and responded several times to market share threats by institutionalizing innovations developed elsewhere. In every instance, they improved on the innovations they borrowed. U.S. firms, although first to perfect interchangeability and mass production, were unable to stabilize the domestic market, and undercut themselves several times through predatory pricing behavior. Due to the size of its domestic market and its protection by trade barriers, the U.S. industry also was more insulated than the Swiss industry. It was not as concerned about the world's vision of the ideal watch. Thus, while the Swiss momentarily lost the U.S. market, they maintained 80 percent of the world watch market. This, and a persistent posture of creating products for world markets, sustained Switzerland as the world leader in watch production for long periods of time.

Distribution and Market Segmentation

One of the Swiss watch industry's unequaled strengths was its continuous export orientation and successful distribution networks. Switzerland's early success in distribution was partially responsible for undermining the British industry through the sale of low-cost Swiss parts to British watchmakers.[2] As time went on, the Swiss used various means to maintain and, in some instances gain, market share through strategic actions—for example, the purchase of U.S. firms, entry into joint-production agreements with U.S. and Japanese firms, and early sale of parts to other watch producers.

Due to timing and institutional factors, the German, French, and British watch industries chose to compete at the high end, in very narrow product markets. Although the reasons for this choice are not completely consistent across the three countries, local demand appears to have influenced the eventual selection of markets. Producers also sank significant capital in this end of the business and appeared unwilling to add additional investment to compete in other markets. In the United States, the

Elgin Watch Company also maintained a narrow product specialization. As time went by, this strategy became a liability.[3]

The Japanese were by far the most effective industry participant in terms of controlling distribution channels. At home, Seiko broke with tradition as rising demand and a shifting retail sector provided an opportunity to redefine channel strategy. In the United States, Seiko made the necessary investment to create a national sales and repair service system that got their product into customers' hands and allayed their fears about repair work. This set of aggressive tactics paid off as the Japanese successfully penetrated both the medium and low-end segments of the U.S. watch industry.

Hong Kong has benefited from distribution channel leadership due to its role as an entrepôt economy. As the gateway to Asia, Hong Kong had a formidable position as the go-between between watch producers and the world's largest and fastest growing market. For the Hong Kong Chinese, as for the Swiss, no market was too small or too underdeveloped. Their marketing expertise and comprehension of market signals continue to allow them to dominate the largest market segment in the industry.

The Role of Skill

The U.S., early British, and Swiss industries deemed the perpetuation of skill critical to the longevity of the watchmaking industry. The British institutionalized the exclusive nature of the product they made by restricting who could be considered a master watchmaker. This gesture, while preserving position for certain employees, prevented the extension of the skill base of the industry. This proved to be a major limitation, as the British watch industry never developed the engineering talent to move into new market niches. The Swiss carried this development farthest, establishing schools, contests, standards, and other activities to perpetuate industry innovation. And although certain U.S. firms followed suit, their tendency to diversify into nonwatchmaking ventures did not reinforce the benefits of such social institutions. Moreover, unlike the Swiss industry, the U.S. watch industry's early application of mass-production techniques and vertically integrated structure resulted in production so routinized that it simply did not yield as many multiskilled individuals.

The importance of human skill to the Japanese industry is only known anecdotally. The industry's relocation to the Suwa region during World War II was purportedly made due to the rural population's skill level, developed as a result of long industrial experience with silk manufacturing. More recent evidence clearly shows a high recognition of the importance of skilled workers to the industry, yet firms within it have maintained relatively flat occupational profiles. This has more to do with the Japanese system of manufacturing than with the watch industry itself. The truly great innovation pioneered by the Japanese was the reduction of direct labor costs, which comes about when products are standardized and organizations are streamlined.

Hong Kong has proved to be a place where flexibility reigns supreme. The country's industry is not built on skilled labor, however. This strategy has reached its limits and the national government has spent considerable sums trying to build a base of skilled workers, understanding that the future of the industry depends on being able to move upmarket.

In contrast with the United States, where watchmaking was spread across several states, the relatively large size of the Swiss industry and its geographic concentration worked in its favor. The large number of people in the industry allowed for extreme specialization, which led to continued technical development and the nurturing of skilled artisans and inventors. This same type of spatial concentration provided initial benefits to the British industry, but as time passed spatial concentration precipitated entrenchment, which functioned as an insurmountable roadblock for needed industrial change.

Institutional Arthritis Sets In

Throughout the history of the watch industry, countries have traded off market leadership. Over time, competitive pressures resulted in a number of responses from watch producers, ranging from the contracting out of manufacturing to the complete automation of production. Government protection of various forms has also been pursued by competitors in the watch industry since its inception.[4] The British used tariff barriers and other restrictive covenants to maintain control over their markets.

The Swiss used government protection at different points in the life of the industry; this protection reinforced myopic tendencies. Since the 1930s, the industry cartelization provided certain protections. As a major stockholder in the largest watchmaking holding company, the government exerted control over the manufacturing and exporting of watches. More important, the government long limited technology transfer from Switzerland, prohibited Swiss acquisitions of foreign companies, and closed the doors to foreign-owned watchmaking firms. Had it not been so protected by the government, the Swiss industry might have initially been more concerned about Japan's entrance into watchmaking. But while the Swiss industry maintained its protectionist policies, U.S. firms (which had Swiss subsidiaries) were selling technology to the Japanese. No doubt these joint production agreements provided Japanese watchmakers with needed technology and access to markets early in the life of their industry.

Today, attempts to limit competition once again haunt the industry. A government law defining what constitutes "Swiss made" bought time for some producers whose product markets were being raided by lower cost Hong Kong producers. But this same law also served to tie the hands of many participants in the industry, making low-cost sourcing difficult for them. This has become a particular problem as exchange rates make Swiss products increasingly expensive compared with those of their competitors.

The Japanese have prospered thanks to the protection of the home market from foreign competition. This protection is especially important during periods of yen appreciation. The Japanese government also encouraged industry concentration by limiting the availability of foreign exchange at certain times, while at other periods it has restricted the importation of watches by distributors and other watch assemblers. As it has done with other industries, the Japanese government has maintained a policy of rationalizing and consolidating firms in the watch industry. The government forced weak competitors out of business through liquidation or merger with stronger partners.

In contrast to extensive intervention by both the Swiss and the Japanese governments, the U.S. government protected the domestic market for its watchmakers, but only indirectly assisted in rationalizing the structure of the domestic watch industry. In many cases, watch producers benefited from economy-wide tariffs enacted on behalf of general manufacturing. Starting at the beginning of this century, tariffs were used to regulate the flow of products into the United States. Through the 1930s, protection fixed the watch industry structure in place. Between the 1920s and 1940s, national defense arguments were used in attempts to save what was by then a dying industry. Special trade programs promoting economic development facilitated the importation of cheap watches from territories such as the Virgin Islands. These trade policies assisted the early offshoring of U.S. watch manufacturing.

Defense contracts are probably the greatest point of connection between the government and U.S. watchmakers, and even this has produced mixed results for U.S. companies. Hamilton, for example, had to pull many workers out of watch production during the Vietnam War to meet fuse production deadlines. On the positive side, some defense-related R&D had applications in watch production.

Industry Structure, Regional Performance, and Enduring Competitive Advantage

Viewing the world watch industry historically reveals the fragility and in some cases the ephemeral nature of market dominance. Inertia often caught market leaders off guard, while more nimble competitors forged ahead by being flexible and capable of change. Under siege, numerous watch producers sought refuge in government protection. In most cases protection only delayed the inevitable. Ultimately, when government restrictions were lifted, these countries' watch industries were moribund and unable to compete internationally.

Failure to size up markets correctly has repeatedly spelled the demise of watch producers. The ability to create new markets continues to define world leadership in the industry. Historically, responsiveness to competition from within a region's industrial complex often meant the difference between success and failure for individual firms. The ability to perpetuate the technological superiority of a region's product provided a resiliency that competitors were hard-pressed to overcome.

Yet the same unity of purpose faltered when major adjustments were warranted in response to changes in technology. The shift to electronics-based watch products exemplifies this development. Watchmakers did not control the pace of technological change because they were simply one of many users. The old strategy based on precision and sophistication no longer assured mass-market dominance. The advent of electronics shifted the entire playing field of watch production to a new and different plane. Now anyone could buy the internal workings of an electronic watch and assemble it into a final product. The skill was no longer in the production of the final product. In place of mechanical prowess, marketing and distribution became the keys to world market leadership.

Endogenous Attributes and the Weight of Exogenous Events

How do we explain the evolution of an industry through different organizational configurations, technology cycles, and corporate cultures? I believe explanations can be found by stepping back from a specific moment in the evolution of an industry and exploring the interactions between factors that are endogenous to it and the effects of exogenous events that help shape possible avenues of response by industry participants. In this book I have invoked such concepts as path dependence, technology lock-in, and institutional inertia to help explain the fortunes of the leading competitors in this industry over time. By so doing, I have attempted to put flesh on the bones of such concepts. At the same time, however, I have suggested that the environment external to the firm and even the industry has much to say about what actually shapes the choices firms have at their command in periods of significant industry upheaval.

From one perspective, the historical evolution of the world watch industry can be seen as a simple problem of increasing returns to specific modes of industrial organization, technology choice, and institutional configuration. The evolution of watch technology and systems of production would then reflect the acceptance over time of superior configurations of technology, social institutions, corporate structures, and ancillary business practices, including marketing and distribution. At different moments in the life of the industry competitive outcomes would be based on successive waves of successful adaptation to evolving problems. Evolutionary economic theory does not require that a superior technology or outcome always be chosen, just that once selection has taken place, lock-in occurs and hence history is made.

While this characterization goes a significant way in helping to explain the evolution of the industry, the various paths that presented themselves as the industry evolved are simply too numerous and complicated to be fully captured by invocation of just a few concepts. This project attempts to account for the interaction of competitors as they come to know one another and react to each others' behaviors, extrafirm constraints that exist in the surrounding industrial environment and form the immediate cultural context, and the still larger environment in which exogenous events limit the effective choice of action. As Brian Arthur (1994) suggests, the effects of technological lock-in and path dependency are not immutable, but rather take many forms depending upon the culture and sector in which they occur. Neither lock-in nor path dependency occurs in isolation; their durability is contingent on changes taking place in a larger context. More fundamentally, it is the operation of these constraints at different spatial scales that helps explain the contingent nature of development. Ultimately, the meaning and significance of concepts such as path dependency and lock-in are best understood when only one level of change is operating; otherwise their meaning becomes indeterminate as more processes are set in motion. To fully appreciate the outcome of change, it is necessary to explore the contingent nature of development at many points in time and at multiple spatial scales.

At the microscale, the watch industry is rich with examples of what commonly lies behind technological lock-in. In this book I look behind the meaning of the term and explore the various ways in which technological and other forms of lock-in occur, interact with, and impinge on the actions of industrial actors. In the case of the British industry's adherence to the "large full-plate" watch, including the maintenance of its design and the production system supporting it, and yet the institutional rigidities that kept the system in place arose from multiple limitations, including the regional labor market and the then-prevailing industrial distribution channel, which relied exclusively upon agents. It wasn't that the need for change was unknown or unknowable; at its base, the fragmented production system that characterized the British industry required many participants to act in concert for change to occur. While in the single-agent case the problem might have been overcome by simply internalizing skill development and the sales and distribution functions, in a disaggregated system no single firm's actions would have been sufficient to change the overall industry tide. Had it not been for extra-regional interventions on the part of the national government, however, the industry might have been forced to change if for no other reason than pressure from the flood of competition. Instead, in this case the invocation of trade restrictions bought time and reinforced the status quo.

At the smaller within-firm scale, a similar logic also applies to the difficulty of reorienting the Waltham and Elgin Corporations on the eve of the explosion of the wristwatch market. In the case of Waltham Watch, the hundreds of small shareholders wanted nothing to do with the capital investments needed to take a serious position in this new market niche. For them, a pocket watch was just fine. A wristwatch was quite out of the realm of immediate need or possibility for most people. Moreover, at the time, there simply wasn't enough capital for such an investment. At Elgin, a lower level of product differentiation, fewer investors, and fewer sunk costs and/or less misdirected capital meant more investment funds were available and the expenditure approval process less cumbersome. Yet this company also hung back, waiting for louder market signals. A stalemate ensued and no real movement occurred on either side. As strictly U.S. firms competing in an uncontested domestic market, Waltham and Elgin's waiting wouldn't have been a problem. But the Swiss were not sitting idle. Instead, they were being encouraged by the Germans to produce wristwatches for the war. Their market was expanding and their capital flow remained strong. In fact, capital flow was actually increasing because capital requirements needed for the switchover were being subsidized by the war effort. Without a war, it is possible that everyone in the industry would have hung back. But this is not what happened. Thus, as this example suggests, variations at the microscale tell only half the story of both the causes of lock-in and path dependency; variations at this scale alone are insufficient to explain the eventual outcome.

Another variant on this theme is seen in the resistance to paradigm-challenging innovations that arise from within an industry, but which are rejected because they threaten the preexisting system of social and economic relations. There are numerous examples in the history of the watch industry where turning a blind eye was momentarily efficient but often had major implications, such as Hetzel's tuning-fork invention and resistance to the creation of a truly cheap watch. Yet, in both cases, everyone knew everything—the importance of such moves were simply the writing on the wall. Institutional rigidities hampered the evolution of the industry despite the availability of solid information and the ability to execute the task successfully. But it was not just resistance for the sake of saving face that limited the introduction of the tuning-fork or the dollar watch. In both cases, the initial product was imperfect, and, in the case of the tuning-fork watch, it was difficult to build and maintain. In the final analysis, however, the critical source of resistance was not the production system, because both products were ultimately produced. Indeed, in the case of the cheap watch, in the 19th century with the Roskopf and dollar watches and in the 20th century with the truly cheap electronic watch, distribution channels acted as bottlenecks. Resistance arises from numerous directions. Its effect very much depends on the overall structure of the industry, the point in the business cycle, the intensity of larger external economic conditions, and also, but not singularly, the production system. In other words, resistance usually extends substantially beyond the production system itself.

Throughout the history of the watch industry the formation of standards has had a profound effect on the rate of technological change and the nimbleness of different competitors. Standards usually benefit some in the industry while hurting others. The creation of benchmarks means that competitors must conform to these standards if they wish to participate. This in turn generates a form of collective lock-in as firms participating in the standard make investments that then impede their ability or willingness to change by moving beyond established boundaries. But a standard is more than specification of a product and its functions. Indeed, the requirement that parts for railroad watches had to be manufactured in the United States acted as a trade barrier, inhibiting the Swiss from selling equally qualified products. The tightening of the overall trade system helped to prop up and reinforce the trend toward railroad watches. Had trade been unburdened, not only might competition have produced greater product differentiation even within railroads watches, but it also could have provided a larger window onto the coming wristwatch innovation. The railroad standard provided U.S. firms with a momentary respite from Swiss competition, but it became an impediment to change when the growth in demand for wristwatches far outstripped the demand for pocket watches of the railroad era. Thus, again, more than one source of rigidity lies behind lock-in.

A second realm to which this book contributes is its greater detail about the meaning and manifestation of path dependency. Like technological lock-in, path dependency is often invoked uncritically as an explanation for a particular industrial experience. Usually lying behind the notion of path dependency is a series of factors that together add up to a directional bias. Just exactly what provokes path dependency is rarely communicated, however; this often erroneously leads to unidimensional invocations of the term.

How is path dependency created? One way is through the creation, dissemination, and preservation of knowledge. The world watch industry presents an exceptional context in which to explore the meaning of and relationship between tacit and codified knowledge and their respective roles in technological change. The Swiss ability to overtake the British in the manufacture of precision products was the result of two interrelated systems of skill development. The primary system provided the basis for the labor market. Learning by doing created a workplace culture that transferred know-how to new labor market entrants through practice and social relations. On the surface, this system of skill development looked quite similar to Britain's guild system, but the British system relied on a rigid hierarchy where knowledge was created and disseminated through replication of tasks specified by masters. It was culturally based and ritualistically reproduced. Knowledge remained almost exclusively tacit (or at least its communication remained of a tacit form), and thus the development of skill was confined to a small group of artisans. The Swiss, in contrast, placed a heavy emphasis on converting tacit knowledge into codified knowledge that could be widely disseminated among workers in the region. It was a more anarchistic society, but also more democratic. This effort to convert intangible knowledge into tangible knowledge created high levels of local competency. The use of contests and records, as well as the establishment of norms and institutions to generate, disseminate, and retain new knowledge, served to strengthen the overall Swiss watchmaking complex and enhance the ability of the Swiss system to move beyond incremental technological change. Thus, within the existing technological paradigm, codification of knowledge became a source of competitive advantage by allowing for a more democratic form of knowledge preservation and dissemination.

Another source of path dependency arises from instabilities that form as foundational elements of a system. Despite the seeming distance between old and new constellations of firms, the knowledge base of new industrial complexes often arises out of the old complexes. The early formation of the U.S. industry was derived in part from knowledge extracted from the British and Swiss industries. The same cross-fertilization can be seen in the proliferation of new firms in the formative days of the U.S. watch industry, where new firms and plants were established as skilled workers left existing firms to join new operations or to start new firms of their own. In the early stage of a new industry, the knowledge base is likely to be very circumscribed, that is, it is shared among few individuals. Thus the proliferation of new businesses is an outgrowth of the individual search for personal gain and the occasional failed enterprise, which frees workers to take positions in new ventures. This proliferation has its downsides inasmuch as labor raiding diminishes the ability of operating businesses to conduct daily operations. The proliferation of new establishments is often erroneously regarded as a sign of industry health when in fact, on an individual business basis, labor mobility can have devastating effects on the operation of existing firms. It also is by its nature a means of constructing path dependency, as mobility locks in place certain practices and erodes the ability of depth of practice to form, which in turn prevents change from being sustained.

Path dependency is often invoked outside of an existing context, disembodied from larger trends. And yet another reoccurring theme in the history of the world watch industry is the degree of foreknowledge that existed among major and emergent competitors even as path dependency set in. Corporate histories often suggest that managers are caught off guard when a new competitor appears seemingly out of nowhere. In fact, new competitors are rarely a complete surprise. This is because cross-fertilization and information transfer often occur among players in a system. Moreover, by the time information is exchanged, the new competitor's competency is usually significant enough to be a threat. In the case of the British and Swiss industries, on the eve of the 19th century, the Swiss were serious, almost equal competitors, even though they did not begin to truly challenge the British for market share until sometime after the 1820s. British and Swiss watchmakers were traveling to see one another's operations. Production plans were sometimes exchanged through purchase, while in other cases they were stolen. Regardless, they were replicated in some way. The same degree of information transfer was evident in the U.S. industry, where machinists traveled to Britain and Switzerland to observe and memorize operations to be implemented back at home. In the case of Japan, the first rounds of machinery were purchased from the United States and set up and run with the help of U.S. engineers. Path dependency occurs not as a result of an isolated event but as the result of a stream of ongoing deeply embedded processes. The beginning and the eventual end point are often blurred in the process.

Path dependence clearly also arises as firms chart a path in the absence of preexisting paradigms. First mover advantages are a source of path dependency that can, in the absence of a dominant and pervasive paradigm, become a liability. Being first is perceived as conveying an advantage. There are countless examples, however, including those of the world watch industry, where being first had serious liabilities, particularly in periods of continuous technological change. No system of production is ever complete upon introduction; rather, it is typically a prototype, and hence subject to change as adoption occurs by others. Great Britain, arguably the most important watch producer in the 18th century, boasted of a system reliant on guild-based skill development. This system *was* the best for its time. But despite the British industry's need to generate a larger pool of skilled workers, the guild system was not flexible enough to expand the supply of trained employees. Waltham Watch Company, as the first to attempt a full factory-based system of watch production, also suffered from the burden of being first in an industry based on a labor-intensive, machine-aided production system that relied heavily on skilled labor. Try as they might, Waltham's initial owners could not purchase or produce a truly turnkey production system. Instead, the system embodied bits and pieces of both old and new models of organization.

A third contribution of this book is its recognition of the importance of a country's technomilitary ideology in the evolution of industrial activities. The Swiss outlook toward war was decidedly mercantile, and their mind-set was to produce for consumption and sell as much as possible. In the United States, war was to be won and it was the duty of all to participate. The economy was bound to come to the aid of the military, regardless of the longer term consequences. The Japanese made few distinctions between war efforts and civilian production for consumption and export. Since the dawn of the Meiji era, Japan has viewed industrialization as if it were a military operation; indeed, there has always been a symbiosis between military and industrial activities.

Over the last 150 years, military engagements have had a significant influence on firm behavior, the nature of industrial competition, and the diffusion of technological change. The effect of military endeavors has been most pronounced on patterns of trade and investment in new technology. The wars of the last century consistently altered the pattern of trade among nations. In most cases, markets were closed off, generating serious instability for exporting countries. At the same time, depending on the prevailing technomilitary ideology of a country, government requirements for war materiel realigned industry production, leaving open domestic markets to competitors. In the case of the United States, exports flooded in; in the case of Japan, consumer demand went unmet. In all cases, war efforts altered the status quo, privileging traders versus manufacturers, depending on the prevailing domestic policy toward production for the war.

The effect of war on the diffusion of technologies is also dependent upon the relative position of the combatants. In some cases, technological change is precipitated by war needs. In this instance a domestic industry could emerge with a far more sophisticated competency than it enjoyed prior to the engagement. In other cases, particularly for neutrals, there may be little stimulus to take up new technologies because their efforts are focused on production for consumption rather than production for strategic value.

Conclusions

Throughout this book I have tried to show that regional fortunes are increasingly intertwined with global events that are largely beyond a single community's control. External economic forces influence the trajectory of a region's development through their effects on locally established firms. Relationships between regions and industry therefore are reciprocal. As an industrial culture comes to define a place, it structures future opportunities for innovation. Given that job creation tends to mirror a region's skills base, and supply constraints govern the availability and mobility of technically trained labor, policy only has a modest effect on reestablishing regional innovation.

The long tradition of watchmaking and precision machining was the technological base for many regions, yet this history is also responsible for limited development. The Swiss experience must be understood within a framework that exposes how technological paradigm shifts challenge previous ways of organizing an industry, a culture, and society. Paradigm shifts present a series of strategic turning points that industrial leaders must navigate during a period of technological change. The British, Swiss, and American experiences were no exception.

A paradigm shift is an extreme event that can precipitate a disjunction between the past and a future means of regional development. Clearly, it is important to distinguish between adjustment costs associated with incremental and discontinuous change. As the Swiss watchmaking industry in all its permutations illustrates, these two types of environmental change often converge. In the face of incremental and then discontinuous change, a system's reaction may be dulled by market signals that mask the need for radical change. The ability to make incremental adjustments often creates a false sense of flexibility and therefore lulls industry members into believing in their invincibility. As the examples in this book show, an industrial system's preeminence often is derived endogenously. Inevitably, internal actors may have difficulty remaining open and sensitive to external signals. The inability to change suggests numerous images, including that of the proverbial ostrich, with its head hidden in the sand.

Notes

CHAPTER TWO

1   There is so much more that can be said about Japan's relative economic position at the end of the 19th century. Clearly, Japan was developing in far more sophisticated ways than was acknowledged by many writers of the day. Moreover, as early as the 1880s, the outlines of Japan's deliberate development trajectory could clearly be seen. Positioning of industry through the use of fiscal and monetary policies laid the foundation for accelerated development even as the rest of the world was suffering through a deep economic recession.
2   Robert Brenner (1998) notes that Gerschekron ignored the writings of several authors writing on the same subject who predated him.

CHAPTER THREE

1   Landes points out the lag in technology adoption that occurred on the Continent: "Later on, when the alternative of concentrated factory production became available, many an entrepreneur, in the continental countries especially, delayed shifting over because of the flexibility of the older arrangements" (1969, p. 57).
2   Hounshell (1984) says of John H. Hall's early recognition of the extraordinary effort required to achieve interchangeability that Hall was the exception: "North and others had already considered this same object and had made progress, but their thought lacked Hall's precision and their commitment to interchangeability of parts was moderated far more by practical or economic considerations than was Hall's. It is important to keep in mind the fact that interchangeability, despite its powerful appeal and seeming rationality, was considered a somewhat irrational pursuit because it continually flew in the face of experience, which had borne out time and again that the system could not survive practical application" (Hounshell, 1984, p. 39).
3   This is not entirely true; the United States, during the Great Depression, required the Japanese to conform to voluntary trade restrictions, particularly in the case of textile exports.

275

CHAPTER FOUR

1. Other more recent treatments of the question of time include Dohrn-van Rossum's book *The History of the Hour* (1996), a historical sociological account of the creation of time increments.
2. The same can be said of the industries in the United States, Hong Kong, and Japan, which all started in major cities.
3. In the early era of watch manufacturing, the industry was off limits to women, which is why I use the term "craftsmen" here. Over time, women were given the opportunity to perform the most tedious, patience-filled tasks in watchmaking. During periods of stress in the industry, women's work was taken back by male watch parts manufacturers. Women's labor was used as an early valve in managing periods of excess capacity and severe price cutting. In the mid-1700s, in England, women and children also were frequently employed because they were seen as docile and willing to follow the strict requirements of foremen in the early factories (Mantoux, 1928).
4. The term "factory system" should not be confused with mechanized production. It was not until the last quarter of the 18th century that machine manufacture became widespread in British industries, and even then the application of machines was concentrated in relatively few industries; watchmaking was not among them.
5. See Mantoux (1928, p. 109) for a description of the rise of Liverpool as a port.
6. Conventionally defined, a *monopsonist* firm is one that, because of its position as the sole market for firms' output, has significant market power. In the case of the watch industry, the term is used to indicate that parts suppliers faced a homogeneous final market that did not motivate them to pursue product variation.
7. See Aldcroft (1968) for a detailed investigation of the effect of trade protection on British access to U.S. markets. Sandberg (1974) makes a similar argument in support of a counterexplanation for the decline of the British textile industry.

CHAPTER FIVE

1. The notion of being slow was mentioned several times during the many interviews I conducted with Swiss watch officials and political representatives.
2. This tendency and the hubris that follows is amply demonstrated in Florida and Kenney's book, *The Breakthrough Illusion: Corporate America's Failure to Move from Innovation to Mass Production* (1990).
3. I appreciate the translation help afforded me in this section on the French watch industry by Stephen Belcher, assistant professor of comparative literature, Pennsylvania State University.
4. In the 1970s, the French industry did not respond promptly to the introduction of the quartz watch (Lechot et al., 1994). Initially, like Swiss manufacturers, the few remaining French watch companies felt that quartz was a fad and chose not to invest in this new technology. Rather, they hid behind tar-

iffs and produced mechanical watches for the French market. This reluctance to follow the new technology was in part a response to the problem of sunk costs. The industry as a whole had considerable investment in the production of mechanical watch components. Unlike the Japanese, or to a lesser extent the Swiss, which had established distribution channels for the sale of watch movements, the French lacked suitable outlets to securely sell parts in order to maintain the components industry. Thus, efforts to guard the industry through protectionist measures became imperative.

But market protection did not afford the stability necessary to ensure the industry's longevity, nor did it secure its ability to retreat further into mechanical movements. With the insistence of and support from the national government, a firm was eventually created to produce electronic watches. This enterprise was short-lived. Major electronics participants dropped out of the arrangement as watch firms bickered among themselves. In an about-face, in the 1980s two of the remaining French companies moved toward electronic watches. Rather than drawing movement supplies from the government-financed manufacturer set up in the late 1970s, the watch manufacturers teamed up with Seiko. As Lechot et al. (1994) note, "The lack of internal coherence within the industry and the lack of a strong lead firm to guide it, resulted in a retreat to domestic protection to save the remaining remnants of the industry" (p. 137).

5   The Edict of Nantes, signed in the 16th century, was a religious compromise between the king and his Protestant subjects signed by Louis the XIV. Its revocation implied that either Protestants would be forced into exile or would have to discretely withdraw from public life (Sandberg, 1974).

6   Perhaps one explanation for Geneva's higher costs was the watch guild's absolute resistance to women working in the industry.

7   The social, economic and human conditions of the pre-Industrial Revolution era experienced in Britain did not evolve in the same way in Switzerland.

CHAPTER SIX

1   The history of the American watch industry is remarkably well documented. This record is the work of a sophisticated collector and repair community. The National Clock and Watch Association maintains an unusually rich museum and library in Pennsylvania where literally thousands of documents and rare watches are on permanent display. The association itself has a long and illustrious history of chronicling the fate of the world watch industry. One author of particular note and reputation is Mike Harrold, an engineer at General Electric in Massachusetts, who wrote one of the most complete watch technology histories ever produced. His monograph is a remarkable treasure in terms of its data coverage and detail. Along with a few other notable books, such as Abbott's (1881) chronology of American factories and Crossman's (1885) depiction of the story of key companies, the intellectual history of the industry has been captured with remarkable detail. In some sense the penchant for chronicling the industry follows in the path of the

early watchmakers who were a small cohort of highly dedicated precisionists, an unusually geographically mobile group of craftsmen who established an industry that was at one time a serious contender for preeminent leadership of the world watch industry.

2. It is actually difficult to say with assurance either how many firms were created during the early period of the watch industry. Many firms were simply reincarnates of firms that for various reasons failed.

3. The first major exchange of watch factories and the establishment of a new business out of existing equipment occurred in 1853. Boston Watch Corporation was sold at auction for $65,000.

4. Some industry representatives argue that true interchangeability was actually not achieved until the 1850s.

5. Waltham's initial strategy was to sell through an agent who serviced retailers. Retailers were sold a finished watch; this had the advantage of controlling prices while also raising costs. Waltham eventually developed its own sales force to eliminate the middleman.

CHAPTER SEVEN

1. Twenty percent of all Swiss exports were associated with watches in the latter part of the last century.

2. Landes (1979) suggests the early emphasis on widespread education both enabled and compelled the Swiss to move to a new model of production.

3. By the turn of the century, U.S. watch firms had become dependent upon the Swiss machine builders for new technology needed to expand their mass-production system.

4. World War I severely disrupted the world watch market. Russia, a major Swiss market, closed its borders to international trade, while other countries raised trade barriers in attempts to preserve domestic industries. Demand for Swiss watches declined precipitously between 1916 and 1921.

5. The larger firms making capital investments in equipment wanted to inject order into the historically anarchistic industry. To recoup capital investments, the more advanced firms had to control the small firms that easily sprang up and produced cheap watches (Jacquet and Chapuis, 1970).

CHAPTER EIGHT

1. See MacIsaac, *Summary Report of the U.S. Strategic Bombing Survey*, vol. 7. The point was made that Japan was not prepared to fight a war with adversaries capable of a long engagement. The Japanese strategy, according to U.S. interests, was to conduct a series of strategic attacks that would debilitate the United States in the Pacific, and which would be complemented by the total devastation and domination of Europe by the Germans. The Japanese assumed that the United States would relinquish its role in Asia once Japan proved insurmountable. As history shows, the outcome was quite different. U.S. air power so devastated Japan that its army could not influence the out-

come. This was a strategic oversight on the part of the Japanese, who expected the United States to invade by land. And, although this was part of the United States's original plan, the war was fought almost entirely by air and sea.
2 The first wall clock factory was located in Nagoya and the first mechanical clock factory was established in the city's environs.
3 The Otay Watch Company was established in 1891 and failed a year later due to insufficient financing.
4 Personal conversation with Noriuki Suguira, professor of geography and president of the Westchester Japanese High School, 1996.
5 Postwar Japanese industrial organization consisted of large business groups and trading companies. Since the 1960s, the Japanese manufacturing system has experienced profound rationalization. The existing system has evolved from high levels of vertical integration to the present situation of high levels of subcontracting (Glasmeier, 1990).

CHAPTER NINE

1 Aside from watches, Bulova—like Hamilton and Timex in their day—produced fuses for artillery and mortar shells. The company consistently emphasized its belief that it would survive, and perhaps be even better off, without the defense business.
2 I have tried to contact Timex, but because it is privately held, it does not have to share its corporate information.
3 Siem was probably the ultimate winner of the conflict; in Lemkuhl's later years he had allowed subordinates to make more decisions. After the shakeup, Siem recentralized control in his own office.
4 The SGT holding company also acquired two U.S. watch companies, Waltham and Elgin.

CHAPTER TEN

1 It has been suggested that engineers were more enamored of LEDs than LCDs. Because the American electronics industry was dominated by engineers, there is speculation that the LED's appeal was driven more by preference than by practicality (Numagami, 1996).
2 An exception to this trend is cotton spinning, which has developed as one of Hong Kong's primary manufacturing industries.
3 Replacement of the reexport industry occurred as the domestic industry built up to supply the world's watchmakers with cases.
4 OEMs are producers who manufacture a product that may ultimately bear the label of another firm.
5 The structure of the watch industry is remarkably similar to the structure of the electronics industry in Hong Kong.
6 FOB stands for free-on-board, a common method of shipping whereby the recipient of a good pays all transportation charges on its receipt.

280 ■ Notes to Chapter Eleven

7   As products achieve market success, they are reevaluated for possible mass production within a vertically integrated setting in Japan.
8   ISO 9000 is a standard in an international certification system that verifies that a firm follows certain quality control, accounting, inventory tracking, and labor training practices.

CHAPTER ELEVEN

1   Another example of reorganization precipitated by the crisis was the introduction of shift work in factories.

CHAPTER TWELVE

1   This policy was later pursued by U.S. industry. It no doubt significantly affected America's long-run competitiveness.
2   This policy also was pursued in U.S. industry, no doubt again affecting long-run competitiveness.
3   Many of the findings on this score have analogies with contemporary high-tech industries. For example, the United States maintains a competitive lead in the most innovative customized semiconductor production technology (analogous to the British and French watch industries).
4   Some authors argue that the eventual pressures to mass-produce watches in Switzerland killed the region's previously great potential to create a constantly expanding skilled labor pool. Thus presumably, when the threat of quartz crystal technology loomed for Swiss producers, rigid institutional relations (built up since the 1940s) limited the industry's ability to adapt and respond quickly.

References

Abbott, H. 1881. *Watch Factories of America*. Exter, NH: Adams Brown.
Aikin, J. 1795. *A Description of the Country from Thirty to Forty Miles Round Manchester* (p. 311). London: Printed for J. Stockdale.
Akimitsu, S. 1995. The Dynamics of Innovation and Learning by Doing: The Case of the Integrated Circuit Industry. In *Acquiring, Adapting and Developing Technologies*, edited by R. Minami, K. Kim, F. Makino, and J. Seo, pp. 165–190. London: St. Martin's Press.
Aldcroft, D. 1968. *The Development of British Industry and Foreign Competition, 1875–1914*. First ed. Toronto: University of Toronto Press.
Alft, E. D. 1984. *Elgin: An American History, 1835–1985*. Elgin, IL: Crossroads Communications.
Allen, G. C. 1928. *Modern Japan and Its Troubles*. New York: Dutton.
Allen, G. C. 1940. *Japanese Industry: Its Recent Development and Present Condition*. New York: Institute of Pacific Relations.
Allen, G. C. 1959. *British Industries and their Organization*. London: Longmans.
Allen, G. C. 1963. *A Short Economic History of Japan*. New York: Praeger.
Amsden, A. 1989. *Asia's Next Giant: South Korea and Late Industrialization*. New York: Oxford University Press.
Arthur, B. 1988. Competing Technologies: An Overview. In *Technical Change and Economic Theory*, edited by G. Dosi, C. Freeman, R. Nelson, G. Silverberg, and L. Soete, pp. 590–607. London: Pinter.
Arthur, B. 1994. *Increasing Returns and Path Dependence in the Economy*. Ann Arbor: University of Michigan Press.
Bagley, J. J. 1961. *A History of Lancashire*. London: Darwen Finlayson.
Bailey, C. H. 1975. *Two Hundred Years of American Clocks and Watches* (p. 209). Englewood Cliffs, NJ: Prentice Hall.
Baillie, G. H. 1929. *Watches: Their History, Decoration, and Mechanism*. London: Methuen.
Barker, T. C., and J. R. Harris. 1959. *A Merseyside Town in the Industrial Revolution: St. Helens, 1750–1900*. First ed. London: Frank Cass.
Barnhart, M. 1987. *Japan Prepares for Total War: The Search for Economic Security 1919–1941*. Ithaca, NY: Cornell University Press.
Barrelet, J. M., and Ramseyer, J. 1990. *La Chaux de Fonds ou le défi d'une cité horlogère*. La Chaux de Fonds, Switzerland: D'en haut.

Berger, S., and R. Lester. 1997. *Made by Hong Kong.* First ed. Oxford, England: Oxford University Press.
Best, M. 1990. *The New Competition.* Cambridge, MA: Harvard University Press.
Blanc, J. F. 1988. *Suisse-Hong Kong: Le défi horloger.* Geneva: La Collection Nord-Sud de l'Institut Universitaire d'Études de Developpement de Genève.
Bolli, J.J. 1957. *The Clock Industry Component of Americano-Swiss Commercial Relations: 1929-1950.* La Chaux de Fonds, Switzerland: La Suisse Horlogère.
Boulianne, L. 1982. Technological Change: Firm and Region: A Case Study. In *Technology: A Key Factor for Regional Development,* edited by. D. Maillat, pp. 39-68. Saint-Saphorin, Switzerland: Georgi.
Boyer, E. 1984, November 12. A Family Rift Roils Seiko. *Fortune,* pp. 44-54.
Brearley, H. 1919. *Time Telling through the Ages.* First ed. New York: Doubleday, Page.
Brenner, R. 1998. The Economics of Global Turbulence. *New Left Review* (Special Issue), pp. 1-265.
Broadbridge, S. 1966. *Industrial Dualism in Japan.* First ed. Chicago: Aldine.
Bruton, E. 1979. *The History of Clocks and Watches.* New York: Rizzoli.
Bulletin Conjoncturel Horloger. 1988. Sommaire exportations horlogères suisses en 1987. No. 211, pp. 22-29.
Bulova Watch. 1968-1978. *Corporate Annual Reports.* New York: Author.
Bumbacher, U. 1992. *The Swiss Watch Industry* (Harvard Business School Case Study 9-792-046). Boston, MA: Harvard Business School Press.
Business Week. 1875, October 27. Digital Watches: Bringing Watchmaking Back to the U.S. P. 78.
Business Week. 1977a, February 28. Dog Days at National Semiconductor. P. 70.
Business Week. 1977b, May 2. The Great Digital Watch Shakeout. P. 78.
Business Week. 1977c, May 5. Hong Kong Moves In. P. 106.
Business Week. 1980, May 5. Technology Takes Selling. P. 168.
Business Week. 1984a, November 26. A Last Minute Comeback for Swiss Watchmakers. P. 139.
Business Week. 1984b, February 20. Can Timex Take a Licking and Keep on Ticking? Pp. 102F-102L.
Business Week 1984c, February 20. Price Wars and a Glut Have the World's Watchmakers in Chaos. Pp. 102D-102F.
Business Week. 1988, July 18. A Wristwatch with a Power Plant in It. P. 70.
Business Month. 1988, March. Up from Swatch. P. 57.
Cardinal, C. 1989. *The Watch: From Its Origins to the XIXth Century.* Secaucus, NJ: Wellfleet Press.
Casio Computer Inc., Japan. 1984-1988. *Corporate Annual Reports.* Tokyo: Author.
Central Federation of Societies of Commerce and Industry, Japan. 1987. *The Shokokai.* Tokyo: Author.
Chamberlain, P. 1941. *It's about Time.* New York: Richard Smith.
Chambre de Commerce et d'Industrie de Genève. 1989. Genève et le marché intérieur de la communauté de 1992. Geneva, Switzerland: Author.
Chandler, A. 1991. Creating Competitive Capability: Innovation and Investment in the United States, Great Britain and Germany from 1860 to World War I.

In *Favorites of Fortune: Technology, Growth, and Economic Development Since the Industrial Revolution*, edited by P. Higonnet, D. S. Landes, and H. Rosovsky, pp. 423–458. Cambridge, MA: Harvard University Press.

Chapuis, A. 1939. *La Suisse dans le monde: Avec neuf croquis.* Reprint. Paris: Payot.

Chapuis, A., and Eugene, J. 1956. *The History of the Self-Winding Watch, 1770–1931.* New York: Central Book.

Child-Hill, R. 1989. Comparing Transnational Production Systems: The Automobile Industry in the USA and Japan. *International Journal of Urban and Regional Research*, *13*(3): 462–480.

Chinitz, B. 1960. Contrasts in Agglomeration: New York and Pittsburgh. *American Economic Association, Papers and Proceedings*, 2: 279–289.

Choi, Y. B. 1993. *Paradigms and Conventions: Uncertainty, Decision Making, and Entrepreneurship.* Ann Arbor: University of Michigan.

Church, C. 1989, July. The Swiss Way of Change: Politics since the 1987 Election. *World Today*, pp. 117–121.

Church, R. A.1975. Nineteenth-Century Clock Technology in Britain, The United States and Switzerland. *Economic History Review*, *28*(24): 616–630.

Citizen Watch Company (Japan). 1985–1988. *Corporate Annual Reports* Tokyo: Author.

Clark, G. L. 1994. Strategy and Structure: Corporate Restructuring and the Scope and Characteristics of Sunk Costs. *Environment and Planning A*, *26*: 9–32.

Conroy, M., and A. Glasmeier. 1995. Industrial Strategies, the Newly Industrializing Economies, and New International Trade Theory in Latin America. *Environment and Planning A: Commentary*, *27*(1): 110.

Corat, P. 1985. *L'incidence de la politique industrielle regionale: Le cas des régions dont l'économie est menacée.* Neuchâtel, Switzerland: Institut de Recherches Economiques et Régionales.

Crevoisier, O., M. Fragomichelakis, F. Hainard, and D. Maillat. 1989, September. *Know-how, Innovation, and Regional Development.* Paper presented at the 29th European Congress, Cambridge, England.

Crossman, C. S. 1885. *The Complete History of Watch Making in America.* Reprint. Exter, NH: Adams Brown.

Cuss, C. 1967. *The Country Life of Watches.* London: Country Life.

Davies, A. 1993. British Watchmaking and the American System. *Business History*, *35*(1): 40–58.

Davies, K. 1996. *Hong Kong after 1997.* London: Economist Press.

de Mestral, A. 1957. *Daniel Jean Richard, Founder of the Jura Watch Industry, 1672–1741.* Zurich: Institute of Economic Research.

Dicken, P. 1998. *Global Shift.* Second ed. New York: Guilford Press.

Director. 1988, November. Switzerland: Alone in a Crowd? P. 27.

Director. 1989, June. Patek Philippe: Is Time on Its Side? Pp. 117–120.

Dohrn-van Rossum, G. 1996. *History of the Hour: Clocks and Modern Temporal Orders.* Chicago: University of Chicago Press.

Dosi, G. 1984. Technological Paradigms and Technological Trajectories. In *Long Waves and the World Economy*, edited by C. Freeman, pp. 27–43. London: Butterworth.

Dosi, G., C. Freeman, R. Nelson, G. Silverberg, and L. Soete, eds. 1988. *Technical Change and Economic Theory*. First ed. London: Pinter.
Dosi, G., and L. Orsenigo. 1988. Coordination and Transformation: An Overview of Structure, Behaviors, and Change in Evolutionary Environments. In *Technical Change and Economic Theory*, edited by G. Dosi, C. Freeman, R. Nelson, G. Silverberg, and L. Soete, pp. 13–37. London: Pinter.
Dowling, J., and J. Hess. 1996. *The Best of Time: Rolex Watches, an Unauthorized History*. Hong Kong: Write Time Partners.
Duncan, B. 1959. *The United States and Switzerland in the Nineteenth Century*. Ph.D. diss., Department of History, Emory University, Atlanta, GA.
Elgin National Industries. 1950–1970. *Corporate Annual Reports*. Chicago, IL: Author.
Elgin National Industries. 1950–1970. *Corporate 10-K Reports*. Chicago, IL: Author.
Elsasser, H. 1982. Sectoral Shifts in Economic Structures: An Overview. In *The Transformation of Swiss Mountain Regions: Problems of Development between Self-Reliance and Dependence in an Economic and Ecological Perspective*, edited by E. Brugger, pp. 79–101. Berne, Switzerland: Haupt.
Enright, M. J. 1995. Organization and Coordination in Geographically Concentrated Industries, In *Coordination and Information*, edited by N. Lamoreaux and D. Raff, pp. 103–146. Boston:. National Bureau of Economic Research.
Enright, M. J., E. Scott, and D. Dodwell. 1997. *The Hong Kong Advantage*. Oxford, England: Oxford University Press.
Fahey, A. 1988, July 18. Timex, Swatch Push Fashion. *Advertising Age*, p. 4.
Fédération de l'Industrie Horlogère Suisse. 1989. *Rapport 1988*. Bienne: Author.
Fédération de l'Industrie Horlogère Suisse. 1991. *Rapport 1990*. Bienne: Author.
Fédération de l'Industrie Horlogère Suisse. 1988. *Bulletin Conjoncturel Horloger*, no. 211.
Fédération de l'Industrie Horlogère Suisse. 1998. Web site: www.fhs.ch.
Fédération Suisse des Travaillers de la Metallurgie et de l'Horlogerie (FTMH). 1988. *Notre avenir a une histoire, 1888–1988*. Berne, Switzerland: Schlatter AG.
Federation of Hong Kong Watch Trades and Industries Ltd. 1988–1989. *Clock and Watch 37* (Annual magazine). Hong Kong: Author.
Federation of Hong Kong Watch Trades and Industries, Ltd. 1989. *Watch Industry Report*. Hong Kong: Author.
Financial Times. 1986a, June 26. Keeping Time with Fashion. P. 6.
Financial Times. 1986b, December 18. Matra Watches Go to Seiko. P. 6.
Financial Times. 1987a, March 10. Hong Kong Challenges for Top Spot in Watchmaking. P. 7.
Financial Times. 1987b, September 3. Hong Kong Retailer Buys Control of Gillette Unit. P. 11.
Financial Times. 1988a, July 28. Watches Mark Time. P. 13.
Financial Times. 1988b, September 8. Hong Kong Overtakes Japan in Value of Watch Exports. P. 12.
Fischer, G. 1984. *Le canton de Neuchâtel dans l'économie suisse: Problèmes et conséquences pour la politique régionale*. Unpublished paper. Institut de Researches Economiques et Régionales.

Fischer, G., and E. Brugger, eds. 1985. *Problèmes règionaux en Suisse*. Lausanne: Presses Polytechniques Romandes.
Fischer, W. 1991. The Choice of Technique: Entrepreneurial Decisions in the Nineteenth-Century European Cotton and Steel Industries. In *Favorites of Fortune: Technology, Growth, and Economic Development Since the Industrial Revolution*, edited by P. Higonnet, D. S. Landes, and H. Rosovsky, pp. 142–157. Cambridge, MA: Harvard University Press.
Florida, R, and M. Kenney. 1990. *The Breakthrough Illusion: Corporate America's Failure to Move from Innovation to Mass Production*. New York: Basic Books.
Forbes. 1987, March 9. A Tough Sell? Pp. 137–138.
Forbes. 1988, September 5. Sweat Chic. P. 78.
Forbes. 1989, April 17. La Mystique Philippe. Pp. 46–31.
Fortune. 1980, January 14. "How Omega and Tissot Got Ticking Again. P. 68.
Fortune. 1984, November 12. A Family Rift Roils Seiko. Pp. 44–54.
Freeman, C., and C. Perez. 1988. Structural Crises of Adjustment: Business Cycles and Investment Behavior. In *Technical Change and Economic Theory*, edited by G. Dosi, C. Freeman, R. Nelson, G. Silverberg and L. Soete, pp. 38–66. London: Pinter.
Fritzsche, B. 1996. Switzerland. In *The Industrial Revolution in National Context: Europe and the U.S.A.*, edited by M. Teich and R. Porter, pp. 126–148. Cambridge, England: Cambridge University Press.
Fröbel, F., Heinrichs, J., and Kreye, O. 1980. *The New International Division of Labor: Structural Unemployment in Industralized Countries and Industrialization in Developing Countries*. New York: Cambridge University Press.
Fuellhart, K. 1998. *Networks, Location and Information Acquisition: An Analysis of Small Manufacturing Establishments*. Unpublished Dissertation, Department of Geography, Pennsylvania State University, University Park.
Fuellhart, K., and A. Glasmeier. 1999. *The Geography of Learning, Knowledge, and Information Acquisition of Small- and Medium-Sized Firms*. Unpublished paper.
Gabher, G. 1993. *The Embedded Firm*. First ed. London: Routledge.
Gertler, M. 1993. Implementing Advanced Manufacturing Technologies in Mature Industrial Regions: Towards a Social Model of Technology Production. *Regional Studies*, 27: 665–680.
Gerschenkron, A. 1962. *Economic Backwardness in Historical Perspective: A Book of Essays*. New York: Praeger.
Gitelman, H. M. 1965. The Labor Force at Waltham Watch during the Civil War Era. *Journal of Economic History*, 25(2): 214–243.
Gitelman, H. M. 1974. *Workingmen of Waltham: Mobility in American Urban Industrial Development, 1850–1890*. Baltimore: Johns Hopkins University Press.
Glasgow, D. 1885. *Watch and Clock Making*. London: Cassell.
Glasmeier, A. K. 1988. Factors Governing the Development of High Technology Complexes: A Tale of Three Cities. *Regional Studies*, 22D: 287–301.
Glasmeier, A. K. 1989a. *The Hong Kong Watch Industry: Final Report to the Tissot Economic Foundation, Le Locle, Switzerland*. Graduate Program in Community and Regional Planning, University of Texas at Austin, Working Paper 14.
Glasmeier, A. K. 1989b. *The Japanese Small Business Sector: Final report to the Tissot Economic Foundation, Le Locle, Switzerland*. Graduate Program in Community and Regional Planning, University of Texas at Austin, Working Paper 16.

Glasmeier, A. K. 1990. A Missing Link: The Relationship between Distribution and Industrial Complex Formation. *Entrepreneurship and Regional Development*, 2: 315-333.

Glasmeier, A. K. 1991. Technological Discontinuities and Flexible Production Networks: The Case of Switzerland and the World Watch Industry. *Research Policy*, 21: 469-485.

Glasmeier, A., and Fuellhart, K. 1996. What Do We Know about Firm Learning? *Innovation and International Business*, pp. 277-312. Stockholm, Sweden: Institute of International Business.

Glasmeier, A., Fuellhart, K., Feller, I., and Mark, M. 1996. The Relevance of Firm Learning to the Design of Manufacturing Modernization Programs. *Economic Development Quarterly*, 12(1): 36-49.

Glasmeier, A., Thompson, J., and Kays, A. 1993. The Geography of Trade Policy: Trade Regulation and Location Decisions in the Textile and Apparel Complex. *Transactions of the British Geographers*, 18: 19-35.

Glasmeier, A. K., and R. Pendall. 1989. *The History of the World Watch Industry: Preliminary Report to the Tissot Economic Foundation, Le Locle, Switzerland*. Graduate Program in Community and Regional Planning, University of Texas at Austin, Working Paper 13.

Gohl, A. 1977, December. The Wristwatch. *Bulletin of the National Association of Watch and Clock Collectors, Inc.*, p. 587.

Gordon, D., and R. Dangerfield. 1947. *The Hidden Weapon: The Story of Economic Warfare*. London: Harper & Brothers.

Gourevitch, P. 1986. *Politics in Hard Times*. Ithaca, NY: Cornell University Press.

Gourevitch, P. 1996. *Politics in Hard Times*. Second ed. Ithaca, NY: Cornell University Press.

Graham, G. 1986, December 18. Matre Watches Go to Seiko. *Financial Times*, p. 34.

Grove, A. 1996. *Only the Paranoid Survive*. New York: Doubleday.

Hainard, F. 1988. *Savoir-faire et culture technique dans l'Arc jurassien*. Geneva, Switzerland: Université de Neuchâtel.

Hamilton Watch Company (After 1972: GMH). 1950-1986. *Corporate Annual Reports*. Lancaster, PA: Author.

Hamilton Watch Company (After 1972: GMH). 1950-1986. *Corporate 10-K Reports*. Lancaster, PA: Author.

Harrold, M. C. 1984. American Watchmaking: A Technical History of the American Watch Industry, 1850-1930. *Bulletin of the National Association of Watch and Clock Collectors, Inc.*, No. 14 (Supplement).

Harrold, M. C. 1989, February. The "Un-rarest" American Watches. *Bulletin of the National Association of Watch and Clock Collectors Inc.*, pp. 14-20.

Harrold, M. C. 1993, June. The Wichita Watch that Wasn't. *Bulletin of the National Association of Watch and Clock Collectors, Inc*, pp. 259-273.

Hart, J. 1994. *Rival Capitalists*. Ithaca, NY: Cornell University Press.

Harvard Business School. 1975. *Timex Corporation* (HBS Case Study 9-376-102). Boston: Harvard Business School Press.

Hattori Seiko, Inc. 1985. *Corporate Annual Report*. Tokyo: Author.

Henderson, J. 1989a. *The Globalization of High Technology Production*. London: Routledge.

Henderson, J. 1989b. Pacific Rim Cities in the World Economy. In *Comparative Urban and Community Research*, Vol. 2, edited by Michael Peter Smith, pp. 95–115. New Brunswick, NJ: Transactions.

Hieronymi, O., and A. Gabus, eds. 1983. *La diffusion de nouvelles technologies en Suisse*. Saint-Saphorin, Switzerland: Georgi.

Higonnet, P., D. S. Landes, and H. Rosovsky, eds. 1991. *Favorites of Fortune: Technology, Growth, and Economic Development Since the Industrial Revolution*. Cambridge, MA: Harvard University Press.

Hirschmeier, J. 1964. *The Origins of Entrepreneurship in Meiji Japan*. Cambridge MA: Harvard University Press.

Hirschmeier, J., and T. Yui. 1975. *The Development of Japanese Business*. Cambridge, MA: Harvard University Press.

Hoff, E. J. 1985. *Hattori-Seiko and the World Watch Industry in 1980*. Boston: Harvard Business School Press.

Hoffman, M. 1987. *Switzerland's Development Policy*. Berlin: German Development Institute.

Hoke, D. 1990. *Ingenious Yankees: The Rise of the American System of Manufactures in the Private Sector*. First ed. New York: Columbia University Press.

Hoke, D. 1987. British and American Horology: Time to Test Factor-Substitution Models. *Journal of Economic History*, 47(2): 321–327.

Hong Kong Government Industry Department. 1985, October. *Industrial Profile: Hong Kong's Watches and Clocks Industry*. Hong Kong: Author.

Hong Kong Government Industry Department. 1989. *Industry Profile: Hong Kong's Watches*. Hong Kong: Author.

Hong Kong Goverment Industry Department. 1993. *Industry Profile: Hong Kong's Watches*. Hong Kong: Author.

Hong Kong Manufacturers Association, 1988. *Annual Report*. Hong Kong.

Hong Kong Productivity Council. 1988. *Clocks and Watches: Information for Design, Marketing, and Manufacturing*. Hong Kong: Author.

Hong Kong Trade and Development Council. 1988, August. *Trade Developments: Brief on Hong Kong Watch and Clock Industry and Export*. Hong Kong: Author.

Hong Kong Trade and Development Council. 1989. *Hong Kong Watches and Clocks Annual Directory*. Hong Kong: Author.

Hong Kong Trade and Development Council. 1993. *Industry and Product Profile*. Hong Kong: Author.

Hong Kong Trade and Development Council. 1994. *Industry and Product Profile*. Hong Kong: Author.

Hong Kong Trade and Development Council. 1998. *Industry and Product Profile* Hong Kong: Author.

Hong Kong Watch Distributors Association. 1989. Briefing. In *Hong Kong Watches and Clocks Annual Directory*. Hong Kong: Hong Kong Trade Development.

Hong Kong Watch Trade Association. 1994. *Hong Kong Watch and Clock Industry, 1993–1994*. Hong Kong: Author.

Hong Kong Watch Trade Association. 1993. *Hong Kong Watch and Clock Industry, 1992–1993*. Hong Kong: Author.

Hounshell, D. A. 1984. *From the American System to Mass Production, 1800–1932:*

The Development of Manufacturing Technology in the United States. Baltimore: Johns Hopkins University Press.
Hsia, R. 1978. *Industrialization, Employment, and Income Distribution* (A Study Prepared for the International Labour Office). London: Croom Helm.
Hyltin, T. 1978. *The Digital Electronic Watch.* New York: Van Nostrand Reinhold.
Inkster, I. 1991. *Science and Technology in History.* First ed. New Brunswick, NJ: Rutgers University Press.
International Management. 1984, July. Hattoris Inspire Seiko to a High-Tech Future. Pp. 20–25.
International Management. 1985, June. Can New Management Team Keep Swiss Watches Ticking? Pp. 57–58.
International Management. 1987, January. How the Smokestack Crowd Discovered Marketing. Pp. 16–19.
International Management. 1989a, January. From Black Sheep to White Knight. Pp. 40–42.
International Management. 1989b, April. The Paranoia Gripping Japanese Business. Pp. 24–28.
Ishihara, M. 1988, April. San Jose Found: History of the Osaka Watch Company, Osaka, Japan. *Bulletin of the National Watch and Clock Association, Inc,* pp. 134–145.
Japan Clock and Watch Association. 1994. *Watches and Clocks: Production, Exports, and Imports.* Tokyo: Author.
Japan Company Handbook. 1988, Autumn. Toyo Keizai Shinposha. Section 1.
Japan Economic Journal. 1982, October 27. Semiconductor Firms See IC Sales for Watches Drop. P. 9.
Japan Economic Journal. 1986, December 6. Casio to Assist India Firm to Produce Digital Watches. P. 19.
Japan Economic Journal. 1987, June 20. Citizen Watch Opens U.S. Research Unit. P. 15.
Japan Economic Journal. 1988, January 16. Output Declines in Nation's Mainstay Industries. P. 22.
Japan Economic Journal. 1988, October 1. Citizen to Move Parts Production to China. P. 17.
Japan Small Business Corporation. 1987a. *Guide to Japan Small Business Corporation.* Tokyo: Author.
Jaquet, E., and A. Chapuis. 1970. *Technique and History of the Swiss Watch.* Second ed. London: Spring Books.
Jenkins, J. 1986. *Jura Separatism in Switzerland.* New York: Clarendon Press.
Jequier, F. 1991. Employment Strategies and Production Structures in the Swiss Watchmaking Industry. In *Favorites of Fortune: Technology, Growth, and Economic Development Since the Industrial Revolution,* edited by P. Higonnet, D. S. Landes, and H. Rosovsky, pp. 322–338. Cambridge, MA: Harvard University Press.
Johnstone, B. 1999. *We Were Burning: Japanese Entrepreneurs and the Forging of the Electronic Age.* First ed. New York: Basic Books.
Kahlert, H., R. Muhe, and G. L. Brunner. 1986. *Wristwatches: A History of a Century's Development.* Chester, PA: Schiffer.

Katzenstein, P. 1984. *Corporatism and Change: Austria, Switzerland, and the Politics of Industry*. Ithaca, NY: Cornell University Press.
Katzenstein, P. 1985. *Small States and World Markets: Industrial Policy in Europe*. Ithaca, NY: Cornell University Press.
Kemp, T. 1993. *Historical Patterns of Industrialization*. Second ed. London: Longmans.
Kenwood, A. G., and A. L. Loughheed. 1992. *The Growth of the International Economy, 1920-1990*. London: Routledge.
Kitamatsu, K. 1988. Rising Wages in Asian NICs Force New Offshore Strategy. *Japan Economic Journal*, 26(13-15): 1-6.
Klinger, L. 1986, May 26. Keeping in Time with Fashion. *Financial Times*, p. S6.
Knickerbocker, A. 1974. *Notes on the Watch Industries of Switzerland, Japan, and the United States*. Boston: Harvard Business School Press.
Knickerbocker, A. 1976. *Notes on the Watch Industries of Switzerland, Japan, and the United States*. Revised ed. Boston: Harvard Business School Press.
Koselka, R. 1987, June 15. Made in the U.S.A. *Forbes*, pp. 80-87.
Kwok, R. Y.-W., and A. Y. So. 1997. *The Hong Kong–Guangdong Link: Partnership in Flux*. London: M. E. Sharp.
Landes, D. S. 1969. *The Unbound Prometheus: Technological Change and Industrial Development in Western Europe from 1750 to the Present*. London: Cambridge University Press.
Landes, D. S. 1979. Watchmaking: A Case Study of Enterprise and Change. *Business History Review*, 53(1): 1-38.
Landes, D. S. 1983. *Revolution in Time: Clocks and the Making of the Modern World*. Cambridge, MA: The Belknap Press of Harvard University Press.
Landes, D. S. 1984, January. Time Runs Out for the Swiss. *Across the Board*, pp. 46-55.
Landes, D. S. 1991. Introduction: On Technology and Growth. In *Favorites of Fortune: Technology, Growth, and Economic Development Since the Industrial Revolution*, edited by P. Higonnet, D. S. Landes, and H. Rosovsky, pp. 1-32. Cambridge, MA: Harvard University Press.
Lap Heng Co. 1986. *Corporate Annual Report*. Hong Kong: Author.
Lau, H., and C. Chan. 1991. *Structural Adaptation: The Response of Hong Kong Garment Manufacturers*. Paper presented at the Symposium on Industrial Policy in Hong Kong, held at the Hong Kong Institute of Asia-Pacific Studies, the Chinese University of Hong Kong, New Territories, Hong Kong.
Lazonick, W. 1991. *Business Organization and the Myth of the Market Economy*. New York: Cambridge University Press.
Learned, E. R., Christensen, and K. Andrews. 1961. *Problems of General Management: Business Policy. A Series Casebook*. Homewood, IL: Irwin.
Lechot, G., B. Lecoq, and M. Pdister. 1994. *Analyse comparative de l'évolution structurelle des millieux: le cas de l'industrie horlogère dans l'Arc jurassien suisse et francais*. Neuchâtel: Université de Neuchâtel Institute de Recherches Economiques et Régionales.
Levitt, B., and J. Mar 1988. Organizational Learning. *Annual Review of Sociology*, 14: 319-340.
Lin, T., V. Mok, and Y. Ho. 1980. *Manufactured Exports and Employment in Hong Kong*. Hong Kong: Chinese University Press.

Lockwood, W. 1954. *The Economic Development of Japan: Growth and Structural Change, 1868–1938*. Princeton, NJ: Princeton University Press.
Lockwood, W. 1965. *The State and Economic Enterprise in Japan: Essays in the Political Economy of Growth*. First ed. Princeton, NJ: Princeton University Press.
Loertscher, R. 1982. Comment on the Small Region and the International Division of Labour: The Swiss Case. In *Technology: A Key Factor for Regional Development*, edited by D. Maillet, pp. 48–74. Saint-Saphorin, Switzerland: Georgi.
Loew's Corporation. 1979–1987. *Corporate Annual Reports*. New York: Author.
Long Term Credit Bank of Japan, Ltd. 1986, May. *Monthly Economic Review*. Tokyo: Author.
Los Angeles Times. 1989, December 12. Swiss Industry's Time Has Come Again. Business section, p. 8.
Luetkens, W. L. 1986, April 28. Fashion Circus Lifts Exports. *Financial Times*, p. 6.
Lui, T., and S. Chiu. 1991. *Industrial Restructuring and Labor Market Adjustment under Positive Non-intervention*. Paper presented at the Workshop on Industrial Restructuring and Regional Adjustment in Asian NIEs, East-West Center, HI.
Lui, T., and S. Chiu. 1994. Restructuring of Two Hong Kong Industries. *Environmental and Planning A, 26*: 53–70.
Lundvall, B. A. 1988. Innovation as an Interactive Process: From User–Producer Interaction to the National System of Innovation. In *Technical Change and Economic Theory*, edited by G. Dosi, C. Freeman, R. Nelson, G. Silverberg, and L. Soete, pp. 349–369. London: Pinter.
MacIssac, D., ed. 1976. *The United States Strategic Bombing Survey*. 31 Volumes. New York: Garland.
Maddison, A. 1991. *Dynamic Forces in Capitalist Development*. First ed. Oxford, England: Oxford University Press.
Maillat, D., ed. 1982. *Technology: A Key Factor for Regional Development*. Saint-Saphorin, Switzerland: Georgi.
Maillat, D. 1984. *De-Industrialization, Tertiary-Type Activities, and Redeployment*. Paper presented at the 24th European Congress of the Regional Science Association, Neuchatel, Switzerland.
Maillat, D. 1986. *New Technologies from the Viewpoint of Regional Economics: The Case of Neuchâtel*. Paper presented at the International High-Tech Forum Seminar, Université de Neuchâtel, Switzerland.
Maillat, D. 1988a. The Case of the Franco–Swiss Border from Geneva to Basel. In *Transatlantic Colloquy on Cross-Border Relations: European and North American Perspectives*, edited by D. Maillat. Zurich: Schulthess Polygraphischer Verlag.
Maillat, D. 1988b. Transfrontier Regionalism: The Jura Arc from Basel to Geneva. In *Perforated Sovereignties and International Relations*, edited by I. D. Duchacek, D. Latouche, and G. Stevenson, pp. 201–207. New York: Greenwood Press.
Maillat, D. 1989, October 2. *Local Dynamism, Milieu, and Innovative Enterprises*. Paper presented at the conference, "L'innovation dans l'entreprise: Le contexte spatial et culturel." Neuchâtel: Université de Neuchatel.
Maillat, D., and J. Vasserot. 1988. Economic and Territorial Conditions for Indig-

enous Revival in Europe's Industrial Regions. In *High Technology Industry and Innovative Environments: The European Experience*, edited by P. Aydalot and D. Keeble. London: Routledge.

Mailloux, K. F. 1961. Labor Problems and Conditions at the Boston Manufacturing Company, 1813–1848. *Cotton History Review*, 2(3): 139–146.

Management Today. 1984, February. Patek Philippe's Better Time. Pp. 78–81.

Management Today. 1987, November. Letter from Switzerland: Settling Matters Swiss Style. P. 26.

Mantoux, P. 1928. *The Industrial Revolution in the Eighteenth Century: An Outline of the Beginnings of the Modern Factory System in England.* London: Cape.

Markusen, A. 1985. *Profit Cycles, Oligopoly, and Regional Development.* Cambridge, MA: MIT Press.

Markusen, A., and J. Yudken. 1992. *Dismantling the Cold War.* New York: Basic Books.

Marsh, E. A. 1896. *The Evolution of Automatic Machinery as Applied to the Manufacture of Watches at Waltham, Mass., by the American Waltham Watch Company.* Chicago: Hazlitt & Co.

Marshall, A. 1997. *Principles of Economics.* New York: Prometheus Books.

Mathews, A. E. 1974, February. The Osaka Watch Company. *Bulletin of the National Association of Watch and Clock Collectors, 168*: 143–146.

Matsuzaka, T. 1992, August 22. Swiss Swatches Set to Be Made in Japan. *Nikkei Weekly*, p. 10.

McCraw, T. 1986. *America versus Japan.* Boston, MA: Harvard Business School Press.

Meier, H. 1970. *Friendship under Stress: U.S.–Swiss Relations, 1900–1950.* First ed. Berne, Switzerland: Herbert Lang and Co.

Milward, A. 1969. The *Development of the Economies of Continental Europe, 1850–1914.* First ed. Cambridge, MA: Harvard University Press.

Mole, D. 1997. *Managing the New Hong Kong Economy.* First ed. Oxford, England: Oxford University Press.

Monroe, W., and E. Sakaibara, eds. 1977. *The Japanese Industrial Society: Its Organizational, Cultural, and Economic Underpinnings.* Austin: University of Texas Press.

Moore, C. W. 1945. *Timing a Century: History of the Waltham Watch Company.* Cambridge, MA: Harvard University Press.

Morais, R. C. 1987, December 14. "Abe Breguet, Meet Christian Dior." *Forbes*, pp. 92–97.

Morris-Suzuki, T. 1994. *The Technological Transformation of Japan: From the Seventeenth to the Twenty-first Century.* First ed. Cambridge, England: Cambridge University Press.

Musée International d'Horlogerie. 1988. *Le mètre et la seconde: Charles-Edouard Guillaume (1861–1938), Prix Nobel de Physique.* La Chaux de Fonds: Author.

Nakane, C., and S. Oishi. 1990. *Tokugawa Japan: The Social and Economic Antecedents of Modern Japan.* Translated by C. Totman. First ed. Tokyo: University of Tokyo Press.

Neff, R. 1984, July. Hattoris Inspire Seiko to a High-Tech Future. *International Management*, pp. 20–25.

North, D. 1990. *Institutions, Institutional Change and Economic Performance.* Cambridge, England: Cambridge University Press.
Norton, R. D., and J. Rees, 1978. The Product Cycle and the Spatial Decentralization of American Manufacturing. *Regional Studies, 13*: 141–151.
Numagami, T. 1996. Flexibility Trap: A Case Analysis of U.S. and Japanese Technological Choice in the Digital Watch Industry. *Research Policy, 25*(1): 133.
Ocampo, R. 1992, September 24. Asia Edges out U.S. in Top Sales Spot for Piaget. *South China Morning Post,* p. 2.
Odagiri, H., and A. Goto. 1993. The Japanese System of Innovation: Past, Present, and Future. In *National Innovation Systems: A Comparative Analysis,* edited by R. Nelson, pp. 76–114. Oxford, England: Oxford University Press.
Ohkawa, K., and Y. Hayami. 1973. *Economic Growth: The Japanese Experience since the Meiji Era.* 2 vols. Tokyo: Japan Economic Research Center.
Ohkawa, K., and H. Rosovsky. 1973. *Japanese Economic Growth: Trend Acceleration in the 20th Century.* Stanford, CA: Stanford University Press.
Palmer, B. 1950. *The Book of American Clocks.* New York: Macmillan.
Parkes, C. 1987, March. Hong Kong Challenges for Top Spot in Watchmaking. *Financial Times,* p. 7.
Parkes, C. 1988, July 28. Watches Mark Time. *Financial Times,* p. 10.
Patrick, H. 1973. *Japanese Industrialization and Its Social Consequences.* First ed. Berkeley and Los Angeles: University of California Press..
Phillips, B., B. Kirchhoff, and S. Brown. 1990, April. *Formation, Growth, and Mobility of Technology Based Firms in the U.S. Economy.* Paper presented at the 10th Annual Babson Entrepreneurship Research Conference, Wellesley, MA.
Phillips, L. E. 1986, October 13. Seiko Winds Up Using Testimonials. *Advertising Age,* p. 101.
Pilarski, L. 1985, June 17. Can New Management Team Keep Swiss Watches Ticking? *International Management, 40*(6): 57–58.
Pilarski, L., and A. Gabor. 1984, November 26. A Last-Minute Comeback for Swiss Watchmakers. *Business Week,* pp. 139–142.
Pinot, R. 1979. *Paysans et horlogers jurassiens.* Geneva, Switzerland: Editions Grounauer.
Piore, M., and C. Sabel. 1984. *The Second Industrial Divide: Possibilities for Prosperity.* New York: Basic Books.
Porter, M. 1990. *The Competitive Advantage of Nations.* Cambridge, MA: Harvard University Press.
President of the United States. 1851. *Friendship, Reciprocal Establishments, Commerce and Extradition Convention.*
Radja, T. 1990. *Unpublished tables.* Bienne, Switzerland: Fédération de l'Industrie Horlogère Suisse.
Rappard, W. E. 1914. *La révolution industrielle et les origines de la protection légale du travail en suisse.* Berne, Switzerland: Staempfli and Ciê.
Redford, L., ed. 1980. *The Occupation of Japan: Economic Policy and Reform.* Norfolk, VA: MacArthur Memorial.
Richardson, K. 1972. *Twentieth Century Coventry.* London: Macmillan.
Riedel, J. 1974. *The Industrialization of Hong Kong.* Tubingen, Germany: J. C. B. Mohr Paul Siebeck.

Rosenberg, N. 1969. *The American System of Manufactures.* Edinburgh, Scotland: University of Edinburgh Press.
Rosovsky, H. 1961. *Capital Formation in Japan, 1868–1940.* First ed. New York: Free Press.
Rosovsky, H. 1966. *Industrialization in Two Systems: Essays in Honor of Alexander Gerschenkron.* First ed. New York: John Wiley.
Rossel, P. 1990. *Translation of Working Note on the Jura Region.* La Locle, Switzerland: Tissot Economic Foundation.
Sabel, C., and J. Zeitlin. 1997. *World of Possibilities: Flexibility and Mass Production in Western Civilization.* First ed. Cambridge, England: Cambridge University Press.
Samuels, R. 1994. *Rich Nation, Strong Army: National Security and the Technological Transformation of Japan.* First ed. Ithaca, NY: Cornell University Press.
Sandberg, L. 1974. *Lancashire in Decline: A Study in Entrepreneurship, Technology, and International Trade.* First ed. Columbus: Ohio State University Press.
Sandholtz, W., M. Borrus, J. Zsyman, K. Conca, J. Stowsky, S. Vogel, and S. Weber. 1992. *The Highest Stakes: The Economic Foundations of the Next Security System.* Oxford, England: Oxford University Press.
Sauer, D. 1992. *Time for America: Hamilton Watch, 1892–1992.* First ed. Lititz, PA: Sutter House.
Saxenian, A. 1990. *The Origins and Dynamics of Production Networks in Silicon Valley* (Working Paper 516). Berkeley: Institute of Urban and Regional Development, University of California at Berkeley.
Saxenian, A. 1994. *Regional Advantage.* Cambridge, MA: Harvard University Press.
Schoenberger, E. 1997. *The Cultural Crisis of the Firm.* London: Blackwell.
Schweizer, B. 1986. The Swiss Watchmaking Industry (UBS Publ. no. 100). Zurich: Union Bank of Switzerland.
Scott, A. 1988. *From Metropolis to Urban Form.* Berkeley and Los Angeles: University of California Press.
Scranton, P. 1997. *Endless Novelty: Speciality Production and American Industrialization, 1865–1925.* Princeton, NJ: Princeton University Press.
Seiko, H. I. J. 1985. *Corporate Annual Report.*
Shannon, F. A. 1967. *The Centennial Years: A Political and Economic History of American from the Late 1870s to the Early 1890s.* New York: Doubleday.
Shibata, Y. 1987, May 20. Hattori Seiko Shows First Pre-Tax Loss. *Financial Times,* p. 5.
SMH. 1985. *Corporate Annual Reports: 1985.* Biel, Switzerland: Author.
SMH. 1986. *Corporate Annual Report: 1986.* Biel, Switzerland: Author.
SMH. 1987. *Corporate Annual Report: 1987.* Biel, Switzerland: Author.
Smith, A. 1973. *The Lancashire Watch Company.* NH: Roberts.
Smith, M. R. 1977. *The Harpers Ferry Armory and the New Technology, the Challenge of Change.* Ithaca, NY: Cornell University Press.
Smitka, M. J. 1991. *Competitive Ties.* First ed. New York: Columbia University Press.
Smitka, M. 1998. *Japan's Ascent: International Trade, Growth, and Postwar Reconstruction.* First ed. New York: Garland.

Sobel, D. 1996. *Longitude*. New York: McGraw-Hill.
South China Morning Post. 1993, November 26. Time pieces. P. 31.
Stanwood, E. 1903. *American Tariff Controversies in the Nineteenth Century*. Boston: Houghton, Mifflin.
Steinberg, J. 1976. *Why Switzerland?* First ed. Cambridge, England: Cambridge University Press.
Storper, M. 1997. *The Regional World: Territorial Development in a Global Economy*. First ed. New York: Guilford Press.
Storper, M., and W. R. Walker. 1989. *The Capitalist Imperative*. London: Blackwell.
Sun, B. 1987, March. Watch Retailer Times It Right in Hong Kong. *Asian Business*, pp. 52–53.
Suzuki, T. M. 1994. *The Technological Transformation of Japan*. Cambridge, England: Cambridge University Press.
Switzerland Federal Statistics Office. 1985. *Recensement Fédéral de la Population 1980* (Vols. 8, 25, 27). Berne, Switzerland: Federal Statistics Office.
Takeshi, T. 1996. A History of Japanese Industry, 4: Japan's Industrial Revolution, 1890–1913. *Journal of Japanese Trade and Industry*, 6: 44–46.
Takeshi, T. 1997. A History of Japanese Industry: Industrial Development Between the Two World Wars (1914–1936). *Journal of Japanese Trade and Industry*, 1: 48–51.
Taylor, W. 1993, June. Message and Muscle: An Interview with Swatch Titan Nicholas Hayek. *Harvard Business Review*, pp. 98–115.
Thompson, E. P. 1967. Time, Word and Discipline, and Industrial Capitalism. *Past and Present*, 38, 57–97.
Tissot Economic Foundation Newsletter. 1989, December. No. 4.
Tomer, J. F. 1992. The Social Causes of Economic Decline: Organizational Failure and Redlining. *Review of Social Economy*, 50(1): 61–81.
Toyo Keizai Shinposha. 1988. *Japan Company Handbook*. Tokyo: Author.
Uchida, H. 1985. *The Osaka Watch Company, Study Note No. 1*. Tokyo, Japan: Hattori Seiko.
Union Bank of Switzerland. 1986. *The Swiss Watchmaking Industry* (UBS Publications on Business, Banking, and Monetary Topics, No. 100). Zurich: Author.
Union Bank of Switzerland. 1987. *Economic Survey of Switzerland, 1987*. Zurich: Author.
Union Bank of Switzerland. 1989. *Economic Trends in Switzerland, 1988–1989*. Zurich: Author.
Union Bank of Switzerland. 1993. *Economic Survey of Switzerland*. Zurich: Author.
Union Bank of Switzerland. 1996. *Sectoral Trends in Switzerland, 1994–1995*. Basel: Author.
Union Bank of Switzerland. 1997. *Economic Survey of Switzerland*. Zurich: Author.
U.S. Department of Commerce. 1950. *Post War Watch Markets*. Washington, DC: U.S. Government Printing Office.
U.S. Senate. 1951. *Committee on Finance, Hearings on the Trade Agreements Extension Act of 1951*. Washington, DC: U.S. Government Printing Office.
Uttinger, W., and D. R. Papera. 1965. Threats to the Swiss Watch Cartel. *Western Economic Journal*, 3(2): 200–216.

Uyeda, T. 1938. *The Small Industries of Japan*. New York: Secretariat Institute of Pacific Relations.

Uyterhoeven, H., and F. Knickerbocker. 1972. *Timex Corporation* (Harvard Business School Case Studies 373-080). Boston: Harvard Business School Press.

Varaldo, R., and Ferrucci, L. 1996. The Evolutionary Nature of the Firm within Industrial Districts. *European Planning Studies*, 4(1): 27–35.

Wadsworth, A. P. , and J. Mann. 1931. *The Cotton Trade and Industrial Lancashire, 1600–1780*. Manchester, England: Manchester University Press.

Waldo, L. 1886, May 21. The Mechanical Art of American Watchmaking. *Journal of the Society of the Arts*, pp. 740–751.

Willatt, N. 1984, February. Patek Philippe's Better Time. *Management Today*, pp. 78–81.

Williams, E. 1982, October 20. World Watch Industry: Swiss Survive at a Price. *Financial Times*, section 1, p. 12.

Williamson, O. 1991. *The Nature of the Firm: Origins, Evolution, and Development*. New York: Oxford University Press.

Wong, S.-L. 1988. *Emigrant Entrepreneurs*. First ed. Oxford, England: Oxford University Press.

Yankelovich, D. 1969. *Preliminary Study of the United States Electronic Watch Market Opportunity*. Report prepared for Fédération Suisse des Associations de Fabricants d'Horlogerie by Daniel Yankelovich, Inc. New York: Author.

Yasuba, Y. 1973. The Evolution of Dualistic Wage Structure. In *Japanese Industrialization and Its Social Consequences*, edited by Hugh Patrick, pp. 249–298. Berkeley and Los Angeles: University of California Press.

Yoffie, D. 1990. *International Trade and Competition: Cases and Notes in Strategy and Management*. New York: McGraw-Hill.

Youngson, A. J. 1972. *Economic Development in the Long Run*. First ed. London: Allen & Unwin.

Index

Accutron watch, 1, 186, 189
Adams, John C., 117, 161, 162
American production system, 46–49
   British watch industry and, 83–85, 86, 87
   Swiss watch industry and, 103, 104
   U.S. watch industry and, 107
   Waltham Watch Company and, 112
Analog quartz watches, 212, 228, 244
Anchor fork, 67
Apprenticeship system, 81
Armaments industry, 48
Astronolon, 245
Astron watch, 206
Avery, Thomas, 120

Balance wheel, 67
Benetton, 189
Boer War, 141
Boston, MA, location of clock manufacturing, 108, 112
Boston Watch Company, 278(n3)
British economy, 24, 25–26, 46, 65, 71
British Horological Institute, 84, 86
British watch industry
   American production system and, 83–85, 86, 87
   apprenticeship system and, 81
   chronometers and, 71, 72
   decline of, 70–71, 79–87
   development of, 38, 71–73
   domestic market and, 79–80, 86
   dual system of production, 73–80
   early failures to mechanize, 42, 44–45

   economic crisis of 1873-1896, 25
   factory production and, 81, 85
   first mover advantages, 272
   French industry and, 90
   geographic focal points, 72–73
   institutional rigidity in, 81–85, 261–262
   knowledge codification and, 269
   metalworking industry and, 73, 75
   putting-out system and, 76–79
   stem-wound watches and, 101
   Swiss imports and competition, 80, 82, 85, 86, 87, 90, 93, 99, 261
   tariff protection and, 79
   technological lock-in, 79–80, 267–268
   U.S. industry competition and, 86, 88
   U.S. tariffs and, 82
   during World War II, 32
Bulova Watch Corporation, 154, 166, 168, 265
   Accutron watch, 1, 186, 189
   decline of, 187–189, 191
   flexible production technology, 186
   growth of, 185–186
   international competitors and, 187
   international markets and, 179
   international production system, 187, 188
   marketing and distribution, 186–187
   quartz technology and, 188
   Swiss mergers and, 200
   tuning-fork technology and, 186
   U.S. tariffs and, 181

297

Business culture. *See* Corporate culture
Business Week (magazine), 207, 214

Calculators, 207
Capital-intensive industry, 51, 52-53
Capital investment
  British watch industry failure and, 82
  Hong Kong manufacturing industry and, 218-220
  postwar reconversion and, 34
Caravelle watch, 186
Cartels, Swiss, 149-150, 198-199
Cartier, 250, 253
Casio, 168, 171, 173, 176, 253, 258
Chandler, Alfred, 50-51, 52
Chicago, Elgin Watch company, 117, 126
China, 195, 218, 230, 232, 234, 238, 258
Chronographs, 99
Chronometers, 70, 71, 72, 201
Citizen Watch Company, 164, 165-166, 168, 171, 172, 173, 176, 186, 187, 258
Civil War (U.S.), 31-32, 114, 115, 116, 133
Clockmaker's Company, 72
Clockmaking, 65-66
Collective capitalism, 53-58
Consensus decision-making, 56, 89
Cornell, Paul, 161, 162
Cornell Watch Company, 161-162
Corporate culture
  change and, 11-12, 61
  conceptualizations of, 58-60
  power and, 60
  regional identity and, 59, 62-63
  See also Organizational culture
Cottage production
  British watch industry and, 76-79
  in Japanese economy, 158-159
  Swiss watch industry and, 97-98, 136-140
  See also Putting-out system
Coventry, England, 72-73, 79, 81, 86

Craft era, 41-45
Culture, 58. *See also* Corporate culture
Currency devaluations, 232

Dailywin Corporation, 229, 234
Daimler Chrysler, 257
Daini Seikosha, 164, 166, 168. *See also* Seiko Corporation
Dennison, Aaron, 107-108, 112-114, 125
Designer watches
  Swiss watch industry and, 249-250
  time-to-market needs, 256
Digital display technology, 206, 210-212, 213, 215, 225-226, 262
Digital watch industry
  display technology and, 206, 210-212, 213, 215
  Hong Kong and, 212, 216, 225-226, 228
  impact of, 206-207
  market size, 209-210
  origins of, 207-208
  plastic watches, 213
  price competition in, 216
  price declines, 208-209, 212, 213, 216, 226
  quartz analog watches and, 228
  strategic decisions of manufacturers, 214-216
  technical problems and, 209
  U.S. semiconductor firms in, 207-208, 209, 210, 211, 212-214, 215, 216
Dollar watches, 121, 122, 123-124, 125, 189. *See also* Pin-lever watches

Ebauches, 67, 98-99
Ebauche SA, 149, 187, 192
Economic crises
  of 1872-1896, 23-25
  of 1929-1949, 27-29
  of the 1950s, 29-30
  of the 1970s, 30-31
  effects of, 22
  Great Depression, 25-27
  industrial change and, 10-11

Economic development theories, 9
Electric watches, 68. *See also* Digital watch industry; Quartz technology
Electronics technology
   impact of, 1–5, 6–7, 67
   labor and, 203–204
Electronic watch industry
   consumer surveys, 1–2
   early watches, 205–206
   global competitors and, 205
   Hamilton watches, 184
   hand assembly production in, 204, 205
   trends governing, 2–3
   See also Digital watch industry; Quartz technology
Elgin Watch Company, 116
   decline of, 179, 182, 191
   dollar watch competition and, 125
   economic crisis of 1893, 118–119
   electric watches and, 205–206
   founding of, 117
   interwar period, 180
   price competition and, 118, 119, 120
   production methods and system, 118, 120
   sales levels, 179
   success of, 118
   technological lock-in, 268
   time-motion studies and, 120
   U.S. tariffs and, 181
   Waltham Watch Company and, 117–118
   wristwatch production and, 145
Enterprise groups, 55
Escapement mechanisms, 67
ETA. *See* SA Fabriques d'Ebauches

Factory-based production
   British watch industry and, 81, 85
   early trends in mechanization, 42–45
   in Japanese economy, 158, 159
   labor and, 45–46, 48
   managerial capitalism and, 49–53
   metal-cutting technology and, 49–50
   Swiss watch industry and, 101–105
   See also American production system
Fairchild Corporation, 208, 212
Family-controlled businesses, 55
Fashion watches, 253–254
Fédération de l'Industrie Horlogère Suisse (FH), 1, 149, 199, 255
Federation of Metal and Watchmaking Workers (Swiss), 247
Federation of the Swiss Watch Industry, 149
Files, 75
Fontainemelon, Switzerland, 99, 139
Fordism, 8
Fordrey-McCumber Tariff Act (U.S.), 181
Fossil, Inc., 249
Free trade, 23–24
French watch industry
   British industry and, 90
   electronic watches and, 277(n4)
   emigration of watchmakers, 94
   impact on Swiss industry, 90–91
   innovation and, 90, 91, 94, 262
   trends to thin watches, 93–94

Garment industry, 204, 219
Gautier, Ralph, 244–245
General Agreement on Tariffs and Trade (GATT), 195
General Watch Company, 200
Geneva, Switzerland, 92–93, 100
Germany
   economic crisis of the 1970s, 30
   Great Depression and, 26
   protectionism and, 24
   Timex expansion into, 191
   trade with Switzerland during World War I, 146, 147
Great Britain. *See* British economy; British watch industry
Great Depression, 25–27
Gruen and Company, 142
Guilds, 41, 72, 93, 261

300 ■ Index

Hairsprings, 67
Hall, John H., 48, 275(n2)
Hamilton Watch Company, 127, 145
    decline of, 179, 185, 191
    defense contracts and, 183, 184, 265
    digital watch production and, 212
    electric watches and, 184, 206
    integrated circuits and, 192
    postwar recovery and growth, 183–184
    Pulsar watch, 185
    reorganization of production system, 182–183
    U.S. tariffs and, 181
Harpers Ferry Armory, 48
Harrison, John, 70, 72
Hattori Seiko, 164–166. *See also* Seiko Corporation
Hayek, Nicholas, 243, 249, 257
Hetzel, Max, 1, 3, 201, 206
Hewitt, T. P., 78–79
Home-work. *See* Cottage production; Putting-out system
Hong Kong, 168, 173
    China and, 218
    description of, 216, 218
    distribution and marketing of Swiss watches, 224, 249, 256
    economy of, 204, 218
    electronics industry in, 225, 281
    international trade policy and, 219
    manufacturing structure, 205, 218–221
Hong Kong watch industry
    China and, 230, 234, 238
    competition in, 237–238
    consumer-oriented product variation, 230–231, 238
    dependence on foreign components, 235
    development of quality reputation and, 235–236
    development of watch brands, 225
    digital display technology and, 215
    digital watch production, 225–226, 228
    distribution channels and problems, 238, 263
    domestic retail market, 239–240
    early history of, 221, 223–224
    electronic watches and, 188
    emergence of, 204–205
    as export center, 221, 223, 224, 225, 230
    future prospects, 238–240, 258–259
    hand assembly in, 205
    international competitors, 232–234
    labor and, 234–235
    markets, 231–232
    mechanical watch production, 226, 228
    organizational characteristics, 240–241
    organization of, 229–234
    original equipment manufacture, 225, 229, 230
    parts exports and, 231–232
    patent disputes, 236–238
    plastic watches, 234
    problems facing, 234–238
    quartz analog watch production, 228
    recent industry trends, 229–231
    reunification with China and, 258
    skilled labor and, 264
    Swiss movements and, 251
    technical innovations and, 228–229
    time-to-market capabilities, 256
    watch assembly operations, 212, 216, 223–224, 225, 226, 230
    watchband manufacture, 224, 225
    watch case manufacturing, 223, 232
    watch prices, 231
Hong Kong Watch Manufacturers Association, 236, 240
Hong Kong Watch Trades Association, 239, 240
Horological schools, 100
Howard, Edward, 107–108, 112, 113
Hughes Aircraft Corporation, 208
Huguenots, 91, 92, 94

Incremental innovation, 17
India, 195
Industrial change
  economic crises and, 10–11, 22–31
  endogenous conditions and, 37–39
  industrial culture and, 11–12
  institutional character and, 11
  long-term outcomes and, 12
  organizational culture and, 58–61
  overview of, 15–17
  preconditions and, 37–38
  regional culture and, 62–63
  responses to crisis, 15
  technological change and, 10, 17–22
  technomilitary ideology and, 12, 34–36
  war and, 12, 31–36
Industrial culture. *See* Corporate culture
Industrial development
  collective capitalism, 53–58
  craft era, 41–45
  factory era, 45–49
  managerial capitalism, 49–53
Industrialization
  Japanese economy and, 156, 157–160
  significance of timekeeping to, 65
Industrial organization, 8–9
Industrial reconversion, postwar, 33–34, 35–36
Industrial Revolution, 71
Ingersoll dollar watch, 123–124, 189
Ingold, Pierre Frédéric, 102
Integrated circuits, 192
Intel Corporation, 208
Interchangeability, 47, 84, 103–104, 275(n2)
ISO 9000 standards, 237, 281(n8)

Japanese economy
  collective capitalism and, 54–57
  dualistic structure of production in, 158–159
  economic crises, 24, 25, 30
  export-led production system, 168, 173, 176
  Great Depression and, 26–27
  Hong Kong and, 219
  independent corporations in, 57
  industrialization and, 54, 156, 157–160, 275(n1)
  Korean War and, 29
  Meiji Restoration and, 157
  national economic policies and, 165
  postwar recovery, 30, 36
  relationship between capital and the state, 57
  silk industry, 158, 166
  technomilitary ideology and, 34–35
  wall clock industry, 279(n2)
  World War II and, 28, 29, 32, 155, 160, 166
Japanese watch industry
  acquisition of clock technology, 160–161
  Bulova Watch Corporation and, 187
  complementary product development, 257–258
  digital display technology and, 210, 211–212
  digital watches and, 208–209, 214–215
  distribution channels, 263
  domestic market, 168–169, 176
  dual production system in, 173
  early history of, 161–164
  electronics technology and, 203–204
  export-led production system, 168, 173, 176
  geographic centers, 161
  Hong Kong and, 168, 173, 223, 224, 232
  interdivisional rivalry, 172
  international markets, 232, 233–234
  labor costs and, 203–204
  machine-based production in, 159
  marketing strategies, 172–173
  micromechanical capability, 169, 171–172
  national economic policies and, 165, 265
  postwar recovery, 168–169

302 ■ Index

Japanese watch industry (cont.)
  quartz technology and, 206
  rise of, 156
  Roskopf watch and, 156
  skilled labor and, 263-264
  superficial production variation, 169
  technological innovation, 22, 172
  U.S. industry and, 160, 162-163, 177
  vertical integration and, 164-165, 168, 173, 176
  world leadership in production, 177
  World War II and, 160, 166
  See also Seiko Corporation
Japy, Frederick, 98
Jeweled watches
  description of, 67-68
  dollar watch competition and, 124
  Swiss watch industry and, 93, 193-195
  Timex and, 191
  U.S. watch industry and, 121, 152-154
Johnson, Lyndon B., 192
Jura watch industry
  mechanization and tool making in, 98-99, 101-102, 132-134
  organizational culture, 99-101
  rise of, 94, 96
  specialization of tasks, 97-98
  support and development of innovation, 99-101
  See also Swiss watch industry

K. Hattori & Company, 164, 168
Kieratsu, 57
Knowledge transfer, 269
Korean War, 29, 184, 191, 218, 223

Labor
  electronics technology and, 203-204
  in Hong Kong manufacturing, 219
  Hong Kong watch industry and, 234-235
  in Japanese collective capitalism, 55, 56-57
  rise of managerial capitalism and, 52

Swiss watch industry and, 104-105, 137-139, 151, 152, 199, 264, 282(n4)
  transition to factory-based production and, 45-46
  See also Skilled labor
Labor-intensive industry, 51-52
La Chaux de Fonds (Switzerland), 97, 98
Lady Hamilton watch, 127
Lancashire, England, 72, 73, 75-79, 86
Lancashire Watch Company, 85
Lancaster and Keystone Club Watch Companies, 127
Landes, D. S., 64
Laurel watch, 164
LCDs. See Liquid crystal displays
LEDs. See Light-emitting diodes
Le Locle, Switzerland, 96, 97
Lemkuhl, Jaokim, 189, 192
Lifetime employment, 56-57
Light-emitting diodes (LEDs), 69, 185, 206, 210-211, 213, 215, 225, 262
Liquid crystal displays (LCDs), 69, 206, 210-211, 213, 214, 215, 225, 226, 262
Liverpool, England, 72-73, 75, 76, 86
Locomotive "940" watch, 127
Loew's Corporation, 188
Longines, 154
Longitude (Sobel), 70

Machine manufacturing
  early trends in, 42-45
  Swiss watch industry and, 98-99, 101-102, 104, 106, 110-111, 132-134, 139-140
Machine tool industry
  British watch industry and, 84
  Japanese watch industry and, 169, 171-172
  Swiss watch industry and, 98-99, 104
  U.S. watch industry and, 110
Managerial capitalism, 49-53

Mass production
  American system, 46–49
  effect on skilled labor, 282(n4)
Mechanical watches, 67–68, 226, 228, 255
Meiji Restoration, 157
Mercedes Benz, 257
Merchants, 41–42
Metalworking industry, 49–50, 73, 75. *See also* Machine tool industry
Micro Compact Car, 257
Microelectronics, 19, 185. *See also* Electronics technology
Monopsonist firms, 276(n6)
Morrill Tariff Act (U.S.), 24
Mostek Corporation, 188

Nagano, Japan, 166
Nagoya, Japan, 161
Nail manufacturing, 75, 79
National Clock and Watch Association (U.S.), 277(n1)
National Semiconductor, 208, 211
Neuchâtel, Switzerland, 94, 96, 100. *See also* Jura watch industry
Newark Watch Company, 162
Newly industrialized countries (NICs), 232, 234

Oligopolies, 53
Olympic Games, 177, 206
Omega Corporation, 149, 244, 245, 250
Organizational culture, 58–61, 99–101. *See also* Corporate culture
Osaka, Japan, 161
Osaka Clock Company, 161
Osaka Watch Company, 163, 165
Otay Watch Company, 161, 162, 279(n3)

Paradigm shifts, 18–22, 268–269, 273
Patek Philippe, 200
Patent disputes, 236–238
Path dependency, 267, 269–272. *See also* Technological lock-in
Patriotic Emulation Society (Switzerland), 100

Philadelphia Centennial Exposition of 1876, 103, 129
Piaget, 200, 250
Pin-lever watches, 68, 121, 122–123, 133, 182. *See also* Dollar watches
Plastic watches, 213, 234, 245. *See also* Swatch watch
Platinum watches, 253–254
Power, corporate culture and, 60
Precision timekeeping, 66, 70
Prescott, England, 76, 77
Price competition
  digital watch industry and, 216
  U.S. watch industry and, 118, 119, 120, 143, 262
Protectionism, 23–24
Protofactories, 101
Pulsar watch, 185, 212
Putting-out system, 41–42, 43, 76–79, 97–98, 136–137. *See also* Cottage production

Quartz analog watches, 212, 228, 244
Quartz chronometers, 201
Quartz technology
  Bulova Watch Corporation and, 188
  impact of, 4–5, 6, 204, 206–207
  precursor technologies, 205–206
  Swiss watch industry and, 3, 4, 6, 201, 206, 207, 242–243
  Timex and, 192
  tuning-fork regulator and, 201
  See also Digital watch industry
Quartz watches, 68–69

Radical innovations, 17
Railroad standards, 121, 125–126, 127, 128, 129, 269
Railroad watches, 127, 128
Reflective innovation, continuous, 56
Regional change, 10–12, 62–63
Regional identity, 59, 62–63
  Jura watch industry and, 97–98
Revolution in Time (Landes), 64
Richard, Daniel Jean, 96
Ricoh, 166, 168

Robbins, Royal, 114
Robbins and Appelton Company, 114
Rock Watch, 246, 253
Rolex, 140, 200
Roskell, Robert, 73
Roskopf, George, 121, 122
Roskopf watch, 115, 121, 122–123, 156, 245. *See also* Pin-lever watch
Rotherham Watch Company, 81
Roxbury, MA, 112, 113
Russia, 278(n4)

SA Fabriques d'Ebauches (ETA), 247, 248–249
San Jose Watch Company, 162
Seiko Corporation
  analog quartz watches and, 212
  assembly-line production, 172
  complementary product development, 257–258
  digital display technology and, 211–212
  digital watch technology and, 215
  distribution channels, 263
  early history of, 164
  export-led production system, 168, 176
  Hong Kong and, 232
  independence of, 57
  interdivisional rivalry and, 172
  labor costs, 173
  marketing strategies, 172–173
  micromechanical capability, 171, 172
  national economic policies and, 165
  postwar recovery, 168
  quartz technology and, 188, 206, 212
  vertical integration and, 164–165, 173
  World War II and, 166
  worldwide reputation, 177
Seikosha, 164, 166. *See also* Seiko Corporation
Self-winding watches, 184

Semiconductor industry
  American firms, 207–208, 209, 210, 211, 212–214, 215, 216
  in Hong Kong, 226
Sherman Anti-Trust Act, 178
Silk industry, 158, 166
Sino-Japanese War, 158
Skilled labor
  American production system and, 48
  British watch industry and, 83–84
  effect of mass production on, 282(n4)
  Hong Kong watch industry and, 234–235
  importance of, 263–264
  transition to factory-based production and, 45–46
  U.S. watch industry and, 110
Small firm clusters, 8–9
Smoot–Hawley Act (U.S.), 181
Smuggling, 194–195
Sobel, Dava, 70
Société des Garde-Temps (SGT), 200
Société General de l'Horlogerie Suisse SA (ASUAG), 1, 149, 151, 200
Société Suisse de Microélectronique et de l'Horlogerie (SMH), 187, 246, 252
Société Suisse pour l'Industrie Horlogerie (SSIH), 149, 187, 200, 242, 244–245, 245
Solid-state watches, 69. *See also* Digital watch industry
Spring-powered watches, 67–68
Standards, 269
  ISO 9000 conventions, 237, 281(n8)
  See also Railroad standards
Statut de l'Horlogerie (Switzerland), 150–151, 179, 195, 198–199, 247
Steinberg, Jonathan, 88, 89
Stem-wound watches, 101
Strategy and Structure (Chandler), 50–51, 52
Suwa Seikosha, 168
Swatchmobile, 257

Swatch watch, 234
  antecedents of, 245
  decline of, 255
  emergence of, 246–247
  joint ventures and, 257
  marketing innovation and, 248–249
  platinum model, 253–254
  revival of Swiss watch industry and, 250
  sales value, 247–248
  weaknesses of, 248
Swiss watch industry
  American production system and, 103, 104
  analog quartz watches and, 244
  British industry and, 80, 82, 85, 86, 87, 90, 93, 99, 261
  Bulova Watch Corporation and, 187
  cartels, 149–150, 198–199
  consumer trends and, 253
  designer watches and, 249–250
  digital watch technology and, 208–209, 211, 213, 215
  distribution networks, 224, 249, 256, 262
  dominance of, 105, 140
  early history of, 92–98
  electronics technology and, 1, 2, 3, 203
  emergence of, 91–92
  factory-based production, 101–105
  fashion watches and, 253–254
  foreign trade and, 139
  French Huguenots and, 91, 92, 94
  French industry and, 90–91, 94
  future of, 256–257
  geographic centers, 92–93, 94, 96
  government regulation, 149–151, 198–199, 264–265
  Great Depression and, 27
  Hong Kong and, 224, 249, 251, 256
  inexpensive watches and, 122–123, 133
  interchangeability and, 103–104
  international competition and, 198
  international markets, 93, 105, 198, 233, 234
  interwar period, 148, 149–150
  Japanese industry and, 177
  knowledge transfer and, 269
  labor and, 104–105, 137–139, 151, 152, 199, 264, 282(n4)
  machine manufacturing and, 98–99, 101–102, 104, 106, 110–111, 132–134, 139–140
  marketing trends, 249–250
  mechanical watch trade, 255
  mergers and industry groups, 200–201
  monetary instability and, 252
  movement sales, 247, 248, 251–252
  partial vertical integration in, 139, 140
  plastic watch production, 245
  postwar conditions, 35
  pricing mismanagement, 199–200
  protofactories, 101
  putting-out system in, 97–98, 136–137
  quartz technology and, 3, 4, 6, 201, 206, 207, 242–243
  reorganization of the 1980s, 244–245, 246–249
  rigidity in, 89, 201
  rise of, 89–90
  Statut de l'Horlogerie and, 179, 195, 198–199, 247
  stem-wound watches, 101
  Swatch watch and, 234, 245–249, 250
  technological innovation and, 21, 22, 99–101
  time-to-market problems, 255–256
  tool making, 98–99, 104
  trade unions and, 138–139
  transformation costs, 136–139
  transformation in response to U.S. competition, 131–134, 139–140
  trends to thin watches, 94
  tuning-fork technology and, 1, 2, 3–4
  unemployment in, 148
  U.S. impact on, 131–134, 139–140

U.S. market and, 132, 144, 147, 152–154, 155, 179, 181–182, 193–195, 194, 198, 201–202
U.S. tariffs and, 29, 102–103, 104, 133, 148–149, 181–182
U.S. underselling of, 129
watch prices, 231
world market volume share, 251–255, 262
World War II and, 28–29, 32–33, 152–153, 155
wristwatch production and, 140, 141, 143, 145
Switzerland
consensus decision-making in, 89
economic crisis of the 1970s, 30–31
economic prosperity of, 88
neutrality and, 88–89
during World War I, 138–139, 146, 147
Systems-transforming innovations, 17–18

Taiwan, 168
Tariffs, 24, 79. *See also* U.S. tariffs
Technoeconomic paradigm shifts, 18–22
Technological change
categories of, 17–18
corporate culture and, 61
effect of war on, 12, 32–34, 272–273
exogenous, effects of, 202
geographic effect and, 18
Hamilton Watch Company and, 184
Hong Kong watch industry and, 228–229
industrial change and, 10
industry rigidity and, 261–262
in Japanese collective capitalism, 55
Japanese interdivisional rivalry and, 172
paradigm shifts, 18–22
resistance to, 267–269
Swiss watch industry and, 21, 22, 99–101
Technological lock-in, 143–144, 267–269. *See also* Path dependency

Technomilitary ideology, 12, 34–36, 272
Territorial development, 9–10
Texas Instruments, 208, 209, 210, 212, 213
Textile industry, 42, 158, 159
Thomke, Ernest, 248
Timekeeping, history of, 65–66
Time–motion studies, 120
Timex Watch Corporation, 2, 154
consumer trends and, 253
decline of, 192
design and production strategies, 189–190
digital watch production and, 209, 212, 213, 214
early history of, 189
expansion of product line, 190–191
international competitors and, 187
international growth and markets, 179, 190, 191
inventory control and, 190
marketing and distribution, 124, 189, 190
quartz technology and, 192
U.S. trade policy and, 191
women's watches and, 189
Tissot Watch Company, 96, 149, 244, 245, 246
Tokugawa regime, 157
Tokyo, 161, 163–164, 164–165, 263–264
Tool industry. *See* Machine tool industry
Trade
effect of war on, 31–32
free trade vs. protectionism, 23–24
Trade unions. *See* Unions
Tresor Magique watch, 253–254
Tuning-fork regulator, 201
Tuning-fork watches, 206
Accutron watch, 1, 186, 189
Bulova Watch Corporation and, 186
description of, 68
quartz technology and, 188
resistance to, 3–4, 269
Swiss watch industry and, 1, 2
Two-Timer watch, 246, 253

Union des Branches Annexes de l'Horlogerie (UBAH), 149
Unions, Swiss, 138-139, 199
U.S. economy
  American production system, 46-49
  during the Civil War, 31-32
  crisis of the 1970s, 30
  Great Depression and, 26
  managerial capitalism and, 50-51
  postwar, 29, 35
  protectionism and, 24
  World War II and, 29
U.S. watch industry
  American production system and, 107
  British industry and, 86, 88
  Civil War and, 114, 115, 116
  decline in, 119-121, 178-180, 192, 193
  defense contracts and, 183, 184, 265, 280(n1)
  digital display technology and, 210-211, 213, 215, 262
  digital watch producers, 207-208, 209, 210, 211, 212-214, 215, 216
  distribution costs and, 144
  early history of, 107-112
  economic crisis of 1873-1896, 25
  electronics technology and, 2, 203
  emergence of, 91-92, 107
  emulation of British and Swiss techniques, 110
  foundational instabilities and, 269
  geography of, 108, 110
  golden age of, 115-119, 128-129
  histories of, 277-278(n1)
  impact on Swiss industry, 131-134, 139-140
  industry expansion, 116-117
  inexpensive watches and, 115-116, 121-122, 123-124, 125
  international competition and, 179
  interwar period, 148-149, 178-180, 178-182
  inventory overhead, 127
  Japanese industry and, 160, 162-163, 177
  jeweled watches, 121
  labor costs and, 203-204
  machine tool imports, 104
  machine tool industry and, 110
  national trade policy and, 191-192, 193-195
  new firms, problems of, 111-112
  postwar period, 36, 194
  price competition in, 118, 119, 120, 124, 143, 262
  problems of the 1890s, 134
  production levels in, 118, 119
  product segmentation in, 126
  railroad standards and, 121, 125-126, 127, 128, 129, 269
  Sherman Anti-Trust Act and, 178
  skilled labor and, 110
  Swiss competition and imports, 132, 134, 144, 147, 152-154, 155, 179, 181-182, 193-195, 194, 198, 201-202
  tariff protection and, 181-182, 265
  technological lock-in and path dependency in, 143-144, 268
  underselling of Swiss industry, 129
  venture capital and, 108, 110
  vertical integration and, 53
  wholesale distributors and, 116
  World War I and, 145-146, 180-181
  World War II and, 32, 152-153, 155, 183
  wristwatch production and, 127-128, 142-144, 145
  See also individual companies
U.S. tariffs
  British watch industry and, 82
  cancellation of, 192
  during the Civil War, 24, 32, 133
  protection of U.S. watch industry, 181-182, 265
  Swiss watch industry and, 29, 102-103, 104, 133, 148-149, 181-182
Universal Genève, 187

Vacheron and Constantin watch company, 110
Vertical integration
  American production system and, 49
  Japanese watch industry and, 164–165, 168, 173, 176
  rise of managerial capitalism and, 52–53
Virgin Islands, 191, 265

Wage inflation, 203
Waltham Land Company, 113
Waltham Watch Company, 164
  American production system and, 49, 112
  decline of, 179, 182, 191
  disintegrated production system, 120
  dollar watch competition and, 125
  dominance of, 114–115
  Elgin Watch Company and, 117–118
  first mover effects, 272
  founding of, 112–113
  mismanagement in, 179–180
  Philadelphia Centennial Exposition of 1876, 129
  price competition and, 119, 120
  reorganization under Robbins, 114
  sales strategies, 278(n5)
  Swiss competition and, 134
  technological lock-ins, 268
  U.S. tariffs and, 181
  vertical integration and, 53
  during World War I, 145–146
  wristwatch production and, 145
War, effects of, 12, 31–36
Watchbands, 224, 225
Watch case manufacturing, 223, 232, 280–281(n3)
Watch faces, 69
Watch industry
  American production system and, 47, 49
  analog quartz watches in, 228
  development in
    collective capitalism and, 53–58
    contingent nature of, 260–262
  craft era, 41–45
  factors affecting, 38–39
  factory era, 45–49
  managerial capitalism and, 49–53
  organizational culture and, 58–61
  overview of, 40
  regional culture and, 62–63
  understanding, 266–267
  digital technology and, 214–216
  distribution channels, 262, 263
  early antecedents of, 69–70
  electronics technology and, 1–5, 6–7, 67, 203–204
  exogenous technical change and, 202
  first mover effects, 271–272
  foreknowledge of competitors, 271
  foundational instabilities and, 269–270
  French innovation and, 90, 91, 94
  government protection and, 264–265
  historical trends, 5–6, 130
  inexpensive watches and, 121–122
  institutional rigidity and, 261
  international competition and, 201–202
  knowledge transfer and, 269
  limits on competition and, 265
  linkage between industries and regions, 6–7
  market dominance instability, 266
  paradigm shifts and, 273
  path dependency and, 269–272
  postwar conditions, 35, 36
  quartz technology and, 4–5, 6, 204
  skilled labor and, 263–264
  standards and, 269
  technological lock-ins and, 267–269
  technomilitary ideology and, 272
  thin watches and, 93–94
  vertical integration and, 53
  war and, 272–273
  watch precision and, 66, 70
  women in, 102, 276(n3)
  world market segmentation, 250–251, 262–263

during World War II, 32–33
See also individual companies and countries
Watch technology
  history of, 66–69
  precision and, 66, 70
Waterbury Clock Company, 123, 189
Waterbury Watch Company, 123
Why Switzerland? (Steinberg), 88, 89
Women
  in watch manufacturing, 102, 276(n3)
  wristwatches and, 142, 145
Women's watches, 127, 189
World War I
  Swiss watch industry and, 146–148
  Switzerland and, 138–139, 146, 147
  U.S. watch industry and, 145–146, 180–181
  world watch market and, 278(n4)
  wristwatches and, 141, 142

World War II
  economic impact of, 28–29
  Japanese economy and, 155, 160, 166
  Japanese watch industry and, 166
  postwar recovery and reconversion, 29–30, 33–34, 35–36
  Swiss watch industry and, 152–153
  U.S. watch industry and, 183
  Waterbury Clock Company and, 189
  world watch industry during, 32–33
Wristwatches
  consumer attitudes toward, 141–142, 145
  military use of, 141, 142
  Swiss watch industry and, 140, 141, 143, 145
  U.S. watch industry and, 127–128, 142–144, 145
  women and, 142, 145

About the Author

Amy K. Glasmeier, PhD, is the director of the Center on Regional Research and Industrial Studies, at the Institute for Policy Research and Evaluation, The Pennsylvania State University. She has published three books on international industrial and economic development—*High Tech America* (1986), *The High-Tech Potential: Economic Development in Rural America* (1991), and *From Combines to Computers: Rural Services Development in the Age of Information Technology* (1995)—and more than 50 scholarly articles. Her popular writings include *Global Squeeze on Rural America: Opportunities, Threats, and Challenges from NAFTA, GATT, and Processes of Globalization* (1994), and *Branch Plants and Rural Development in the Age of Globalization* (1995). In 1996–1998, Dr. Glasmeier was the John D. Whisman Appalachian Scholar for the Appalachian Regional Commission. She has served as a consultant with the Economic Development Administration of the U.S. Department of Commerce, the National Aeronautics and Space Administration, the U.S. Department of Housing and Urban Development, the Congressional Office of Technology Assessment, the Organization for Economic Cooperation and Development, the U.S. Department of Transportation, The Economic Policy Institute, and the Regional Government of Emilia Romagna, Italy. She is a member of the National Academy of Sciences, National Research Council Board on the Constructed Environment. Her current research focuses on community impacts of globalization, regional development, poverty alleviation, and industrial change.